A Gentle Introduction to Stata

Second Edition

A Gentle Introduction to Stata

Second Edition

ALAN C. ACOCK
Oregon State University

A Stata Press Publication
StataCorp LP
College Station, Texas

Second edition 2008

Published by Stata Press, 4905 Lakeway Drive, College Station, Texas 77845
Typeset in $\LaTeX\,2_\varepsilon$
Printed in the United States of America
10 9 8 7 6 5 4 3 2 1

ISBN-10: 1-59718-043-2
ISBN-13: 978-1-59718-043-6

Acknowledgments

I acknowledge the support of the Stata staff who have worked with me on this project. Special thanks goes to Lisa Gilmore, the Stata Press production manager, and Jennifer Neve and Deirdre Patterson, the Stata Press technical editors. I also thank my students who have tested my ideas for the book. They are too numerous to mention, but Shauna Tominey deserves special recognition for going through the entire draft of the second edition to find errors.

Bennet Fauber, during the time he was affiliated with StataCorp, provided hours and hours of support on all aspects of developing the first edition. He taught me the LaTeX 2_{ε} document preparation system used by Stata Press, and his patience with many of my problems and mistakes inspired me to have more patience with my own students. Bennet also had major input on the topical coverage and organization of the first edition.

Finally, I thank my wife, Toni Acock, for her support and for her tolerance of my endless excuses for why I could not do things. She had to pick up many tasks I should have done, and she usually smiled when told it was because I had to finish this book.

Contents

Tables

Figures

Preface

This book was written with a particular reader in mind. This reader is learning social statistics and needs to learn Stata but has no prior experience with other statistical software packages. When I learned Stata, I found there were no books written explicitly for this type of reader. There are certainly excellent books on Stata, but they assume extensive prior experience with other packages, such as SAS or SPSS; they also assume a fairly advanced working knowledge of statistics. These books moved quickly to advanced topics and left my intended reader in the dust. Readers who have more background in statistical software and statistics will be able to read chapters quickly and even skip sections. The goal is to move the true beginner to a level of competence using Stata.

With this target reader in mind, I make far more use of the Stata dialog system than do any other books about Stata. Advanced users may not see the value in using the dialogs, and the more people learn about Stata, the less they will rely on the dialogs. Also, even when you are using the dialog system, it is still important to save a record of the sequence of commands you run. Although I rely on the commands much more than the dialogs in my own work, I still find value in the dialogs. They include many options that I might not have known or might have forgotten.

To illustrate the dialog system as well as graphics, I have included more than 80 figures, many of which show dialog boxes. I present many tables and extensive Stata "results" as they appear on the screen and interpret them substantively in the belief that beginning Stata users need to learn more than just how to produce the results—users also need to be able to interpret them.

I have tried to use real data. There are a few examples where it is much easier to illustrate a point with hypothetical data, but for the most part, I use data that are in the public domain. For example, I use the General Social Surveys for 2002 and 2006 in many chapters, as well as the National Survey of Youth, 1997. I have simplified the files by dropping many of the variables in the original datasets, but I have kept all the observations. I have tried to use examples from several social science fields, and I have included a few extra variables in several datasets so that instructors, as well as readers, can make additional examples and exercises that are tailored to their disciplines. People who are used to working with statistics books that have contrived data with just a few observations, presumably so work can be done by hand, may be surprised to see more than 1,000 observations in our datasets. Working with these files provides better experience for other real-world data analysis. If you have your own data and the dataset has a variety of variables, you may want to use your data instead of the data provided with this book.

The exercises use the same datasets as the rest of the book. Several of the exercises require some data management prior to fitting a model because I believe that learning data management requires practice and cannot be isolated in a single chapter or single set of exercises.

This book takes the student through much of what is done in introductory and intermediate statistics courses. It covers descriptive statistics, charts, graphs, tests of significance for simple tables, tests for one and two variables, correlation and regression, analysis of variance, multiple regression, logistic regression, reliability, and factor analysis. There is also a chapter on constructing scales to measure variables. By combining this coverage with an introduction to creating and managing a dataset, the book will prepare students to go even further. More advanced statistical analysis using Stata is often even simpler from a programming point of view than what we cover. If an intermediate course goes beyond what we do with logistic regression to multinomial logistic regression, for example, the programming is simple enough. The command `logit` can simply be replaced with the command `mlogit`. The added complexity of these advanced statistics is the statistics themselves and not the Stata commands that implement them. Therefore, although more advanced statistics are not included in this book, the reader who learns these statistics will be more than able to learn the corresponding Stata commands from the Stata documentation and help system.

I assume that the reader is running Stata 10, or a later version, on a Windows-based PC. Stata works as well or better on Mac and on Unix systems. Readers who are running Stata on one of those systems will have to make a few minor adjustments. I will note some Mac-specific differences when they are important. In preparing this book, I have used both a Windows-based PC and a Mac.

Corvallis, Oregon
July 2008

Alan C. Acock

Support materials for the book

All the datasets and do-files for this book are freely available for you to download. At the Stata dot prompt, type

```
. net from http://www.stata-press.com/data/agis2/
. net describe agis2
. net get agis2
```

We will use the default Stata data `C:\data` directory for the examples, but you may want to create a new directory and copy the materials there. `C:\data` is the default directory for storing data, but if you have several projects, it may be useful to have a separate folder for each project.

In a net-aware Stata, you can also load the dataset by specifying the complete URL of the dataset. For example,

```
. use http://www.stata-press.com/data/agis2/firstsurvey
```

This text complements the material in the Stata manuals but does not replace it. For example, chapters 5 and 6, respectively, show how to generate graphs and tables, but these are only a few of the possibilities described in the *Stata Reference Manuals*. My hope is to give you sufficient background so that you can use the manuals effectively.

1 Getting started

1.1 Conventions

Listed below are the conventions that are used throughout the book. I thought it might be convenient to list them all in one place should you want to refer to them quickly.

`Typewriter font` I use this font when something would be meaningful to Stata as input. I also use it to indicate Stata output.

I use a typewriter font to indicate the text to type in the Command window. Because Stata commands do not have any special characters at the end, any punctuation mark at the end of a command in this book is not part of the command. Sometimes, I will put a command on a line by itself with the dot preceding it to be consistent with Stata manuals, as in

```
. sysuse cancer, clear
```

All Stata's dialog boxes generate commands, which will be displayed in the Review window and in the Results window. In the Results window, each command will be preceded by the dot prompt. If you make a point of looking at the command Stata prints each time you use the dialog boxes, you will quickly learn the commands. I may include the equivalent command in the text after explaining how to navigate to it through the dialog boxes.

When you type a Stata command in the Command window, you execute the command when you press the Enter key. The command may wrap onto more than one line, but if you press the Enter key in the middle of a command, Stata will interpret that as the end of the command and will probably generate an error. The rule is that you should just keep typing when entering a command in the Command window, no matter how long the command is. Press Enter only when you want to execute the command.

I also use the typewriter font for variable names, for the names of datasets, and to show Stata's output. In general, I use the typewriter font whenever the text is something that can be typed into Stata or when the text is something that Stata might print as output. This approach may seem cumbersome now, but you will catch on quickly.

Folder names, filenames, and filename extensions, as in "The file `survey.dta` is in the `C:\data` directory (or folder)", are also denoted in the typewriter font. Stata assumes that `.dta` will be the extension, so you can use just the filename without an extension, if you prefer.

Sans serif font We use this font to indicate menu items (in conjunction with the ▷ symbol), button names, dialog box tab names, and particular keys:

- Menu items, such as "Select the Data ▷ Variable utilities ▷ Rename variable menu".
- Buttons that can be clicked, as in "Remember, if you are working on a dialog box, it will now be up to you to click on OK or Submit, whichever you prefer".
- Keys on your keyboard, as in "The Page Up and Page Down keys will move you backward and forward through the commands in the Review window".

  Some functions require the use of the Shift, Ctrl, or Alt keys, which are held down while the second key is pressed. For example, Alt+F will open the File menu.

Slant font I use this font when I talk about labeled elements of a dialog box with the label capitalized as it is on the dialog box.

Italics font I use this font when I refer to a word that is to be replaced.

Quotes I use double quotes when I am talking about labels in a general way, but I will use the typewriter font to indicate a specific label in a dataset. For example, if we decided to label the variable `age` "Age at first birth", we would enter `Age at first birth` in the text box.

Capitalization Stata is case sensitive, so `summarize` is a Stata command, whereas `Summarize` is not and will generate an error if you use it. Stata also recognizes capitalization in variable names, so `agegroup`, `Agegroup`, and `AgeGroup` will be three different variables. Although you can certainly use capital letters in variable names, you will probably find yourself making more typographical errors if you do. I have found using all lowercase letters combined with creating good variable names and value labels to be the best practice.

I will also capitalize the names of the various Stata windows, but I do not set them off by using a different font. For example, we will type commands in the Command window and look at the output in the Results window.

1.2 Introduction

The best way to learn data analysis is to actually do it with real data. These days, doing statistics means doing statistics with a computer and a software package. There is no other software package that can match the internal consistency of Stata, which makes it easy to learn and a joy to use. Stata empowers users more effectively than any other statistical package.

Work along with the book

> Although it is not necessary, you will probably find it helpful to have Stata running while you read this book so that you can follow along and experiment for yourself when you have a question about something. Having your hands on a keyboard and replicating the instructions in this book will make the lessons that much more effective, but more importantly, you will get in the habit of just trying something new when you think of it and seeing what happens. In the end, that is how you will really learn how Stata works. The other great advantage to following along is that you can save the examples we do for future use.

Stata is a powerful tool for analyzing data. Stata makes statistics and data analysis fun because it does so much of the tedious work for you. A new Stata user should start by using the dialog boxes. As you learn more about Stata, you will be able to do more sophisticated analyses with Stata commands. You can save commands in files that Stata calls *do-files*, so you can run a series of commands all at once, saving time and providing a record of what you have done. Learning Stata well is an investment that will pay off in saved time later. Stata is constantly being extended with new capabilities, which you can install using the Internet from within Stata. Stata is a program that grows with you.

Stata is a command-driven program. It has a remarkably simple command structure that you use to tell it what you want it to do. You can use a dialog box to generate the commands (this is a great way to learn the commands or prompt yourself if you do not remember one exactly), or you can enter commands directly. If you enter the `summarize` command, you will get a summary of all the variables in your dataset (mean, standard deviation, number of observations, minimum value, and maximum value). Enter the command `tabulate gender`, and Stata will make a frequency distribution of the variable called `gender`, showing you the number and percentage of men and women in your dataset.

After you have used Stata for a while, you may want to skip the dialog box and enter these commands directly. When you are just beginning, however, it is easy to be overwhelmed by all the commands available in Stata. If you were learning a foreign language, you would have no choice but to memorize hundreds of common words right away. This is not necessary when you are learning Stata because the dialog boxes are so easy to use.

Searching for help

Stata can help when you want to quickly find out how to do something. You use the `search` command along with a keyword. Let's assume you already know that a *t* test is for comparing two means. Enter `search t test`; Stata searches its own resources and others that it finds on the Internet. The second entry of the results is

```
[R] ttest . . . . . . . . . . . . . . . . . . . . . Mean-comparison tests
(help ttest)
```

The `[R]` at the beginning of the line means that details and examples can be found in the *Base Reference Manual*. Clicking on the blue `ttest` will take you to the help file for the `ttest` command. If you think this help is too cryptic, repeat the `search t test` command and look further down the list where you will find lines starting with `FAQ` (frequently asked questions). One of these is "What statistical analysis should I use?" Click on the blue URL, and the FAQ will be opened in your web browser. When using the `search` command, you need to pick a keyword that Stata knows. You might have to try different keywords before you get one that works. Searching these Internet locations is a remarkable capability of Stata. If you are reading this book and want to know more about a command, the online help is the first place to start. Suppose we are discussing the command `summarize` and you want to know more options for this command. Type `help summarize` and you will get an informative help screen.

Stata has done a lot to make the dialog boxes as friendly as possible so that you feel confident using them. The dialog boxes often show many options, which control the results that are shown and how they are displayed. You will discover that the dialog boxes have default values that are often all you need, so you may be able to do a great deal of work without specifying any options.

As we progress, you will be doing more complex analyses. You can do these using the dialog boxes, but Stata lets you create files that contain a series of commands you can run all at once. These files, called do-files, are essential once you have many commands to run. You can reopen the do-file a week or even several months later and repeat exactly what you did. Keeping a record of what you do is essential; otherwise, you will not be able to replicate results of elaborate analyses. Fortunately, Stata makes this easy. You will learn more about this in chapter 4. The do-files that reproduce most of the tables, graphs, and statistics for each chapter are available on the web page for this book (http://www.stata-press.com/data/agis2/).

Because Stata is so powerful and easy to use, we may include some analyses that are not covered in your statistics textbook. If you come to a procedure that you have not already learned in your statistics text, give it a try. If it seems too daunting, you can skip that section and move on. On the other hand, if your statistics textbook covers a procedure that we omit, you might search the dialog boxes yourself. Chances are you will find it there.

Depending on your needs, you might want to skip around in the book. Most people tend to learn best when they need to know something, so skipping around to the things you do not know may be the best use of the book and your time. Some topics, though, require prior knowledge of other topics, so if you are new to Stata, you may find it best to work through the first four chapters carefully and in order. After that, you will be able to skip around more freely as your needs or interests demand.

1.3 The Stata screen

When you open Stata, you will see a screen that looks something like figure 1.1.

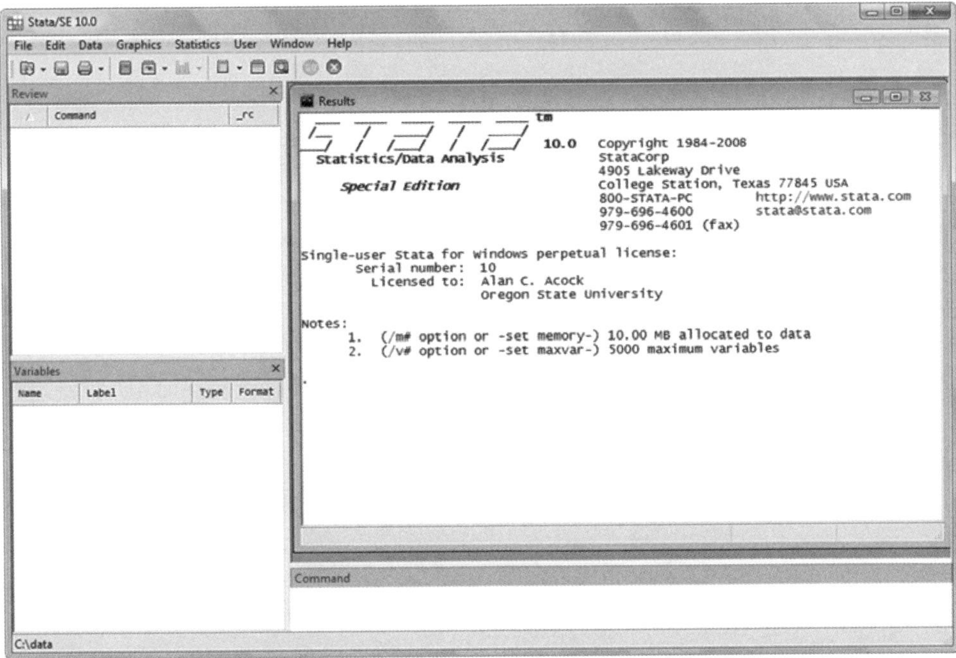

Figure 1.1. Stata's opening screen

You can rearrange the windows to look the way you want them, although many users are happy with the default layout. If you are satisfied with the defaults, you might skip the next couple paragraphs and come back to them if you change your mind later. Many experienced Stata users have particular ways of arranging these screens. Feel free to experiment with the layout. If you have Stata installed on your computer, use the Prefs menu to change your windowing preferences.

Selecting Edit ▷ Preferences ▷ Manage Preferences ▷ Load Preferences ▷ Compact Window Settings will allow you to rearrange your screen so it is more compact. Next select Edit ▷ Preferences ▷ General Preferences... and then click on the tab labeled Windowing. On this tab, you can change the buffer size of the Results window scrollback from the default value of 32,000 to a much bigger value (depending on how much memory your computer has, say, 128,000). Also, check the boxes for *Enable ability to pin or unpin windows*; *Enable ability to dock, undock, or tab windows*; and *Use docking guides*. You might also want to check *Make undocked Viewers persistent*. Once these are checked, click on Apply and then click on the Result Colors tab. There is a drop-down menu for the color scheme where you can change from a *Black background* to a *Blue background*. These are all the changes we will make now, so click on OK.

You can move the sections of the Stata screen around. You may want the section of the screen labeled Command to run the width of the screen. When you click-and-hold the bar across the top of the Command window and move the cursor, several docking icons appear. Drag your cursor to the docking icon on the bottom of the screen and let go. Now your screen should appear as shown in figure 1.2. If you are using a Mac, each window (Results, Review, Variables, and Command) is independent, and you can arrange them any way you like. On a Mac, there are several preferences you get by going to Stata/SE 10.0 ▷ Preferences. To change the scrollback buffer size in the Results window, however, you need to enter the command `set scrollbufsize 128000`, and this takes effect the next time you open Stata.

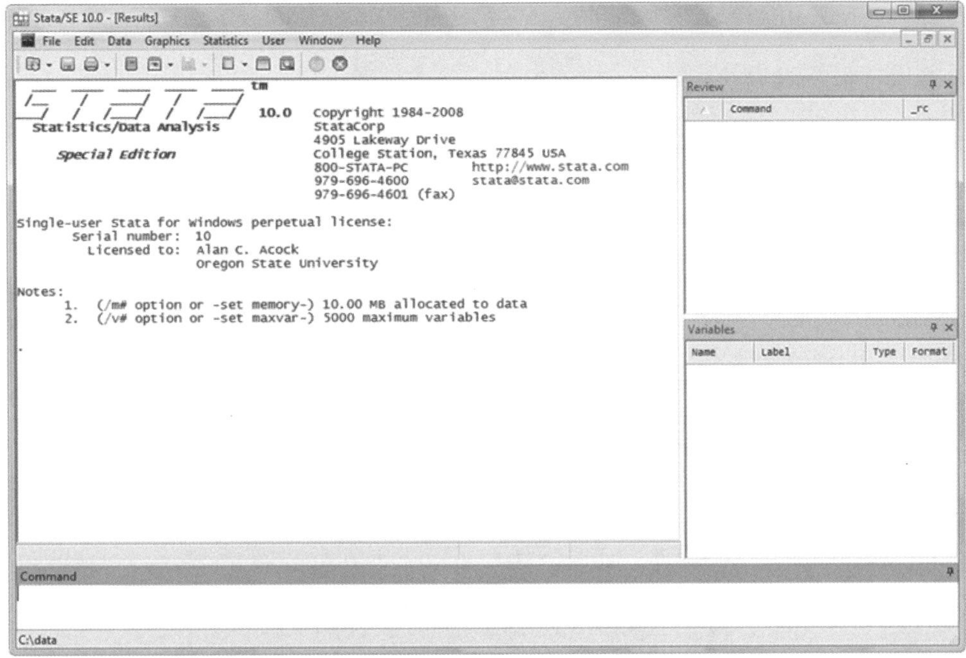

Figure 1.2. Your customized Stata screen

If you are in a computer lab, the computers may be set up to erase all your changes when the computer is restarted. If so, you will not be able to preserve your preferences, but you can rearrange the screen the way you like it when you return to Stata.

When you open a file that contains Stata data, which we will call a Stata dataset, a list of the variables will appear in the Variables window. The Variables window reports the name of the variable (e.g., `abortion`), a label for the variable (e.g., `Attitude toward abortion`), the type of variable (e.g., `float`), and the format of the variable (e.g., `%8.0g`). For now, just consider the name and label. You can vary the width of each column in the Variables window by placing your cursor on the vertical line between the name and label, clicking on it, and then dragging your cursor to the right or left. You can do the same thing to make each window wider.

When Stata executes a command, it prints the results or output in the Results window. First, it prints the command preceded by a `.` (dot) prompt, and then it prints the output. The commands you run are also listed in the Review window. If you click on one of the commands listed in the Review window, it will appear in the Command window. If you double-click on one of the commands listed in the Review window, it will be executed. You will then see the command and its output, if any, in the Results window.

When you are not using the dialog system, you enter commands in the Command window. You can use the Page Up and Page Down keys on your keyboard to recall commands from the Review window. You can also edit commands that appear in the Command window. I will illustrate all of these methods in the coming chapters.

The gray bar at the bottom of the screen, called the status bar, displays the current working directory (folder). This directory will most likely be `C:\data` and is the active folder where Stata will save data, but you can change this folder and store data wherever you wish. For example, if you have a project and want to store all your data in a folder called `\My Documents\thesis`, you could enter the command `cd C:\My Documents\thesis` on the command line. This command assumes this folder already exists on your computer.

Stata has the usual Windows title bar across the top, on the right side of which are the three buttons (in order from left to right) to minimize, to expand to full-screen mode, and to close the program. Immediately below the Stata title bar is the menu bar, where the names of the menus appear. Some of the menu items (File, Edit, and Window) will look familiar because they are used in other programs. The Data, Graphics, and Statistics menus are specific to Stata, but their names provide a good idea of what you will find under them.

Below the menu bar is the toolbar, shown in figure 1.3, which contains icons (or tool buttons) that provide alternate ways to perform some of the actions you would normally do with the menus. If you hold the cursor over any of these icons for a couple of seconds, a brief description of the function appears. The most useful of these are the four leftmost icons. From left to right, you can use these icons to open a dataset, to save a dataset, to print the contents of the Results window (or a Viewer window), and to start or stop a log of your session. The fifth icon from the right opens an editor where you can enter a set of commands. Just to the right of that icon is a spreadsheet view of your data where you can enter data. At the extreme right is a red circle with an X in the middle. This icon is referred to as the Break button because it stops whatever command is being processed. For a complete list of the toolbar icons and their functions, see the *Getting Started with Stata* manual.

Figure 1.3. Stata's toolbar

1.4 Using an existing dataset

Chapter 2 discusses how to create your own dataset, save it, and use it again. You will also learn how to use datasets that are on the Internet. For now, we will use a simple dataset that came with Stata. Although we could use the dialog box to do this, we will enter a simple command. Click once in the Command window to put the cursor there, and then type the command `sysuse cancer, clear`; the Command window should look like the one in figure 1.4.

```
Command
sysuse cancer, clear
```

Figure 1.4. Stata command to open the cancer dataset

The `sysuse` command we just used will find the sample datasets on your computer by name alone, without the extension; in this case, the dataset name is `cancer`, and the file that is actually found is called `cancer.dta`. The `cancer` dataset was installed with Stata. This particular dataset has 48 observations and 4 variables related to a cancer treatment.

What if you forget the command `sysuse`? You could open a file that comes with Stata using the menu File ▷ Example Datasets.... A new window opens in which you click on *Example datasets installed with Stata*. The next window then lists all the datasets that come with Stata. You can click on *use* to open the dataset.

Now that we have some data read into Stata, type `describe` in the Command window. That is it: just type `describe` and press the Enter key, which will yield a brief description of the contents of the dataset.

```
. describe

Contains data from C:\Program Files\Stata10\ado\base/c/cancer.dta
  obs:            48                          Patient Survival in Drug Trial
 vars:             4                          3 Mar 2007 16:09
 size:           576 (99.9% of memory free)
-------------------------------------------------------------------------------
              storage  display     value
variable name   type   format      label      variable label
-------------------------------------------------------------------------------
studytime       int    %8.0g                  Months to death or end of exp.
died            int    %8.0g                  1 if patient died
drug            int    %8.0g                  Drug type (1=placebo)
age             int    %8.0g                  Patient's age at start of exp.
-------------------------------------------------------------------------------
Sorted by:
```

The description includes a lot of information: the full name of the file, `cancer.dta` (including the path entered to read the file), the number of observations (48), the number of variables (4), the size of the file (576 bytes) and how much of Stata's memory is still available, a brief description of the dataset (*Patient Survival in Drug Trial*), and the date the file was last saved. The body of the table displayed shows the names of the variables on the far left and the labels attached to them on the far right. We will discuss the middle columns later.

Now that you have opened the `cancer` dataset, note that the Variables window lists the four variables `studytime`, `died`, `drug`, and `age`.

Internet access to datasets

Stata can use data stored on the Internet just as easily as data stored on your computer. If you did not have the file `cancer` installed on your computer, you could read it by entering `webuse cancer`. However, you are not limited to data stored at the Stata site. Typing `use http://www.ats.ucla.edu/stat/stata/notes/hsb2` will open a dataset stored at UCLA.

1.5 An example of a short Stata session

Since you have loaded the `cancer` dataset, we will execute a basic Stata analysis command. Type `summarize` in the Command window and then press Enter.

Rather than typing in the command directly, you could use the dialog box by selecting Data ▷ Describe data ▷ Summary statistics to open the corresponding dialog box. Simply clicking on the OK button located at the bottom of the dialog box will produce the `summarize` command we just entered. Because we did not enter any variables in the dialog box, Stata assumed we wanted to summarize all the variables in the dataset.

You might want to select specific variables to summarize instead of summarizing them all. Open the dialog box again and click on the pulldown menu within the *Variables* box, located at the top of the dialog box, to display a list of variables. Clicking on a variable name will add it to the list in the box. Dialog boxes allow you to enter a variable more than once, in which case the variable will appear in the output more than once. You can also type variable names in the *Variables* box. A last alternative is to click on the variable name in the Variables window.

If you enter the `summarize` command directly in the Command window, simply follow it with the names of the variables for which you want summary statistics. For example, typing `summarize studytime age` will display only statistics for the two variables named.

In the Results window, the `summarize` command will display the number of observations (also called cases or N), the mean, the standard deviation, the minimum value, and the maximum value for each variable.

```
. summarize

    Variable |       Obs        Mean    Std. Dev.       Min        Max
-------------+--------------------------------------------------------
   studytime |        48        15.5    10.25629          1         39
        died |        48    .6458333    .4833211          0          1
        drug |        48       1.875    .8410986          1          3
         age |        48      55.875    5.659205         47         67
```

The first line of output displays the dot prompt followed by the command. After that, the output appears as a table. As you can see, there were 48 observations in this dataset. *Observations* is a generic term. These could be called participants, patients, or subjects, depending on your field of study. In Stata, each row of data in a dataset is called an observation. The average, or mean, age is 55.875 years with a standard deviation of 5.659, and the subjects are all between 47 (the minimum) and 67 (the maximum) years old. I may round numbers in the text to fewer digits than shown in the output unless it would make finding the corresponding number in the output difficult.

If you have computed means and standard deviations by hand, you know how long this can take. Stata's virtually instant statistical analysis is what makes Stata so valuable. It takes time and skill to set up a dataset so that you can use Stata to analyze it, but once you learn how to set up a dataset in chapter 2, you will be able to compute a wide variety of statistics in little time.

We will do one more thing in this Stata session: we will make the histogram for the variable `age`, shown in figure 1.5.

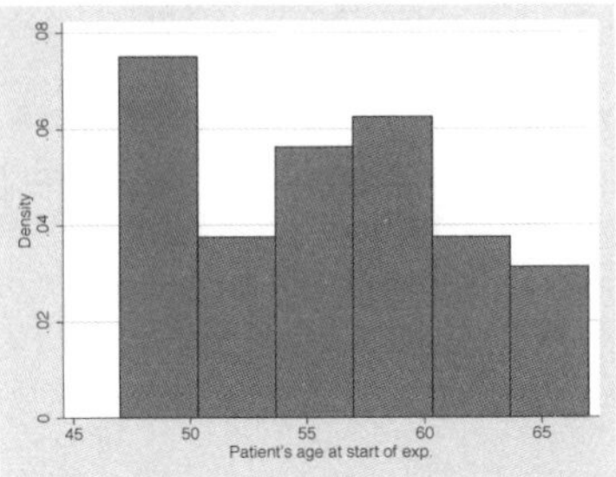

Figure 1.5. Histogram of `age`

A histogram is just a graph that shows the distribution of a variable, such as `age`, that takes on many values.

Simple graphs are simple to create. Just type the command `histogram age` in the Command window, and Stata will produce a histogram using reasonable assumptions. We will show you how to use the dialog boxes for more complicated graphs shortly.

At first glance, you may be happy with this graph. Stata used a formula to determine that six bars should be displayed, and this is reasonable. However, Stata starts the lowest bar (called a bin) at 47 years old, and each bin is 3.33 years wide (this information is displayed in the Results window) although we are not accustomed to measuring years in thirds of a year. Also notice that the vertical axis measures density, but we might prefer that it measure the frequency, that is, the number of people represented by each bar.

Using the dialog box can help us customize our histogram. Let's open the `histogram` dialog box shown in figure 1.6 by selecting Graphics ▷ Histogram from the menu bar.

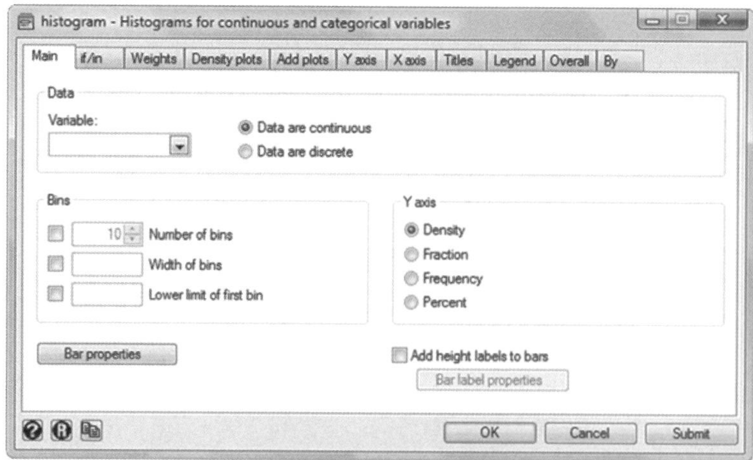

Figure 1.6. The **histogram** dialog box

Let's quickly go over the parts of the dialog box. There is a text box labeled *Variable* with a pulldown menu. As we saw on the **summarize** dialog, you can pull down the list of variables and click on a variable name to enter it in the box, or you can type the variable's name yourself. Only one variable can be used for a histogram, and here we want to use age. If we stop here and click on OK, we will have recreated the histogram shown in figure 1.5.

There are two radio buttons visible to the right of the *Variable* box, one labeled *Data are continuous* (which is shown selected in figure 1.6), and one labeled *Data are discrete*. Radio buttons indicate mutually exclusive items—you can choose only one of them. In this case, we are treating age as if it were continuous, so make sure that is the radio button we select. On the right side of the Main tab is a section labeled *Y axis*. Click on the radio button for *Frequency* so that the histogram shows the frequency of each interval. In the section labeled *Bins*, check the box labeled *Width of bins* and type `2.5` in the text box that becomes active (because the variable is age, the 2.5 indicates 2.5 years). Also check the box labeled *Lower limit of first bin* and type 45, which will be the smallest age represented by the bar on the left.

The dialog box shows a sequence of tabs just under its title bar, as shown in figure 1.7. Different categories of options will be grouped together, and you make a different set of options visible by clicking on each tab. The options you have set on the current tab will not be canceled by clicking on another tab.

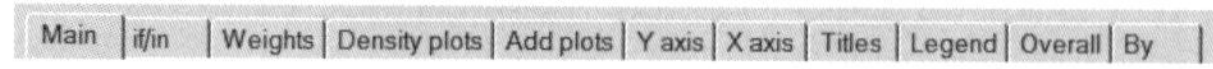

Figure 1.7. The tabs on the **histogram** dialog box

Graphs are usually clearer when there is a title of some sort, so click on the Titles tab and add one. Here we type `Age Distribution of Participants in Cancer Study` in the *Title* box. Let's add the text `Data: Sample cancer dataset` to the *Note* box so that we know which dataset we used for this graph. Your dialog box should look like figure 1.8.

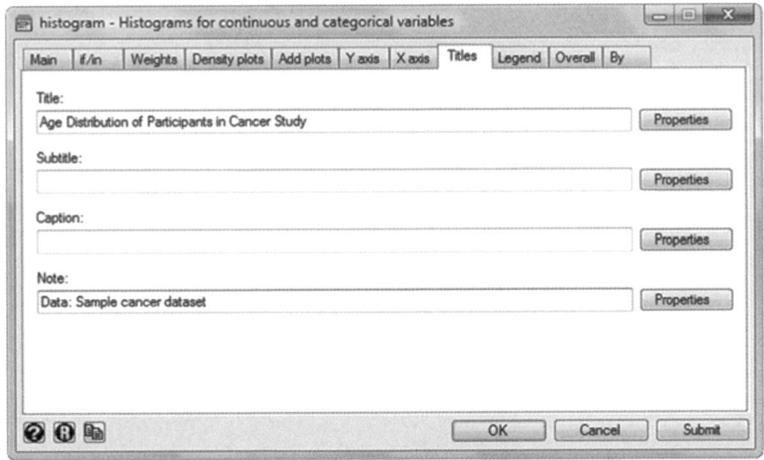

Figure 1.8. The Titles tab of the `histogram` dialog box

Now click on the Overall tab. Let's select *s1 monochrome* from the pulldown menu on the *Scheme* box. Schemes are basically templates that determine the standard attributes of a graph, such as colors, fonts, and size; which elements will be shown; and more.

From the Legend tab, under the *Legend behavior* section, click on the radio button for *Show legend*. Whether a legend will be displayed is determined by the scheme that is being used, and if we were to leave *Default* checked, our histogram might have a legend or it might not, depending on the scheme. Choosing *Show legend* or *Hide legend* overrides the scheme, and our selection will always be honored.

Now that we have made these changes, click on Submit instead of OK to generate the histogram shown in figure 1.9. The dialog box does not close. To close the dialog box, click on the X (close) button in the upper right corner, but we are not ready to do that yet.

(*Continued on next page*)

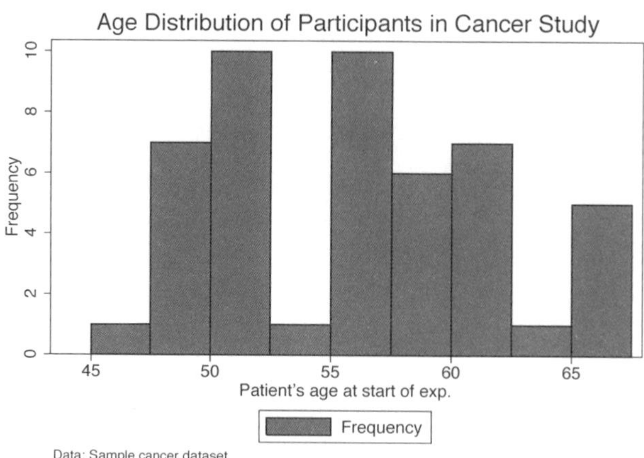

Figure 1.9. First attempt at an improved histogram

If you look at the complex command that the dialog box generated, you will see why even experienced Stata programmers will often rely on the dialog box to create `graph` commands. In reading this command, you will want to ignore the opening dot (Stata prints this in front of commands in the Results window, but the dot is not part of the command and you do not type it). Stata prints the > sign at the start of the second and third line and this might be confusing. Stata uses the Enter key to submit a command. Because of this, Stata sees the entire command as one line. To print the entire line in the confines of the Results window, Stata inserts the > for a line break. If you wanted to enter this command in the Command window, you would simply type the entire thing without the > and let Stata do the wrapping as needed in the Command window. Never press the Enter key until you have entered the entire command.

```
. histogram age, width(2.5) start(45) frequency
> title(Age Distribution of Participants in Cancer Study)
> note(Data: Sample cancer dataset) legend(on) scheme(s1mono)
```

It is much more convenient to use the dialog box to generate that command than to try to remember all its parts and the rules of their use. If you do want to enter a long command in the Command window, remember to type it as one line. Whenever you press Enter, Stata assumes that you have finished the command and are ready to submit it for processing.

When to use Submit and when to use OK

> Stata's dialogs give you two ways to run a command: by clicking on OK or by clicking on Submit. If you click on OK, Stata creates the command from your selections, runs the command, and closes the dialog box. This is just what you want for most tasks. At times, though, you know you will want to make minor adjustments to get things just right, so Stata provides the Submit button, which still runs the command but leaves the dialog open. This way, you can go back to the dialog box and make changes without having to reopen the dialog box.

The resulting histogram in figure 1.9 is an improvement, but we might want fewer bins. Here we are making small changes to a Stata command, then looking at the results, and then trying again. The Submit button is useful for this kind of interactive, iterative work. If the dialog box is hidden, we can use the Alt+Tab key combination to move through Stata's windows until the one we want is on top again.

Instead of a width of 2.5 years, let's use 5 years, which is a more common way to group ages. If you clicked on OK instead of on Submit, you need to reopen the `histogram` dialog box as you did before. When you return to a dialog that you have already used in the current Stata session, the dialog box reappears with the last values still there. So all you need to do is change `2.5` to `5` in the *Width of bins* box on the Main tab and click on Submit. The result is shown in figure 1.10.

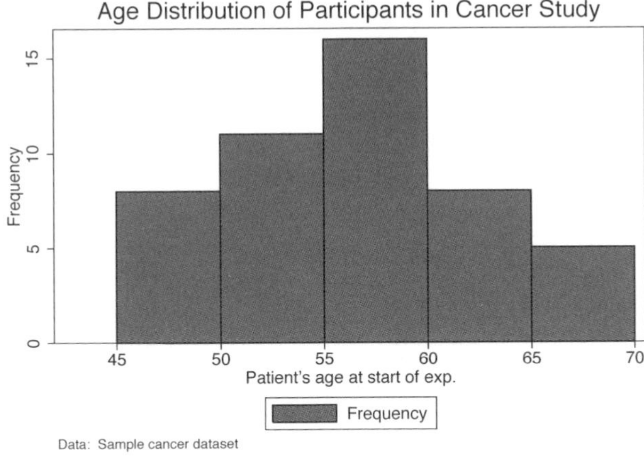

Figure 1.10. Final histogram of age

Notice that these graphs are different, so you need to use judgment to pick the best combination and avoid using a graph that misrepresents a distribution. Can you think of any more improvements? The legend at the bottom, center of the graph is unnecessary.

You might want to go back to the dialog box, click on the Legend tab, and click on *Hide legend* to turn off the legend.

To finish our first Stata session, we need to close Stata. Do this with File ▷ Exit.

1.6 Summary

We covered the following topics in this chapter:

- The font and punctuation conventions we will use throughout the book
- The Stata interface and how you can customize it
- How to open a sample Stata dataset
- The parts of a dialog box and the use of the OK and Submit buttons
- How to summarize the variables
- How to create and modify a simple histogram

1.7 Exercises

Some of these exercises involve little or no writing; they are simply things you can do to make sure you understand the material. Other exercises may require a written answer.

1. You can copy and paste text to and from Stata as you wish. You should try highlighting some text in Stata's Results window, copying it to the clipboard, and pasting it into another program, such as your word processor. To copy highlighted text, you can use the Edit ▷ Copy menu or, as indicated on the menu, Ctrl+C. You will probably need to change the font to a monospaced font (e.g., Courier), and you may need to reduce its font size (e.g., to 9 point) after pasting it to prevent the lines from wrapping. You may wish to experiment with copying Stata output into your word processor now so that you know which font size and typeface work best. It may help to use a wider margin, such as 1 inch, on each side.

2. After you highlight material in the Results window, right-click on it. You can save this output in several formats, including Copy Text, Copy Table (only works with some commands), Copy Table as HTML, and Copy as Picture (copies a graphic image of what you highlighted). The Copy Text option works nicely, but you will need to use a monospaced font, such as Courier, and may need to use a smaller font size when you paste it into your word processing document. The Copy Table option is limited because it only works with a few commands. The Copy Table as HTML option will create a table that looks like what you would see on a web page. Using Word, you can edit the table by making columns wider or less wide and by aligning the columns so each number has the same number of decimal places. Just

copy the tabular results and not the command when using the HTML option. The Copy as Picture option works nicely in Windows, but you cannot edit it in Word because it is a graphic image, so the columns may not line up correctly and you may have a different number of decimal places for some variables. In Word, you can resize the image.

Run the `summarize` command and copy the results to a Word document by using each of the options: Highlight the table. Right-click on it and then select the option you want. Switch to your word processor. Press Ctrl+V to paste what you copied. In your word processor, make the table as nice as you can by adjusting the font, font size, spacing, etc.

3. Stata has posted all the datasets from its manuals that were used to illustrate how to do procedures. You can access the manual datasets from within Stata by going to the File ▷ Example Datasets... menu, which will open a Viewer window. Click on *Stata 10 manual datasets* and then click on *User's Guide [U]*.

   The Viewer works much like a web browser, so you can click on any of the links in the list of datasets. Scroll down to chapter 25, and select the `use` link for `states3.dta`, which opens a dataset that is used for chapter 25 of the *User's Guide*. Run two commands, `describe` and `summarize`. What is the variable `divorce_rate` and what is the mean (average) divorce rate for the 50 states?

4. Open the `cancer` dataset. Create histograms for `age` using bin widths of 1, 3, and 5. Use the right mouse button to copy each graph to the clipboard, and then paste it into your word processor. Does the overall shape of the histogram change as the bins get wider? How?

5. UCLA has a Stata portal containing a great deal of helpful material about Stata. You might want to browse the collection now just to get an idea of which topics are covered there. The URL for the main UCLA Stata page is

   http://www.ats.ucla.edu/stat/stata/

   In particular, you might want to look at the links listed under Learning Stata. On the Stata Starter Kit page, you will find a link to Class notes with movies. These movies demonstrate using Stata's commands rather than the dialog box. The topics we will cover in the first few chapters of this book are also covered on the UCLA web page using the commands. Each movie is about 25 minutes long. Some of these are for older versions of Stata, but they are still useful.

2 Entering data

2.1 Creating a dataset

In this chapter, you will learn how to create a dataset. Data entry and verification can be tedious, but these tasks are essential for you to get accurate data. More than one analysis has turned out wrong because of errors in the data.

In the first chapter, you worked with some sample datasets. Most of the analyses in this book use datasets that have already been created for you and are available on the book's web page, or they may already be installed on your computer. If you cannot wait to get started analyzing data, you can come back to this chapter when you need to create a dataset. You should look through the sections in this chapter that deal with documenting and labeling your data because not all prepared datasets will be well documented or labeled.

The ability to create and manage a dataset is valuable. People who know how to create a dataset are valued members of any research team. For small projects, collecting, entering, and managing the data can be straightforward. For large, complex studies that extend over many years, managing the data and the documentation can be as complex as any of the analyses that may be conducted.

Stata, like most other statistical software, almost always requires data that are set in a grid, like a table, with rows and columns. Each row contains data for an observation (which is often a subject or a participant in your study), and each column contains the measurement of some variable of interest about the observation, such as age. This system will be familiar to you if you have used a spreadsheet.

Variables and Items

In working with questionnaire data, an item from the questionnaire almost always corresponds to a variable in the dataset. So if you ask for a respondent's age, then you will have a variable called `age`. Some questionnaire items are designed to be combined to make scales or composite measures of some sort, and new variables will be created to contain those items, but there is no single item corresponding to the scale or construct on the questionnaire. A questionnaire may also have an item where the respondent is asked to "mark all that apply", and commonly there is a variable for each category that could be marked under that item. The terms *item* and *variable* are often used interchangeably, but they are not synonyms. An item always refers to one question. A variable may be the score on an item or a composite score based on several items.

There is more to a dataset than just a table of numbers. Datasets usually contain labels that help the researcher use the data more easily and efficiently. In a dataset, the columns correspond to variables, and variables must be named. We can also attach to each variable a descriptive label, which will often appear in the output of statistical procedures. Since most data will be numbers, we can also attach value labels to the numbers to clarify what the numbers mean.

It is extremely helpful to pick descriptive names for each item. For example, you might call a variable that contains people's responses to a question about their state's schools `q23`, but it is often better to use a more descriptive name, such as `schools`. If there were two questions about schools, you might want to name them `schools1` and `schools2` rather than `q23a` and `q23b`. If a set of items are to be combined into scales and are not intended to be used alone, you may want to use names that correspond to the questionnaire items for the original variables and reserve the descriptive names for the composite scores. This is useful with complex datasets where several people will be using the same dataset. Each user will know that `q23a` refers to question `23a` whereas "friendly" names like `schools1` may make sense to one user but not to another user. No matter what the logic of your naming, try to keep names short. You will probably have to type them often, and shorter names offer fewer opportunities for typing errors. A respondent's age can be named `age` or `age_of_respondent` (notice no blank spaces are allowed in a name). One problem with the longer name is that some of Stata's tabular output is designed for short names and the longer names will be truncated.

Even with relatively descriptive variable names, it is usually helpful to attach a longer, and one hopes more descriptive, label to each variable. We call these *variable labels* to distinguish them from *variable names*. For example, you might want to label the variable `schools` with the label "Public school rating". The variable label gives us a clearer understanding of the data stored in that variable.

For some variables, the meaning of the values in the data is obvious. If you measure people's height in inches, then when you see the values, it is clear what they mean. For other variables, the meaning of the values needs to be specified with value labels. Most of us have run across questions that ask us if we "Strongly agree", "Agree", "Disagree", or "Strongly disagree", or that ask us to answer some question with "Yes" or "No". This sort of data is usually coded as numbers, such as 1 for "Yes" and 2 for "No", and it can make understanding tables of data much easier if you create *value labels* to be displayed in the output in addition to (or instead of) the numbers.

If your project is just a short homework assignment, you could consider skipping the labeling (though your instructor would no doubt appreciate anything that makes your work easier to understand). For any serious work though, clear labeling will make your work much easier in the long run. Remember, we need a variable name, a descriptive label for the variable, and sometimes labels for the values the variable can have.

2.2 An example questionnaire

We have discussed datasets in general, so now let's create one. Suppose that we conducted a survey of 20 people and asked each of them six questions, which are shown in the example questionnaire in figure 2.1.

(*Continued on next page*)

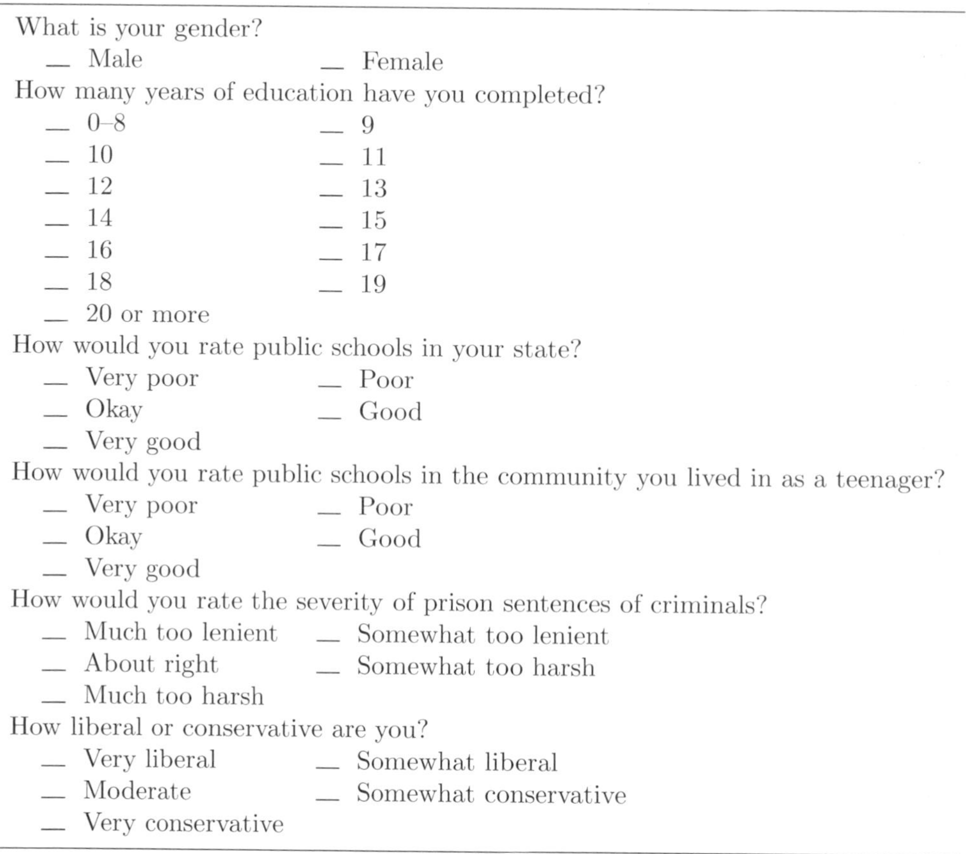

What is your gender?
- __ Male __ Female

How many years of education have you completed?
- __ 0–8 __ 9
- __ 10 __ 11
- __ 12 __ 13
- __ 14 __ 15
- __ 16 __ 17
- __ 18 __ 19
- __ 20 or more

How would you rate public schools in your state?
- __ Very poor __ Poor
- __ Okay __ Good
- __ Very good

How would you rate public schools in the community you lived in as a teenager?
- __ Very poor __ Poor
- __ Okay __ Good
- __ Very good

How would you rate the severity of prison sentences of criminals?
- __ Much too lenient __ Somewhat too lenient
- __ About right __ Somewhat too harsh
- __ Much too harsh

How liberal or conservative are you?
- __ Very liberal __ Somewhat liberal
- __ Moderate __ Somewhat conservative
- __ Very conservative

Figure 2.1. Example questionnaire

Our task is to convert the questionnaire's answers into a dataset that we can use with Stata.

2.3 Developing a coding system

Statistics is most often done with numbers, so we need to have a numeric coding system for the answers to the questions. Stata can use numbers or words. For example, we could type `Female` if a respondent checked that box. However, it is usually better to use some sort of numeric coding, so you might type `1` if the respondent checked the Male box on the questionnaire and 2 if the respondent checked Female. We will need to assign a number to enter for each possible response for each of the items on the survey.

You will also need a short variable name for the variable that will contain the data for each item. Variable names can contain uppercase and lowercase letters, numerals, and the underscore character, and they can be up to 32 characters long. No blank spaces are allowed for variable names. The variable name `mother age` would be interpreted as two variables, `mother` and `age`. Generally, you should keep your variable names to 10 characters or fewer, but 8 or fewer is best. Variable names should start with a letter.

If appropriate, you should explain the relationship between any numeric codes and the responses as they appeared on the questionnaire. For an example, see the example codebook (not to be confused with the Stata command `codebook`, which we will use later) for our questionnaire that appears in table 2.1.

(Continued on next page)

Table 2.1. Example codebook

Question	Variable name	Value labels	Code
Identification number			
	id	Record in order	1 to 20
What is your gender?			
	gender	Male	1
		Female	2
		No answer	−9
How many years of education have you completed?			
	education	0–8	8
		9–20	9 to 20
		No answer	−9
Rate public schools in your state.			
	sch_st	Very poor	1
		Poor	2
		Okay	3
		Good	4
		Very good	5
		No answer	−9
Rate public schools in the community you lived in as a teenager.			
	sch_com	Very poor	1
		Poor	2
		Okay	3
		Good	4
		Very good	5
		No answer	−9
Rate severity of prison sentences of criminals. The are ...			
	prison	Much too lenient	1
		Too lenient	2
		About right	3
		Too harsh	4
		Much too harsh	5
		No answer	−9
How liberal or conservative are you?			
	conserv	Very liberal	1
		Liberal	2
		Moderate	3
		Conservative	4
		Very conservative	5
		No answer	−9

A codebook translates the numeric codes in your dataset back into the questions you asked your participants and the choices you gave them for answers. Regardless of whether you gather data with a computer-aided interviewing system or with paper questionnaires, the codebook is essential to help make sense of your data and your analyses. If you do not have a codebook, you might not realize that everyone with eight or fewer years of education is coded the same way, with an 8. That may have an impact on how you use that variable in later analyses.

We have added an `id` variable to identify each respondent. In simple cases, we just number the questionnaires sequentially. If we have a sample of 5,000 people, we will number the questionnaires from 1 to 5,000, write the identification number on the original questionnaire, and record it in the dataset. If we discover a problem in our dataset (e.g., somebody with a coded value of `3` for `gender`), we can go back to the questionnaire and determine what the correct value should be. Some researchers will eventually destroy the link between the ID and the questionnaire as a way to protect human participants. It is good to keep the original questionnaires in a safe place that is locked and accessible only by members of the research team.

Some data will be missing—people may refuse to answer certain questions, interviewers forget to ask questions, equipment fails; the reasons are many, and we need a code to indicate that data are missing. If we know why the answer is missing, we will record the reason too, so we may want to use different codes that correspond to the different reasons the data are missing.

On surveys, respondents may refuse to answer, may not express an opinion, or may not have been asked a particular question because of the answer to an earlier question. Here we might code "invalid skip" (interviewer error) as `-5`, "valid skip" (not applicable to this respondent) as `-4`, "refused to answer" as `-3`, "don't know" as `-2`, and "missing for any other reason" as `-1`. For example, adolescent boys should not be asked when they had their first menstrual period, so we would enter a `-4` for that question if the respondent is male. We should pick values that can never be a valid response. In chapter 3, we will redefine these values to Stata's missing value codes. In this chapter, we will use only one missing value code, `-9`.

We will be entering the data ourselves, so after we administered the questionnaire to our sample of 20 people, we prepared a coding sheet that will be used to enter the data. The coding sheet originates from the days when data were entered by professional keypunch operators, but it can still be useful or necessary. When you create a coding sheet, you are converting the data from the format used on the questionnaire to the format that will actually be stored in the computer (a table of numbers). The more the format of the questionnaire differs from a table of numbers, the more likely it is that a coding sheet will help prevent errors.

There are other reasons you may want to use a coding sheet. For example, your questionnaires may include confidential information that should be shown to as few people as possible; also, if there are many open-ended questions for which extended answers were collected but will not be entered, working from the original questionnaires can be unwieldy.

In general, if you transcribe the data from the questionnaires to the coding sheet, you will need to decide which responses go in which columns. Deciding this will reduce errors from those who perform the data entry, who may not have the information needed to make those decisions properly.

Whether you enter the data directly from the questionnaires or create a coding sheet will depend largely on the study and on the resources that are available to you. Some experience with data entry is valuable because it will give you a better sense of the problems you may encounter, whether you or someone else enters the data; so, for our example questionnaire, we have created a coding sheet shown in table 2.2.

Table 2.2. Example coding sheet

id	gender	education	sch_st	sch_com	prison	conserv
1	2	15	4	5	4	2
2	1	12	2	3	1	5
3	1	16	3	4	3	2
4	1	8	-9	1	-9	5
5	2	12	3	3	3	3
6	2	18	4	5	5	1
7	1	17	3	4	2	4
8	2	14	2	3	1	5
9	2	16	5	5	4	1
10	1	20	4	4	5	2
11	1	12	2	2	1	5
12	2	11	-9	1	1	3
13	2	18	5	5	-9	-9
14	1	16	5	5	5	1
15	2	16	5	5	4	2
16	1	17	4	3	4	3
17	1	12	2	2	-9	1
18	2	12	2	2	2	2
19	2	14	4	5	4	4
20	1	13	3	3	5	5

Because we are not reproducing the 20 questionnaires in this book, it may be helpful to examine how we entered the data from one of them. We will use the ninth questionnaire. We have assigned an `id` of `9`, as shown in the first column. Reading from left to right, in the second column, for `gender`, we have recorded a `2` to indicate a woman, and for `education`, we have recorded a response of `16` years. This woman rates schools in her state and in the community in which she lived as a teenager as very good, which we can see from the 5s in the fourth and fifth columns. She thinks prison sentences are too harsh and she considers herself very liberal (4 and 1 in the sixth and seventh columns, respectively).

2.4 Entering data

We will use Stata's Data Editor to enter our data from the coding sheet in table 2.2. The Data Editor provides an interface similar to that of a spreadsheet, but it has some features that are particularly suited to creating Stata datasets. Before opening the Data Editor, you might want to save any open files and then enter the command `clear` in the Command window. This step will give you a fresh Data Editor in which to enter data. Enter the `clear` command only if you want to start with a new dataset that has nothing in it. To open the Data Editor, click on Data ▷ Data Editor. There are several other ways you can open the Data Editor: use the shortcut on the toolbar (it looks like a spreadsheet), press Ctrl+7 on your keyboard, or type `edit` in the Command window. The Data Editor window is shown in figure 2.2.

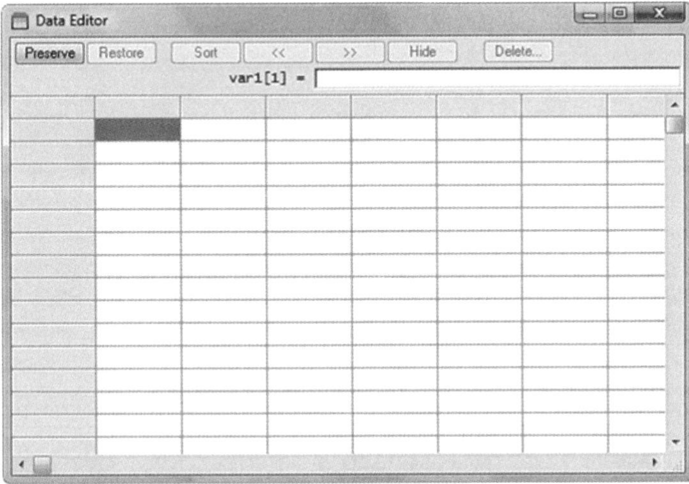

Figure 2.2. Data Editor window

Data are entered in the white columns, just as in other spreadsheets. Here we will only enter the data for the first respondent to the example survey. In the first white cell under the first column, enter the identification number of the first respondent, which is a `1`. Press the Tab key to move across to the next cell, and enter the value from the second column of the coding sheet. This value is 2 because the first case is a woman. Keep entering the values and pressing Tab until we have entered all the values for the first participant. After we have entered the number from the last number in the first row of the coding sheet, press Enter, which will move the cursor to the second row of the Data Editor. Press the Home key to move to the first column of this row.

Let's interrupt the data entry here, after we have the data entered for the first respondent. When we create a new dataset by entering data into the Data Editor, Stata assigns each column a default variable name. As you have probably noticed, the columns are called `var1`, `var2`, `var3`, ..., `var7`. Say that we want to change these to the names recorded in the codebook. We also composed a label for each variable as part of our codebook, and we should enter those, too.

Double-click on the gray cell at the top of the first column, which now contains the variable name `var1`. Double-clicking on this cell opens the *Variable Properties* dialog box. Figure 2.3 shows how we can use this dialog box to change the variable name to `id` and the label to `Identification number`.

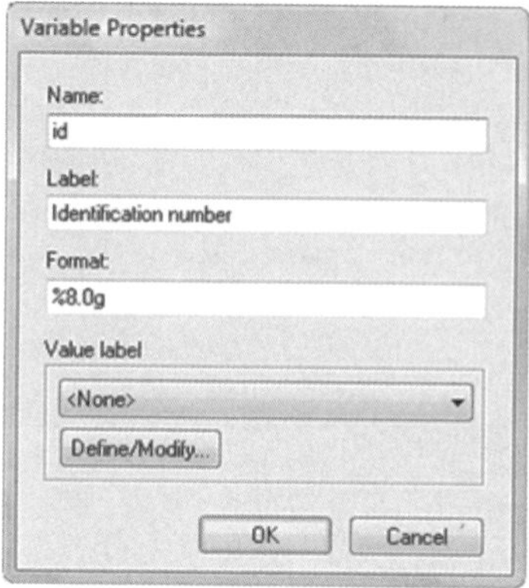

Figure 2.3. Variable name and variable label

Click OK and then close the dialog box. Double-click on each of the remaining generic variable names to rename and label them as listed in table 2.3.

Table 2.3. New variable names and labels

Generic variable name	New variable name	Variable label
`var2`	`gender`	Participant's gender
`var3`	`education`	Years of education
`var4`	`sch_st`	Ratings of schools in your state
`var5`	`sch_com`	Ratings of schools in your community of origin
`var6`	`prison`	Rating of prison sentences
`var7`	`conserv`	Conservatism/liberalism

As well as allowing us to specify the variable name and variable label, the dialog box includes a box where we can specify the format. The current format is `%8.0g`, which is the default format for numeric data in the Data Editor. This format instructs Stata to try to print numbers within eight columns. Stata will adjust the format if you enter a number wider than eight columns in the Data Editor.

I entered the letter `l` for the number `1`

A common mistake is to enter a letter instead of a numeric value, such as typing the letter `l` rather than the number `1` or simply pressing the wrong key. When this happens for the first value you enter for a variable, Stata will make the variable a "string" variable, meaning that it is no longer a numeric variable. When this happens for a subsequent value you enter for a variable, Stata will give you an error message so you can make the correction. If you made a mistake of making a variable a string variable when it should be a numeric variable, you need to correct this. Simply changing the letter to a number will not work because a string variable may include numbers such as a variable for a street address. Stata will still think it is a string variable. There is a simple command to force Stata to change the variable from a string variable back to a numeric variable: `destring` *varlist*`, replace`. For example, to change the variable `id` from string to numeric, you would type `destring id, replace`.

If you have string data (that is, data that contain letters, numbers, or symbols), you will see a format that looks like `%9s`, which indicates a variable with nine characters that are strings (the `s`). For more detailed information about formats, see the *Stata Data Management Reference Manual*.

All our data here are numeric, none of the data contain decimals, and none are wider than eight digits, so we can leave the format alone. Once you have defined a variable as numeric, Stata will warn you if you try to enter data that are not numeric, which can help reduce errors. Be careful to always enter the number `1` and not the lowercase letter `l`.

When you learn how to write Stata programs in do-files, you might want to enter the labels by direct commands without using the dialog box. For example, if we wanted to change the variable named `prison` to `prisons` and then change the label for the variable to "Prison harshness", here are the commands you would use:

```
. rename prison prisons
. label variable prisons "Prison Harshness"
```

The examples in this chapter do not reflect the changes that these commands produce.

If you want to rename a variable, change a variable label, or add a variable label, you can do this directly from the Variables window by right-clicking on the variable. For example, if we right-click on `gender` under the Variables window (not the Data Editor), we get the menu shown in figure 2.4. This menu lists several options including renaming `gender`, editing the variable label for `gender`, dropping `gender` from the dataset, or adding a note about `gender`.

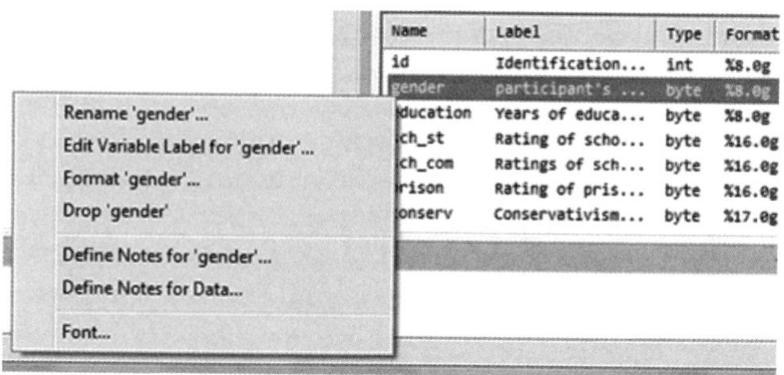

Figure 2.4. Renaming a variable or editing its variable label

2.4.1 Labeling values

It is a good idea to enter the value labels described in table 2.1. Some people do not do this and try to remember what the values mean. You can remember that 1 is the code for a male and 2 the code for a female, but if the questionnaire contains hundreds of items, remembering all the values would be impossible. You could rely on the codebook, but checking the codebook each time you come across a variable you are not sure of will take valuable time.

Creating value labels pays off if you are going to use the dataset more than a few times or if you share a dataset with others. Output will be nicely labeled, making it much easier for you to read and understand your results. Value labels are especially helpful for others on the team who may not work with the dataset regularly or who work only with some parts of the dataset regularly.

Value labels are used when the numbers have some language equivalent, such as a scale from "strongly disagree" to "strongly agree". If we record people's ages, there is no label we could sensibly use for 16. At times, the number of distinct values to be labeled can be daunting (e.g., diagnostic codes).

Labeling values in Stata is done in two stages. In the first stage, we create the label mapping where the text and the value to which it is attached are saved. In the second stage, we assign the label mapping to the appropriate variable. Label mappings are given names, which can be the same as variable names or might describe the type of scale or measurement being labeled. The first step involves creating a mapping for each set of labels in your dataset. In the following example, we will create a mapping called `sex` using 1 for male, 2 for female, and −9 for missing.

At the bottom of figure 2.3—just above OK—you see Define/Modify.... Click on Define/Modify..., which opens the *Define value labels* dialog box in which we enter the name for the new label mapping, `sex`. The name can contain letters, numerals, and the underscore—the same characters that can be used for variable names. Type the name `sex` because we will use this mapping for the gender variable. Click on OK, and a new dialog box opens.

Enter the first value, `1`, and text for the label, `male`. Click on OK, and enter the second value, `2`, and the text for its label, `female`. After clicking on OK, enter the third value, `-9`, and its label, `missing`, and click on OK. Because this is the last value for this mapping, click on Cancel to go back to the *Define value labels* dialog box. Figure 2.5 shows the results.

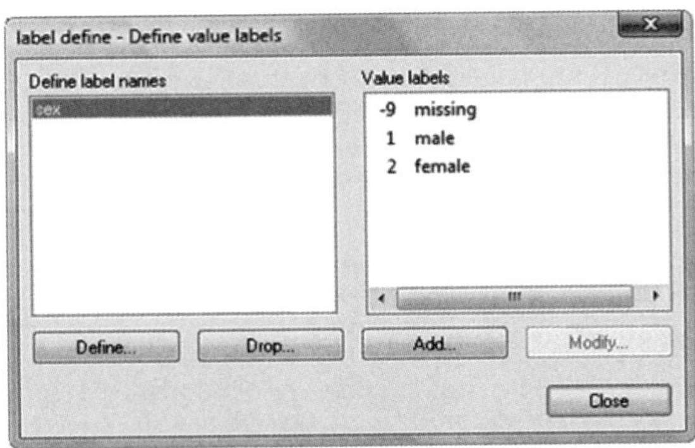

Figure 2.5. Define value labels dialog box

Repeat this process to define the other mappings: Create a mapping called `harsh` using 1 for much too lenient, 2 for too lenient, 3 for about right, 4 for too harsh, 5 for much too harsh, and −9 for missing. Create a mapping called `rating` using 1 for very poor, 2 for poor, 3 for okay, 4 for good, 5 for very good, and −9 for missing. Finally, create a mapping called `conserv` using 1 for very liberal, 2 for liberal, 3 for moderate, 4 for conservative, 5 for very conservative, and −9 for missing. When you are finished, figure 2.5 becomes figure 2.6.

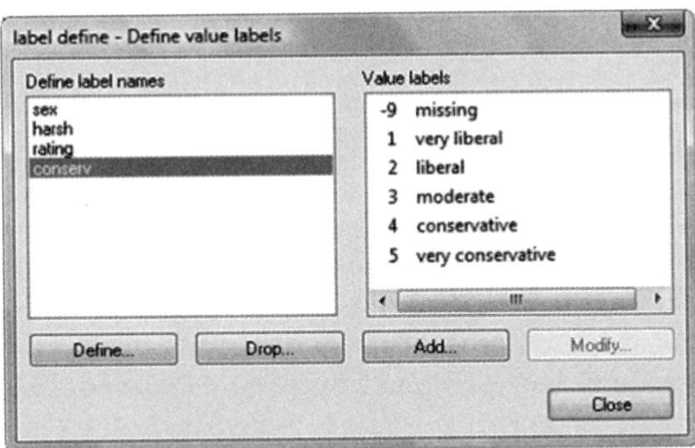

Figure 2.6. Define value labels dialog box

Close this dialog box, and we will return to the original dialog in figure 2.3. This time, click on OK to close this dialog box.

Double-click on the variable name `gender` to open the *Variable Properties* dialog box for this variable. Click on the down-arrow to show the list of value-label mappings; click on `sex` to assign that mapping to the variable `gender`. This is the second step in which we are assigning the value-label mappings to the variables. The dialog box for doing this mapping to `gender` is in figure 2.5. Repeat this process for each variable to assign the mapping `rating` to both `sch_st` and `sch_com`, the mapping `harsh` to the variable `prison`, and the mapping `conserv` to the variable `conserv`. You can now close the Data Editor window.

Going through this series of dialog boxes can be confusing, so it may be easier to simply enter the commands. Fortunately, this is one of the most confusing data management tasks we will cover. The key thing to remember is that assigning label mappings is a two-step process. The first step defines a different mapping for each set of labels. The second step assigns these to the appropriate variables. Here is how we would enter the first step using a series of Stata commands:

```
. label define sex 1 "male" 2 "female" -9 "missing"
. label define harsh 1 "much too lenient" 2 "too lenient"
. label define harsh 3 "about right" 4 "too harsh", add
. label define harsh 5 "much too harsh" -9 "missing", add
. label define rating 1 "very poor" 2 "Poor" 3 "okay"
. label define rating 4 "good" 5 "very good ", add
. label define rating -9 "missing", add
. label define conserv 1 "very liberal" 2 "liberal" 3 "moderate"
. label define conserv 4 "conservative" 5 "very conservative", add
```

Why do we show the dot prompt with these commands?

When we show a listing of Stata commands, we place a dot and a space in front of each command. When you enter these commands in the Command window, you enter the command itself and not the dot prompt or space. We include these because Stata always shows commands this way in the Results window. Also, Stata manuals and many other books about Stata follow this convention.

The example above included the `add` option to the end of some of the `label define` commands, which indicates that we are adding labels to the labeling mappings and not trying to redefine them. What happens if we make a mistake in how we enter a label? The definitions above assigned the text "Poor" to the `rating` mapping value of 2. Because we have not capitalized any of the other labels, we might want to change this to lowercase "poor". To do this, we can enter the following command with the option `modify`. Stata tries to protect us from ourselves and will not let us change the definition without explicitly including the `modify` option.

```
. label define rating 2 "poor", modify
```

The second step is to assign these value labels to the appropriate variables. If we do not want to go through the dialog boxes, we can enter these commands directly:

```
. label values gender sex
. label values sch_st rating
. label values sch_com rating
. label values prison harsh
. label values conserv conserv
```

The command is `label values`. This is followed by the name of the variable, e.g., `gender`, and then the label, e.g., `sex`.

You may have used other statistical packages that do not use this two-step process. Assigning value labels to each variable, one at a time, would be reasonable for a dataset that has just a few variables, such as the example dataset. When you are dealing with complex datasets, dozens of variables may have the same mapping, such as "strongly agree" to "strongly disagree" or yes/no options. A complex dataset might have just 10 mappings but it might need to assign those 10 mappings to 200 or more items. Once you have defined each mapping, you can assign them to all the variables for which they apply. This is an efficient approach for all but the simplest datasets. Stata has an advanced command, `foreach`, that can be used to assign a mapping such as `rating` to a long list of variables. This is covered in the *Programming Reference Manual*.

Many Stata users have written custom commands that may simplify things. One such user is Jeroen Weesie, who wrote a command called `labelvar`. This command can be used the first time you create variable labels and value labels. It does not work if there are already value labels. To install this command on your computer, enter the command `findit labelvar`, click on the highlighted link, and then click on the `click here to install` link. If we have a variable named `abort` that is unlabeled, we could run the command `labelvar abort "abortion should be legal" 1 "strongly agree" 2 "agree" 3 "disagree" 4 "strongly disagree" -9 "missing"`. This one-line command will label the variable and each of the possible values. It is a one-step process rather than a two-step process.

2.5 Saving your dataset

If you look at the Results window in Stata, you can see that the dialog box has done a lot of the work for you. The Results window shows that a lot of commands have been run to label the variables. These also appear in the Review window. The Variables window lists all of your variables. We have now created our first dataset.

Until now, Stata has been working with our data in memory, and we still need to save the dataset as a file, which we can do from the File ▷ Save As... menu. Doing this will open the familiar dialog box to choose where we would like to save our file. Once the data have been saved to a file, we can exit Stata or use another dataset and not lose our changes.

Notice that Stata will give the file the `.dta` extension if you do not specify one. You must use this extension, as it identifies the file as a Stata dataset. You can give it any first name you want, but do not change the extension. Let's call this file `firstsurvey` and let Stata add the `.dta` extension for us. If you look in the Results window, you will see that Stata has printed something like this:

```
. save "C:\data\firstsurvey.dta"
file C:\data\firstsurvey.dta saved
```

It is a good idea to make sure that you spelled the name correctly and that you saved the file into the directory you intended. Note that Stata added quotation marks around the filename. If you type a `save` command in the Command window, you must use the quotes around the filename if the name contains a space; however, if you always use the quotes around the filename, you will never get an error.

Stata works with one dataset at a time. When you open a new dataset, Stata will first clear the current one from memory. Stata knows when you have changed your data and will prompt you if you attempt to replace unsaved data with new data. It is, however, a good idea to get in the habit of saving before reading new data because Stata will not prompt you to save data if the only things that have changed are labels or notes.

A Stata dataset contains more than just the numbers, as we have seen in the work we have done so far. A Stata dataset contains the following information:

- Numerical data
- Variable names
- Variable labels
- Missing values
- Formats for printing data
- Dataset label
- Notes
- Sort order and the name of the sort variables

In addition to the labeling we have already seen, you can store notes in the dataset, and you can attach notes to particular variables, which can be a convenient way to document changes, particularly if more than one person is working on the dataset. To see all the things you can insert into a dataset, type the command `help label`. If the dataset has been sorted (observations arranged in ascending order based on the values of a variable), that information is also stored.

We have now created our first dataset and saved it to disk, so let's close Stata. This is done with the File ▷ Exit menu.

2.6 Checking the data

We have created a dataset, and now we want to check the work we did defining the dataset. Checking for the accuracy of our data entry is also our first statistical look at the data. To open the dataset, use the File ▷ Open... menu. Locate `firstsurvey.dta`, which we saved in the preceding section, and click on Open.

Let's run a couple of commands that will characterize the dataset and the data stored in it in slightly different ways. We created a codebook to use when creating our dataset. Stata provides a convenient way to reproduce much of that information, which is useful if you want to check that you entered the information correctly. It is also useful if you share the dataset with someone who does not have access to the original codebook. Use the Data ▷ Describe data ▷ Describe data contents (codebook) menu to generate Stata's `codebook` from the current dataset. You can specify variables for the codebook entry you want, or you can get a complete codebook by not specifying any variables. Try it first without specifying any.

Scrolling the results

When Stata displays your result, it does so one screen at a time. When there is more output than will fit in the Results window, Stata prints what will fit and displays —`more`— in the lower left corner of the Results window. Stata calls this a *more condition* in its documentation. To continue, you can click on —`more`— in the Results window, but most users find it easier to just press the Spacebar. If you do not want to see all the output—that is, you want to cancel displaying the rest of it—you can press the q key. If you do not like having the output pause this way, you can turn this paging off by typing the command `set more off` in the Command window. All the output generated will print without pause until you either change this setting back or quit and restart Stata. To restore the paging, use the `set more on` command. To see what else you can set, enter the command `help set` and you will get a long list of things you can customize.

Let's look at the codebook entry for `gender` as Stata produced it. For this example, you can either scroll back in the Results window, or you can use the Data ▷ Describe data ▷ Describe data contents (codebook) menu and put `gender` in the *Variables* box, as I have done in the examples that follow.

(*Continued on next page*)

```
. codebook gender

-------------------------------------------------------------------------------
gender                                                     Participant's gender
-------------------------------------------------------------------------------

                  type:  numeric (byte)
                 label:  sex

                 range:  [1,2]                        units:  1
         unique values:  2                        missing .:  0/20

            tabulation:  Freq.   Numeric  Label
                            10         1  male
                            10         2  female
```

Let's go over the display for `gender`. The first line lists the variable name, `gender`, and the variable label, `Participant's gender`. Next the type of the variable, which is `numeric (byte)`, is shown. The value-label mapping associated with this variable is `sex`. The range of this variable (shown as the lowest value, then the highest value) is from 1 to 2, there are two unique values, and there are 0 missing values in the 20 cases. This information is followed by a table showing the frequencies, values, and labels. We have 10 cases with a value of 1, labeled `male`, and 10 cases with a value of 2, labeled `female`.

Using `codebook` is an excellent way to check your categorical variables, such as `gender` and `prison`. If, in the tabulation, you saw three values for `gender` or six values for `prison`, you would know that there are errors in the data. Looking at these kinds of summaries of your variables will often help you detect data-entry errors, which is why it is crucial to review your data after entry.

Let's also look at the display for `education`:

```
. codebook education

-------------------------------------------------------------------------------
education                                                    Years of education
-------------------------------------------------------------------------------

                  type:  numeric (byte)

                 range:  [8,20]                       units:  1
         unique values:  10                       missing .:  0/20

                  mean:     14.45
              std. dev:   2.94645

           percentiles:        10%       25%        50%       75%       90%
                              11.5        12       14.5      16.5        18
```

Because this variable takes on more values, Stata does not tabulate the individual values. Instead, Stata displays the range (8 to 20) along with the mean, which is 14.45, and the standard deviation, which is 2.95. It also gives the percentiles. The 50th percentile is usually called the median, and here the median is 14.5.

With a dataset that has many variables, the complete codebook may give more information than you need. For a quicker, condensed overview, there are two alternative commands, `codebook, compact` and `describe`.

You can obtain the compact form of the codebook by clicking on *Display compact report on the variables* on the Options tab of the `codebook` dialog box. However, it is simpler to enter the command directly: `codebook, compact`. This gives you seven columns of information including the variable name, the number of observations, the number of unique values the variable has in your data, the mean for the item, the minimum and maximum values, and finally the variable label. A variation of the `codebook, compact` command adds one or more variable names to the command, for example, `codebook gender, compact`. This gives you the same information as the `codebook, compact` command, but only for the variable(s) you specify.

The second alternative is the command `describe`. Enter the command in the Command window, or use the Data ▷ Describe data ▷ Describe data in memory menu, which will display the dialog box in figure 2.7.

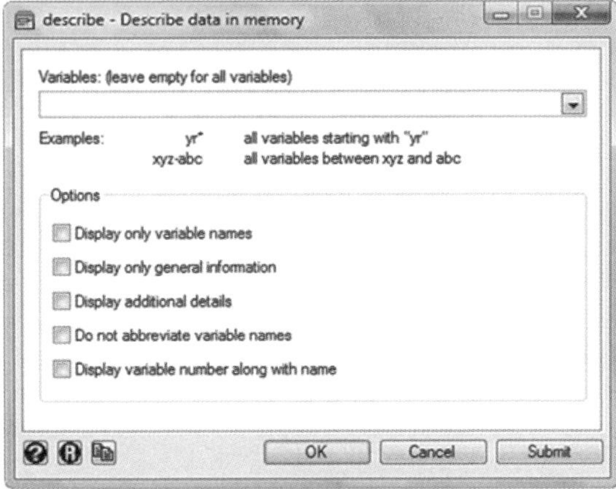

Figure 2.7. The `describe` dialog box

If you do not list any variables in the *Variables* box, all the variables will be included, as shown in the following output:

(*Continued on next page*)

```
. describe
Contains data from C:\data\firstsurvey.dta
  obs:            20
 vars:             7                          27 Oct 2007 11:43
 size:           240 (99.9% of memory free)
-------------------------------------------------------------------------------
              storage  display     value
variable name   type   format      label      variable label
-------------------------------------------------------------------------------
id              int    %8.0g                  Identification number
gender          byte   %8.0g       sex        Participant's gender
education       byte   %8.0g                  Years of education
sch_st          byte   %16.0g      rating     Rating of schools in your state
sch_com         byte   %16.0g      rating     Ratings of schools in your
                                                community of origin
prison          byte   %16.0g      harsh      Rating of prison sentences
conserv         byte   %17.0g      conserv    Conservativism/liberalism
-------------------------------------------------------------------------------
Sorted by:
```

As we can see by examining the output, the path to the data file is shown (directory and filename), as are the number of observations, the number of variables, the size of the file, and the last date and time the file was written (called a *time stamp*). We can attach a label to a dataset, too, and if we had done so, it would have been shown above the time stamp. Below the information that summarizes the dataset is information on each variable: its name, its storage type (there are several types of numbers and strings), the format Stata uses when printing values for the variable, the value-label mapping associated with it, and the variable label. This is clearly not as complete a description of each variable as that provided by `codebook`, but it is often just what you want.

2.7 Summary

We covered the following topics in this chapter:

- How to create a codebook for a questionnaire
- How to develop a system for coding the data
- How to enter the data into a Stata dataset
- How to add variable names and variable labels
- How to create value-label mappings and associate them with variables

2.8 Exercises

1. From the coding sheet shown in table 2.2, translate the numeric codes back into the answers given on the questionnaire by using the codebook in table 2.1.

2. Using the dialog box, attach a descriptive label to the dataset you created in this chapter. What command is printed in the Results window by the dialog box? You will find this dialog box under Data ▷ Labels.

3. Using the dialog box, experiment with Stata's notes feature. You can attach notes to the dataset and to specific variables. See if you can attach notes, display them, and remove them. What might be the advantage of keeping notes inside the dataset?

4. Administer your own small questionnaire to students in your class. Enter the data into a Stata dataset, and label the variables and values. Be sure to have at least 15 people complete your questionnaire, but do not try to enter data for a very large group of people.

3 Preparing data for analysis

3.1 Introduction

Most of the time spent on any research project is spent preparing for analysis. Exactly how much time will depend on the data. If you collect and enter your own data, most of the actual time you spend on the project will not be analyzing the data; it will be getting the data ready to analyze. If you are performing secondary analysis and are using well-prepared data, you may spend substantially less time preparing the data.

Stata has an entire reference manual on data management, the *Data Management Reference Manual*. We will cover only a few of Stata's capabilities for managing data. Some of the data management tasks you perform you will be able to anticipate and plan, but others will just arise as you progress through a project, and you will need to deal with them on the spot. If your data are always going to be prepared for you and you will use Stata for only statistics and graphs, then you might skip this chapter.

3.2 Planning your work

The data we will be using are from the U.S. Department of Labor, Bureau of Labor Statistics, National Longitudinal Survey of Youth, 1997 (NLSY97). I selected a set of four items to measure how adolescents feel about their mothers and another set of four parallel items to measure how adolescents feel about their fathers. I used an extraction program available from http://www.bls.gov/nls/nlsdata.htm and extracted a Stata dictionary that contained a series of commands for constructing the Stata dataset, which we will use along with the raw data in ASCII format. I created a dataset called `relate.dta`. In this chapter, you will learn how to work with this dataset to create two variables, Youth Perception of Mother and Youth Perception of Father.

What is a Stata dictionary file?

A Stata dictionary file is a special file that contains Stata commands and raw data in a plain-text format called ASCII. We will not cover this type of file here, but we use this format to import the raw data into a Stata dataset. Selecting File ▷ Import ▷ ASCII data in fixed format with a dictionary opens a window in which we browse for the directory where we stored the dictionary file. Here we find `relate.dct`. The extension `.dct` is always used for a dictionary file. Sometimes the dictionary will be in one file and the raw ASCII data will be in a separate file. Here they are both in the same file, so we do not need to point to an ASCII data file. Clicking on OK creates the Stata dataset. We can then save it using the name `relate.dta`. The dictionary provides the variable names and variable labels. Had the variable labels not been included in the dictionary, we would have created them prior to saving the dataset. If you need to create a dictionary file yourself, you can learn how to do that in the *Data Management Reference Manual*.

It is a good idea to make an outline of the steps we need to take to go from data collection to analysis. The outline should include what needs to be done to collect or obtain the data, enter (or read) and label the data, make any necessary changes to the data prior to analysis, create any composite variables, and create an analysis-ready version (or versions) of the dataset.

An outline serves two vital functions: it provides direction so we do not lose our way in what can often be a complicated process with many details, and it provides us with a set of benchmarks for measuring progress. For a large project, do not underestimate this latter function as it is the key to successful project management.

Our project outline, which is for a simple project on which only one person is working, is shown in table 3.1.

Table 3.1. Sample project task list

- Consult NLSY97 documentation (see appendix 9 [PDF], page 23, available at http://www.nlsinfo.org/ordering/display_db.php3#NLSY97), to determine which variables are needed.
- Download the data that is in the `dictionary` file and the codebook. (This was done for you as `relate.dct` and `relate.cdb`.)
- Create a basic Stata dataset. (This was done for you as `relate.dta`.)
- Create variable and value labels.
- Generate tables for variables to compare against the codebook to check for errors.
- Convert missing-value codes to Stata missing values.
- Reverse code those variables that need it; verify.
- Copy variables not reversed to named variables; verify.
- Create the scale variable.
- Save the analysis-ready copy of the dataset.

This is a pretty skeletal outline, but it is a fine start for a small project like this. The larger the project, the more detail needed for the plan. Now that we have the NLSY97-supplied codebook to look at, we can fill in some more details of our plan.

First, we will run a `describe` command on the dataset we created, `relate.dta`. Remember you can download `relate.dta` and other datasets used in this book from the book's web page. Alternatively, if you are using Stata 10 or later, you can open the file `relate.dta` by running the command

```
. use http://www.stata-press.com/data/agis2/relate
```

Let's show the results for just the four items measuring the adolescents' perception of their mother. From the documentation on the dataset, we can determine that these items are `R3483600`, `R3483700`, `R3483800`, and `R3483900`. First, we can run the command `describe R3483600 R3483700 R3483800 R3483900`, and then we can get more information by running the command `codebook, compact`. For the codebook, we do not list the variables because the dataset includes only a few variables. Thus we use the following commands to produce a description of the data:

```
. describe R3483600 R3483700 R3483800 R3483900

              storage  display     value
variable name   type   format      label      variable label

R3483600        float  %9.0g                  MOTH PRAISES R DOING WELL 1999
R3483700        float  %9.0g                  MOTH CRITICIZE RS IDEAS 1999
R3483800        float  %9.0g                  MOTH HELPS R WITH WHAT IMPT TO R
R3483900        float  %9.0g                  MOTH BLAMES R FOR PROBS 1999
. codebook, compact
Variable    Obs Unique       Mean  Min   Max  Label

R0000100   8984   8984   4504.302    1  9022  PUBID - YTH ID CODE 1997
R3483600   8984      9  -.5638914   -5     4  MOTH PRAISES R DOING WELL 1999
R3483700   8984      9  -1.425312   -5     4  MOTH CRITICIZE RS IDEAS 1999
R3483800   8984      9  -.5893811   -5     4  MOTH HELPS R WITH WHAT IMPT TO R
R3483900   8984      9  -1.769813   -5     4  MOTH BLAMES R FOR PROBS 1999
R3485200   8984      9  -1.535508   -5     4  FATH PRAISES R DO WELL 1999
R3485300   8984      9  -2.143143   -5     4  FATH CRITICIZE IDEAS 1999
R3485400   8984      9  -1.586487   -5     4  FATH HELPS R WITH WHAT IMPT TO R
R3485500   8984      9  -2.388023   -5     4  FATH BLAME R FOR PROBS 1999
R3828100   8984      8   15.02538   -5    20  SYMBOL!KEY!AGE 1999
R3828700   8984      3   .9319902   -5     2  SYMBOL!KEY!SEX 1999
```

These variable names are not helpful, and we might want to change them later. For example, we may rename `R3483600` to `mopraise`. The variable labels make sense, but notice that there are no value labels listed under the column labeled value label in the results of the `describe` command. `codebook, compact` gives us a bit more information. We see there are 8,984 observations, we see a mean for each item, and we see the minimum (-5) and maximum value (4) reported for the items. Still, without value labels, we do not know what these values mean.

Let's read the codebook that we downloaded at the same time we downloaded this dataset. So you do not need to go through the process of getting the codebook from the NLS web site, the part we use is called `relate.cdb`. You can open this file by pointing your web browser to http://www.stata-press.com/data/agis2/relate.cdb. Table 3.2 shows what the downloaded codebook looks like for the `R3483600` variable.

Table 3.2. NLSY97 sample codebook entries

```
R34836.00      [YSAQ-022]                                   Survey Year: 1999

               MOTHER PRAISES R FOR DOING WELL

How often does she praise you for doing well?

     118       0 NEVER
     235       1 RARELY
     917       2 SOMETIMES
    1546       3 USUALLY
    1701       4 ALWAYS
  -------
    4517

Refusal(-1)           20
Don't Know(-2)         3
TOTAL =========>    4540   VALID SKIP(-4)    3669     NON-INTERVIEW(-5)     775
```

We can see that a code of 0 means `Never`, a code of 1 means `Rarely`, etc. We need to make value labels so that the data will have nice value labels like these. Also notice that there are different missing values. The codebook uses `-1` for `Refusal`, `-2` for `Don't know`, `-4` for `Valid skip`, and `-5` for `Noninterview`. But there is no `-3` code. Searching the documentation, we learn that this code was used when there was an interviewer mistake or an `Invalid skip`. We did not have any of these for this item, but we might for others. The `Don't know` responses are people who should have answered the question but did not answer or did not know what they thought. By contrast, notice there are 3,669 valid skips representing people who were not asked the question for some reason. We need to check this out, and reading the documentation, we learn that these people were not asked because they were more than 14 years old, and only youth 12–14 were asked this series of questions. Finally, 775 youth were not interviewed. We need to check this out, as well. These are 1999 data, the third year the data were collected, and the researchers were unable to locate these youth, or they refused to participate in the third year of data collection. In any event, we need to define missing values in a way that keeps track of these distinctions.

How does the Stata dataset look compared with the codebook that we downloaded? We can run the command `codebook` to see. Let's restrict it to just the variable `R3483600` by entering the command `codebook R3483600` in the Command window. The problem with the data is apparent when we look at the codebook: the absence of value labels makes our actual data uninterpretable.

(Continued on next page)

```
. codebook R3483600

R3483600                                          MOTH PRAISES R DOING WELL 1999

                  type:  numeric (float)

                 range:  [-5,4]                       units:  1
         unique values:  9                        missing .:  0/8984

           tabulation:  Freq.  Value
                          775  -5
                         3669  -4
                            3  -2
                           20  -1
                          118  0
                          235  1
                          917  2
                         1546  3
                         1701  4
```

Before we add value labels to make our codebook look more like the codebook we downloaded, we need to replace the numeric missing values (-5, -4, -3, -2, and -1) with values that Stata recognizes as representing missing values. Stata recognizes up to 27 different missing values for each variable: . (a dot), .a, .b, ..., .z. When there is only one type of missing value, as discussed in chapter 2, we could change all numeric codes of -9 to . (a dot). Because we have five different types of missing values, we will replace -5 with .a (read as dot-a), -4 with .b, -3 with .c, -2 with .d, and -1 with .e. Stata uses a command called mvdecode (missing values decode) to do this. Because all items involve the same decoding replacements, we can do this with the following command:

```
. mvdecode _all, mv(-5=.a\-4=.b\-3=.c\-2=.d\-1=.e)
    R3483600: 4467 missing values generated
    R3483700: 4469 missing values generated
    R3483800: 4468 missing values generated
    R3483900: 4470 missing values generated
    R3485200: 5608 missing values generated
    R3485300: 5611 missing values generated
    R3485400: 5610 missing values generated
    R3485500: 5611 missing values generated
    R3828100: 775 missing values generated
    R3828700: 775 missing values generated
```

We use the _all that is just before the comma to tell Stata to do the missing value decoding for all variables. If we just wanted to do it for variables R3483600 and R3828700, the mvdecode _all would be replaced by mvdecode R3483600 R3828700. Some datasets might use different missing values for different variables and this would require several mvdecode commands. For example, a value of -1 might be a legitimate answer for some items and for these items a value of 999 might be used for Refusal. Notice that each replacement is separated by a backslash (\). The results appear right below the mvdecode command and tell us how many missing values are generated for each variable. For example, R3483600 has 4,467 missing values generated (775 replace the −5, 3,669 replace the −4, 3 replace the −2, and 20 replace the −1).

3.3 Creating value labels

Now let's add value labels, including labels for the missing values. We know from the dataset documentation that we will be reverse coding some of the variables because some items are stated positively (praises) and some are stated negatively (blames). When we create value labels, we want labels for the original coding and labels for the reversed coding. Let's call the labels for the parental relations items `often` and those for the reversed items `often_r`. The text of the labels is taken from the codebook file. Here are the commands to add the labels. You can, if you wish, use the dialog box to do this.

```
. label define often  0 "Never" 1 "Rarely" 2 "Sometimes" 3 "Usually" 4 "Always"
. label define often .a "Noninterview" .b "Valid skip" .c "Invalid skip", add
. label define often .d "Don't know" .e "Refusal", add
. label define often_r  4 "Never" 3 "Rarely" 2 "Sometimes" 1 "Usually" 0 "Always"
. label define often_r .a "Noninterview" .b "Valid skip" .c "Invalid skip", add
. label define often_r .d "Don't know" .e "Refusal", add
```

By doing this, we have entered value labels, but we have not yet attached them to any variables. The next task is to assign the value labels to the variables, and we can do this in several ways. We can open the Data Editor, right-click in any cell in each of the columns to bring up the menu, select *Assign Value Label to Variable*, and choose the appropriate value label from the list. We have done this for `R3483600`, and the results appear in figure 3.1. Notice that after we do this for `R3483600`, the value labels appear in the Data Editor for this variable.

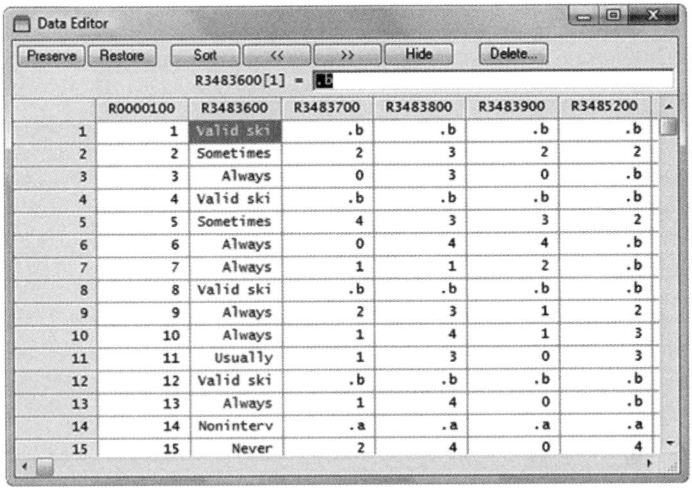

Data Editor

Preserve Restore Sort << >> Hide Delete...

R3483600[1] = .b

	R0000100	R3483600	R3483700	R3483800	R3483900	R3485200
1	1	Valid ski	.b	.b	.b	.b
2	2	Sometimes	2	3	2	2
3	3	Always	0	3	0	.b
4	4	Valid ski	.b	.b	.b	.b
5	5	Sometimes	4	3	3	2
6	6	Always	0	4	4	.b
7	7	Always	1	1	2	.b
8	8	Valid ski	.b	.b	.b	.b
9	9	Always	2	3	1	2
10	10	Always	1	4	1	3
11	11	Usually	1	3	0	3
12	12	Valid ski	.b	.b	.b	.b
13	13	Always	1	4	0	.b
14	14	Noninterv	.a	.a	.a	.a
15	15	Never	2	4	0	4

Figure 3.1. Assign value labels to variables

Sometimes these value labels are what we want to see in the Data Editor, and sometimes we need to see the actual numbers the value labels represent. To see the numbers instead of the labels, we can enter a simple command in the Command window:

```
. edit, nolabel
```

This command opens the Data Editor and shows the values and the missing-value codes.

Another way we can assign the value labels is through the dialog system. Select Data ▷ Labels ▷ Label values ▷ Assign value labels to variable. If we wanted to enter the commands directly, we could type the following in the Command window:

```
. label values R3483600 often
. label values R3483800 often
. label values R3485200 often
. label values R3485400 often
```

We have used the value label called `often` for the positively stated items. You can assign the `often_r` scheme to the other items this same way.

3.4 Reverse-code variables

We have two items for the mother and two for the father that are stated negatively, and we should reverse code these. R3483700 refers to the mother criticizing the adolescent, and R3485300 refers to the father criticizing the adolescent. R3483900 and R3485500 refer to the mother and father blaming the adolescent for problems. For these items, an adolescent who reports that this always happens would mean that the adolescent had a low rating of his or her parent. We always want a higher score to signify more of the variable. A score of 0 for `never` blames the child on this pair of items should be the highest score on these items (4), and a response of `always` blames the child should be the worst response and would have the lowest score (0).

It is good to organize all the variables that need reverse coding into groups based on their coding scheme. It is also a good idea to create new variables instead of reversing the originals. This step ensures that we have the original information if any questions arise about what we have done. I recommend that you write things out before you start working, as I have done in table 3.3.

Table 3.3. Reverse-coding plan

Old value	New value
0	4
1	3
2	2
3	1
4	0

For a small dataset like the one used as the example, writing it out may seem like overkill. When you are involved in a large-scale project, the experience you gain from working out how to organize these matters on small datasets will serve you well, particularly if you need to assign tasks to several people and try to keep them straight.

There are several ways that this recoding can be done, but let's use `recode` right now. Select Data ▷ Create or change variables ▷ Other variable transformation commands ▷ Recode categorical variable.

The `recode` command and its dialog box are straightforward. We provide the name of the variable to be recoded and then give at least one rule for how it should be recoded. Recoding rules are enclosed in parentheses, with an old value (or values) listed on the left, then an equal-sign, and then the new value. The simplest kind of rule lists one value on either side of the equal-sign. So, for example, the first recoding rule we would use to reverse code our variables would be `(0=4)`.

More on recoding rules

More than one value can be listed on the left side of the equal-sign, as in `(1 2 3=0)`, which would recode values of `1`, `2`, and `3` as `0`. Occasionally, you may want to collapse just one end of a scale, as with `(5/max=5)`, which recodes everything from `5` up, but not including the missing values, as `5`. You might use this to collapse the highest income categories into one category, for example. There is also a corresponding version to recode from the smallest value up to some value. For example, you might use `(min/8=8)` to recode the values for highest attained grade in school if you wanted everybody with fewer than 9 years of education coded the same way.

When you use `recode`, you can choose to recode the existing variables or create new variables that have the new coding scheme. I strongly recommend creating new variables. The dialog boxes shown in figures 3.2 and 3.3 show how to do this for all four variables being recoded. Figure 3.2 shows the original variables on the Main tab.

(Continued on next page)

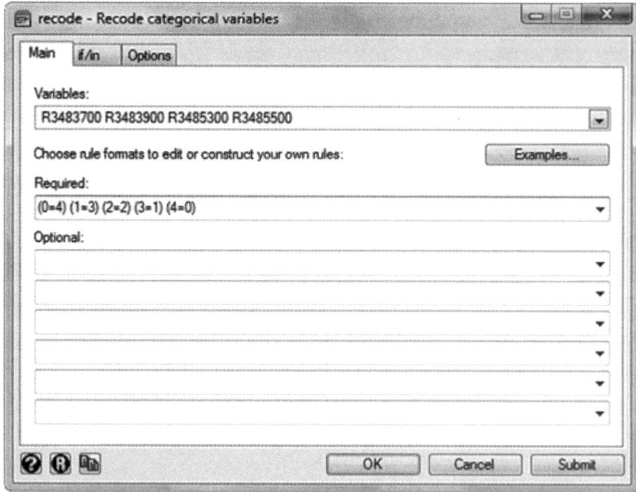

Figure 3.2. `recode`: specifying recode rules on the Main tab

Figure 3.3 shows the names of the new variables to be created on the Options tab.

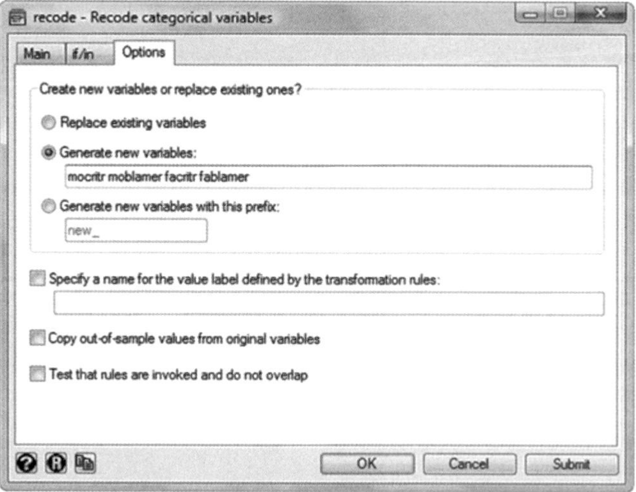

Figure 3.3. `recode`: specifying new variable names on the Options tab

The variables for mothers begin with `mo`, and the variables for fathers begin with `fa`. Adding an `r` at the end of each variable reminds you that they are reverse coded. Stata will pair up the names, starting with the first name in each list, and recode according to the rules stated. We must have the same number of variable names in each list. After you click on OK, the output from the selections shown in figures 3.2 and 3.3 is

```
. recode R3483700 R3483900 R3485300 R3485500 (0=4) (1=3) (2=2) (3=1) (4=0),
> generate(mocritr moblamer facritr fablamer)
(3230 differences between R3483700 and mocritr)
(4037 differences between R3483900 and moblamer)
(2562 differences between R3485300 and facritr)
(3070 differences between R3485500 and fablamer)
```

The output above shows how Stata presents output that is too long to fit on one line: it moves down a line and inserts a > character at the beginning of the new line. This differentiates command-continuation text from the output of the command. If you are trying to use the command in output as a model for typing a new command, you should not type the > character.

The `recode` command labels the variables it creates. The text is generic, so we may wish to replace it. For example, the `mocritr` variable was labeled with the text

```
RECODE of R3483700 (MOTH CRITICIZE RS IDEAS 1999)
```

This text tells us the original variable name (and its label in parentheses), but it does not tell us how it was recoded. It may be worth the additional effort to create more explicit variable labels, so we will do that here. To label the variable, type

```
. label variable mocritr "Mother criticizes R, R3483700 reversed"
. label variable moblamer "Mother blames R, R3483900 reversed"
. label variable facritr "Father criticizes R, R3485300 reversed"
. label variable fablamer "Father blames R, R3485500 reversed"
```

The `recode` command is useful. You can specify missing values either as original values or as new values, and you can create a value label as part of the `recode` command. You can also recode a range (say, from 1 to 5) easily, which is useful for recoding things like ages into age groups. If the resulting values are easily listed (most often integers, but they could be noninteger values), it is worth thinking about using the `recode` command. See the *Data Management Reference Manual* entry for `recode` or type `help recode` in Stata for more information.

3.5 Creating and modifying variables

There are three primary commands for creating and modifying variables: `generate` and `egen` (short for extended generation) are used to create new variables, and `replace` is used to modify the values of existing variables. We will use each of these commands in this section and the next, but they can do much more than we show.

The simplest kind of variable creation is when you assign the same value to every observation in the new variable, which is called creating a *constant* variable. Almost as simple is copying the values from one variable into a new variable. One of the tasks in our project task list (table 3.1) was to create copies of some of the original variables that have more descriptive names than those supplied with NLSY97, and I will show you how to do that now.

To create a new variable using the dialog box, select Data ▷ Create or change variables ▷ Create new variable. The dialog is pretty simple: enter the name of the new variable, and enter what the new variable should equal. We want to create a variable called `id`, and it should contain whatever is in the variable `R0000100`. Our dialog box should look like figure 3.4.

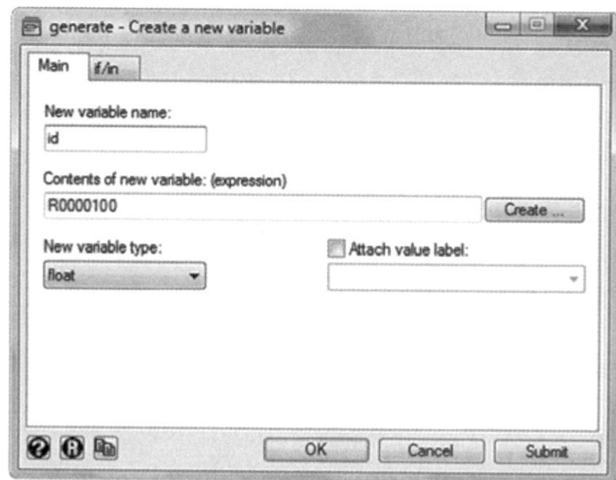

Figure 3.4. Create a new variable dialog box

We could enter the `generate` command directly in the Command window to do the same thing, as shown by the command that the dialog box generates:

```
. generate float id = R0000100
```

Generally, the way to assign values to a new variable with the `generate` command is simple: type `generate`, then the name of the new variable, then an equal-sign, and finally type an expression (or a rule or formula) that tells `generate` the value to put in the new variable. Here is what the commands look like to rename the remaining items that did not need to have their coding reversed:

```
. generate id = R0000100
. generate sex = R3828700
. generate age = R3828100
. generate mopraise = R3483600
. generate mohelp = R3483800
. generate fapraise = R3485200
. generate fahelp = R3485400
```

The `generate` expression can be as simple as a number or the name of another variable, as shown above, or it can be complicated. We can put variable names, arithmetic signs, and many functions into expressions. Table 3.4 shows the arithmetic symbols that can be used in expressions.

Table 3.4. Arithmetic symbols

Symbol	Operation	Example
+	Addition	`mscore + fscore + sibscore`
-	Subtraction	`balance - expenses - penalty`
*	Multiplication	`income * .75`
/	Division	`expenses/income`
^	Exponentiation (x^2)	`x^2`

Note that attempts to do arithmetic with missing values will lead to missing values. So, in the addition example in table 3.4, if `sibscore` were missing (say, it was a single-child household), the whole sum in the example would be set to missing for that observation. From the Command window, we can type `help generate` to see several more examples of what can be done. If you ever have trouble finding a dialog box, but you know the command name, you can open the help file first. In the upper right corner of a help file opened in the Viewer window, it will say `dialog:` *command_name*. Click on *command_name* and the dialog box for that command will open.

For more complicated expressions, order of operations can be important, and you can use parentheses to control the order in which things are done. Parentheses contain expressions too, and those are calculated before the expressions outside of parentheses. Parentheses are never wrong. They might be unnecessary to get Stata to calculate the correct value, but they are not wrong. If you think they make an expression easier to read or understand, use as many as you need.

Fortunately, the rules are pretty simple. Stata reads expressions from left to right, and the order in which Stata calculates things inside expressions is to

1. Do everything in parentheses. If one set of parentheses contains another set, do the inside set first.
2. Exponentiate (raise to a power).
3. Multiply and divide.
4. Add and subtract.

Let's go step-by-step through an example:

```
. generate example = weight/.45*(5+1/age^2)
```

When Stata looks at the expression to the right of the equal-sign, it notices the parentheses (priority #1) and looks inside them. There it sees that it first has to square `age` (priority #2), then divide 1 by the result (priority #3), and then add 5 to that result (priority #4). Once it is done with all the stuff in parentheses, it starts reading from left to right, so it divides `weight` by .45 and then multiplies the result by the final value it got from calculating what was inside parentheses. That final value is put into a variable called `example`.

Stata does not care about spaces in expressions, but they can help readability. So, for example, instead of typing something like we just did, we can use some spaces to make the meaning clearer, as in

```
. generate example = weight/.45 * (5 + 1/age^2)
```

If we wanted to be even more explicit, it would not be wrong to type

```
. generate example = (weight/.45) * (5 + 1/(age^2))
```

Let's take another look at reverse coding. We already reverse-coded variables by using a set of explicit rules like `(0=4)`, but we could accomplish the same thing using arithmetic. Because this is a relatively simple problem, we will use it to introduce some of the things you may need to be concerned about with more complex problems.

Reversing a scale is swapping the ends of the scale around. The scale is 0 to 4, so we can swap the ends around by subtracting the current value from 4. If the original value is 4 and we subtract it from 4, we have made it a 0, which is the rule we specified with `recode`. If the original value is 0 and we subtract it from 4, we have made it a 4, which is the rule we specified with `recode`.

This scale starts at 0, so to reverse it, you just subtract each value from the largest value in the scale, in this case, 4. So, if our scale were 0 to 6, we would subtract each value from 6; if it were 0 to 3, we would subtract from 3. If the scale started at 1 instead of 0, we would need to add 1 to the largest value before subtracting. So, for a 1 to 5 scale, we would subtract from 6 ($6 - 1 = 5$ and $6 - 5 = 1$); for a 1 to 3 scale, we would subtract from 4 ($4 - 3 = 1$ and $4 - 1 = 3$).

What we have said so far is correct, as far as it goes, but we are not taking into account missing values or their codes. The missing-value codes are -1 to -5, and if you subtract those from 4 along with the item responses that are not missing-value codes, we will end up with $4-(-1) = 5$ to $4-(-5) = 9$. So, we must first convert the missing values and then do the arithmetic to reverse the scale.

Many researchers would code the values 1–5 rather than 0–4. To reverse a scale that starts at 1 and goes to 5, you need to subtract the current value from 1 more than the maximum value. Thus you would use $6 -$ *variable*. If the variable's current value is 5, then $6 - 5 = 1$.

Let's try this example. We have already ran the `mvdecode` command on our variables, so let's reverse code `R3485300` and call it `Facritr` (remember that Stata is case sensitive, so `Facritr` and `facritr` are different names). We can use the same dialog that is shown in figure 3.4, except that we enter `4 - R3485300` in the *Contents of new variable* box instead of just `R3485300`. The output is

```
. generate float Facritr = 4 - R3485300
(5611 missing values generated)
```

Whenever you see that missing values are generated (there are 5,611 of them in this example!), it is a good idea to make sure you know why they are missing. These variables have only a small set of values they can take, so we can compare the original variable with the new variable in a table and see what got turned into what. Select Statistics ▷ Summaries, tables, and tests ▷ Tables ▷ Two-way tables with measures of association, which will bring up the dialog shown in figure 3.5.

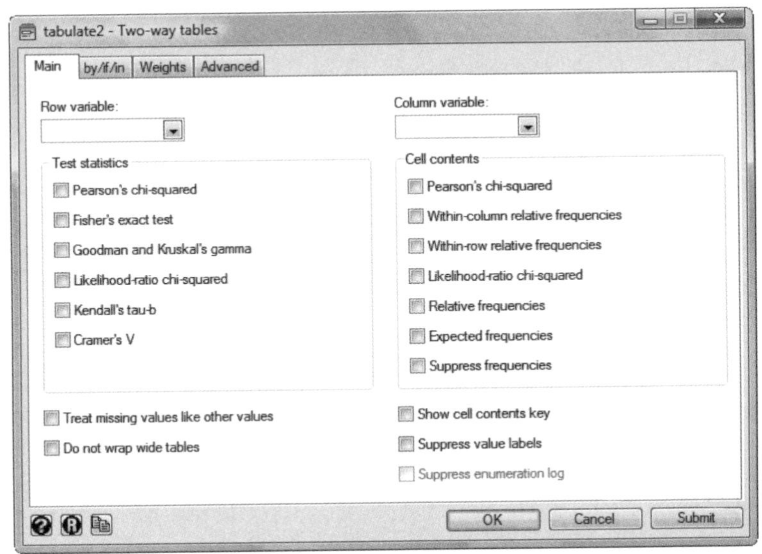

Figure 3.5. Two-way tabulation dialog box

Here we select `Facritr` as the *Row variable* and `R3485300` as the *Column variable*. We also check two of the options boxes: Because we are interested in the actual values the variables take, select the option to suppress the value labels (*Suppress value labels*). We are also interested in the missing values, so we check the box to have them included in the table (*Treat missing values like other values*).

```
. tabulate Facritr R3485300, miss nolabel

           |              FATH CRITICIZE IDEAS 1999
   Facritr |         0          1          2          3          4 |     Total
-----------+-------------------------------------------------------+----------
         0 |         0          0          0          0        117 |       117
         1 |         0          0          0        247          0 |       247
         2 |         0          0        811          0          0 |       811
         3 |         0      1,078          0          0          0 |     1,078
         4 |     1,120          0          0          0          0 |     1,120
         . |         0          0          0          0          0 |     5,611
-----------+-------------------------------------------------------+----------
     Total |     1,120      1,078        811        247        117 |     8,984
```

Facritr	FATH CRITICIZE IDEAS 1999 .a	.b	.d	.e	Total
0	0	0	0	0	117
1	0	0	0	0	247
2	0	0	0	0	811
3	0	0	0	0	1,078
4	0	0	0	0	1,120
.	775	4,816	4	16	5,611
Total	775	4,816	4	16	8,984

From the table generated, we can see that everything happened as anticipated. Those adolescents who had a score of 0 on the original variable (column with a 0 at the top) now have a score of 4 on the new variable. There are 1,120 of these observations. We need to check the other combinations, as well. Also the missing-value codes were all transferred to the new variable, but we lost the distinctions between the different reasons an answer is missing. The new variable Facritr is a reverse coding of our old variable R3485300.

3.6 Creating scales

We are finally ready to calculate our scales. We will construct two scale variables: one for the adolescent's perception of his or her mother and one for the adolescent's perception of his or her father. With four items, each scored from 0 to 4, the scores could range from 0 to 16 points. Higher scores indicate a more positive relationship with the parent.

At first glance, this assertion is straightforward. We just add the variables together:

```
. generate ymorelate = mopraise + mocritr + mohelp + moblamer
. generate yfarelate = fapraise + facritr + fahelp + fablamer
```

Before we settle on this, we should understand what will happen for missing values, and we should check through the documentation to find out what StataCorp does about missing values. The first thing to do is to determine how many observations have one, two, three, or all four of the items answered, i.e., not missing. This need introduces a new command, egen (short for extended generation). To display the egen dialog box, select Data ▷ Create or change variables ▷ Create new variable (extended), which produces figure 3.6.

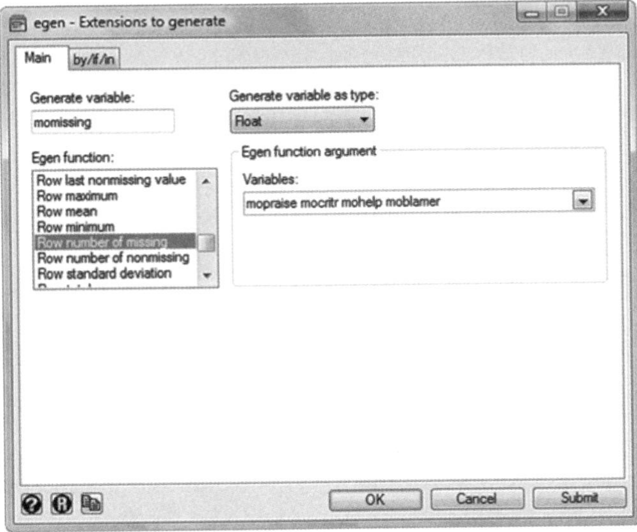

Figure 3.6. The Main tab for the egen dialog box

As you can see, the dialog box allows us to enter the name of the variable to be created, which we will call `momissing` because this will tell us how many of the items about mothers that each adolescent missed answering. The next step is to select the function from the alphabetical list. In this case, we want to look among the row functions for one that will count missing values, which appears as *Row number of missing*. Last, we enter the items about the mother in the box provided. These selections produce

```
. egen float momissing = rowmiss(mopraise mocritr mohelp moblamer)
```

Beyond egen

We have seen how `generate` and `egen` cover a wide variety of functions. You can always type `help generate` or `help egen` to see a description of all the available functions. Sometimes this is not enough. Nicholas J. Cox has written and continues to update a command called `egenmore`, which adds many more functions that are helpful for data management. You can type `ssc install egenmore, replace` in your Command window to install these added capabilities. The option `replace` will check for updates he has added if `egenmore` is already installed on your machine. If you now enter `help egenmore`, you will get a description of all the capabilities Dr. Cox has added.

If we look at a tabulation of the `momissing` variable, the rows will be numbered 0–4, and each entry will indicate how many observations have that many missing values. For example, there are 4,510 observations for which all four items were answered (no missing values), 6 for which there are three items answered (one missing value), 2 for which there are two answered items (two missing values), and 4,466 for which there are no answered items. Our table does not include a row for 3 because there are no observations in our example that have three missing values.

```
. tab momissing

  momissing |      Freq.     Percent        Cum.
------------+-----------------------------------
          0 |      4,510       50.20       50.20
          1 |          6        0.07       50.27
          2 |          2        0.02       50.29
          4 |      4,466       49.71      100.00
------------+-----------------------------------
      Total |      8,984      100.00
```

We can see that there are a total of eight observations for which there are some, but not complete, data. There should be little doubt about what to do for the 4,510 cases with complete information, as well as for the 4,466 with no information (all missing). The problem with using the sum of the items as our score on `ymorelate` is that the eight observations with partial data will be dropped, i.e., given a value of missing.

There are several solutions. We might decide to compute the mean of the items that are answered. We can do this with the `egen` dialog box; select Data ▷ Create or change variables ▷ Create new variable (extended) to display the `egen` dialog displayed in figure 3.6. This time, call the scale `momeana` and pick *Row mean* from the list of `egen` functions. This function generates a mean of the items that were answered, regardless of how many were answered, as long as at least one item was answered. If we tabulate `momeana`, we have 4,518 observations, meaning that everybody who answered at least one item has a mean of the items they answered.

Another solution is to have some minimum number of items, say, 75% or 80% of them. We might include only people who answered at least three of the items. These people would have fewer than two missing items. We can go back to the dialog box for `egen` and rename the variable we are computing to `momeanb`. The by/if/in tab allows us to stipulate the condition that `momissing < 2`, as shown in figure 3.7.

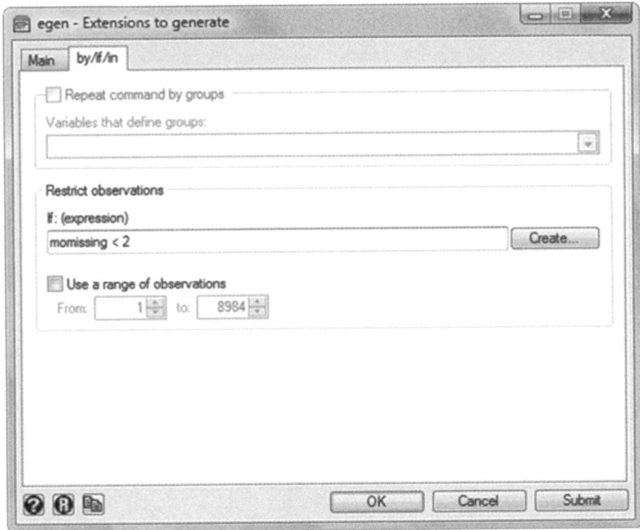

Figure 3.7. The by/if/in tab for the egen dialog box

The Stata command this dialog box generates is

```
. egen float momeanb = rowmean(mopraise mocritr mohelp moblamer) if momissing < 2
```

This command creates a variable, `momeanb`, which is the mean of the items for each observation that has at least three of the four items answered. Doing a tabulation on this shows that there are 4,516 observations with a valid score. This command allows us to keep the six cases that answered all but one of the items and drop the two cases that were missing more than one item.

Setting how much output is in the Results window

If you are working in a class and doing homework is the extent of your Stata usage, you might not run out of space in the Results window. If you are working on a research project or need to generate more than a few models and see their results, you will almost certainly want to increase the number of lines that you can scroll back in the Results window. Select Edit ▷ Preferences ▷ General Preferences.... Click on the Windowing tab. The default size for the scrollback buffer is 32,000 characters. Make it at least 128,000. You may also want to turn the `—more—` message off using the command `set more off`. If you enter the command `help set`, you will find many other ways to personalize the Stata interface. On a Mac, to change the buffer size you have to use the `set` command. You would enter `set scrollbufsize 128000`, and this would take effect the next time you open Stata—not the current session.

Deciding among different ways to do something

When you need to decide among different ways of performing a data management task—which is what we have been doing this whole chapter—it is usually better to choose based on how easy it will be for someone to read and understand the record, rather than based on the number of commands it takes or based on some notion of computer efficiency. For example, computing the row mean has a major advantage over computing the sum of items. The mean is on the same 0–4 scale that the items have, and this often makes more sense than a scale that has a different range. For example, if you had nine items that ranged from `1` for `strongly agree` to `5` for `strongly disagree`, a mean of 4 would tell you that the person has averaged `agree` over the set of items. This might be easier for users to understand than a total or sum score of 36 on a scale that ranges from 9 (one on each item) to 45 (five on each item).

3.7 Saving some of your data

Occasionally, you will want to save only part of your data. You may wish to save only some variables, or you may wish to save only some observations. Perhaps you have created variables that you needed as part of a calculation but do not need to save, or perhaps you wish to work only with the created scales and not the original items. You might want to drop some observations based on some characteristic, such as gender or parent's educational level or geographic area.

To drop variables from your dataset, select Data ▷ Variable utilities ▷ Keep or drop variables, which will display the dialog box shown in figure 3.8.

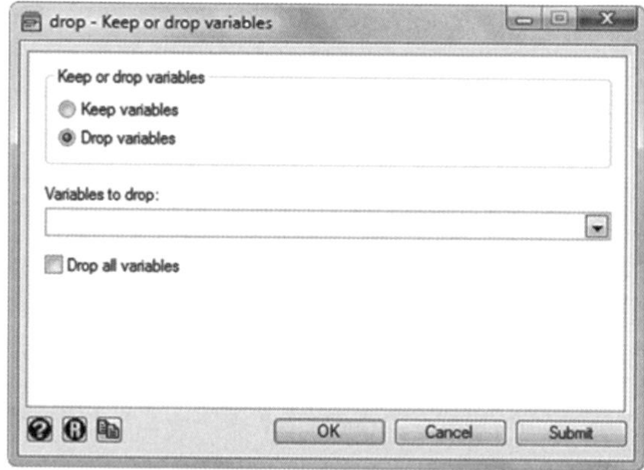

Figure 3.8. Selecting variables to drop

Sometimes it is easier to list the variables you wish to keep instead of those you wish to drop. As you can see from figure 3.8, both tasks are done from the same dialog box. Select the *Keep variables* radio button, and fill in the list of variables you want to keep.

Often the criterion for selecting what to save to a file is not the variable, but the observation. For example, you might have a large master dataset, and you wish to work only with participants 14 years old and younger. You have the choice to specify which observations meet your criteria by specifying which to keep or which to drop. To drop or keep selected observations from your dataset, select Data ▷ Variable utilities ▷ Keep or drop observations, which will display the dialog box shown in figure 3.9. Notice the Keep or drop observations dialog box has two tabs.

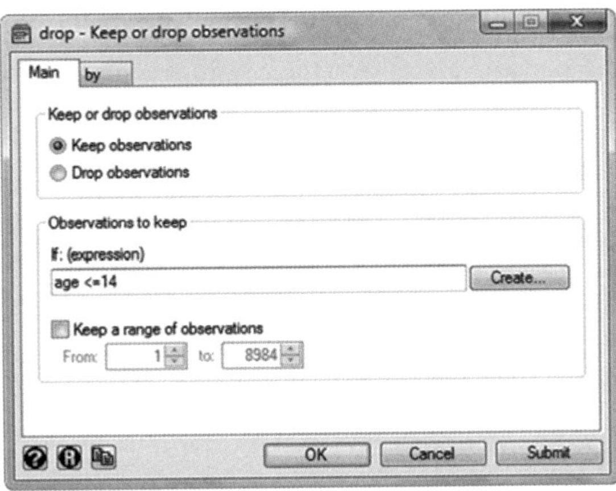

Figure 3.9. Selecting observations to keep

To select by observation, select the *Keep observations* or *Drop observations* radio button, and then fill in the expression. Figure 3.9 shows the dialog filled in to keep respondents age 14 and younger.

Now let's save two versions of the dataset: one that contains all the variables and one from which we have dropped the original R* variables. The commands to do this are

```
. * Save the master dataset before dropping variables
. save nlsy97_mstr, replace
. * Drop the original items because we no longer need them
. drop R3483600-R3828700
. * Reorder the variables when saving to make them easier to read
. order id sex age ymorelate yfarelate m* f*
. save nlsy97_fp, replace
```

Note four things about the save command. First, we give the saved dataset a new name, nlsy97_fp. Second, we leave off the .dta extension and use the replace option. Stata automatically appends .dta to the filename, so we can leave it off. Third, we use the replace option so that, if we rerun these commands, we will not get an error because nlsy97_fp.dta already exists. Finally, there are comments that start with an asterisk (*). These comments are printed but otherwise ignored by Stata. They help explain what we did. Comments like these are helpful when you come back to a project after working on other projects for a few weeks.

Just before the save command is the order command. If you just entered order sex, Stata would put sex first in the list of variables. Notice that we end the line with m* f*. This puts the variables that begin with m and f at the end of the dataset.

3.8 Summary

This chapter has covered a lot of ground. You may need to refer to this chapter when you are creating your own datasets. We have covered how to label variables and create value labels for each possible response to an item. In explaining how to create a scale, we covered reverse-coding items that were stated negatively and creating and modifying items. We also covered how to create a scale and work with missing values, especially where some people who have missing values are included in the scale and some are excluded. Finally, we covered how to save parts of a file, whether the parts were selected items or selected observations.

In the next chapter, we will look at how to create a file containing Stata commands that can be run as a group. The book's web page has such a command file for chapter 3 (chapter3.do). By recording all the commands into a program, we have a record of what we did, and we can rerun it or modify it in the future. We will also see how to record both output and commands in files.

3.9 Exercises

1. Open the `relate.dta` dataset. The variable `R3828700` represents the gender of the adolescent and is coded as a `1` for males and `2` for females. There are 775 people who are missing because they dropped out of the study after a previous wave of data collection. These people have a missing value on `R3828700` of `-5`. Run a tabulation on `R3828700`. Then modify the variable so that the code of `-5` will be recognized by Stata as a missing value. Label the variable so that a `1` is male and a `2` is female. Finally, run another tabulation, and compare this with your first tabulation.

2. Open the `relate.dta` dataset. Using variable `R3483600`, repeat the process you did for the first exercise. Go to the web page for the book and examine the dictionary file `relate.dct` to see how this variable is coded. Modify the missing values so that `-5` is `.a`, `-4` is `.b`, `-3` is `.c`, `-2` is `.d`, and `-1` is `.e`. Then label the values for the variable and run a tabulation.

3. Using the result of the second exercise, run the command `numlabel, add`. Repeat the tabulation of `R3483600`, and compare it with the tabulation for the second exercise. Next run the command `numlabel, remove`. Finally, repeat the tabulation, but add the `missing` option; that is, insert a comma and the word "missing" at the end of the command. Why is it good to include this option when you are screening your data?

4. The `relate.dta` dataset is data from the third year of a long-term study. Because of this, some of the youth who were adolescents the first year (1997) are more than 18 years old. Assign a missing value on `R3828100` (age) for those with a code of `-5`. Drop observations that are 18 or more. Keep only `R0000100`, `R3483600`, `R3483800`, `R3485200`, `R3485400`, and `R3828700`. Save this as a dataset called `positive.dta`.

5. Using the `positive.dta` dataset, assign missing values to the four items `R3483600`, `R3483800`, `R3485200`, and `R3485400`. Copy these four items to four new items called `mompraise`, `momhelp`, `dadpraise`, and `dadhelp`, respectively.

6. Create a scale called `parents` of how youth relate to their parents using the four items (`R3483600`, `R3483800`, `R3485200`, and `R3485400`). Do this separately for boys (`if R3828700 == 1`) and girls (`if R3828700 == 2`). Use the `rowmean()` function to create your scale. Do a tabulation of `parents`.

4 Working with commands, do-files, and results

4.1 Introduction

Throughout this book, I am illustrating how to use the menus and dialog boxes, but underneath the menus and dialogs is a set of commands. Learning to work with the commands lets you get the most out of Stata, and this is true of any other statistical software. Even the official Stata documentation is organized by command name, as illustrated by the three-volume *Stata 10 Base Reference Manual*, which has more than 1,800 pages explaining Stata commands.

With the logical organization of the menu system, you may wonder why you need to even think about the underlying commands. There are several reasons: Entering commands can be quicker than going through the menus. More importantly, you can put the commands into files that are called *do-files* and run repeatedly. These do-files allow you to replicate your work, something you should always ensure you can do. When you collaborate with co-workers, they can use your do-file as a way to follow exactly what you did. It is hard enough to remember all the commands you create in a session, and if there is a delay between work sessions, it is impossible to remember all those commands. Even when you are using the menu system, it is useful to save the commands generated from the menus into a do-file, and Stata has a way to facilitate doing this that you will soon learn.

What is a command? What is a program?

A command instructs Stata to do something, such as construct a graph, a frequency tabulation, or a table of correlations. Once you know how Stata does this, you will be able to understand a program—even if you have not yet learned all the procedures used in the program. A program is a collection of commands. Stata has a special name for programs; it calls them do-files. This is a good name because it is so descriptive. The program in the file tells Stata what to "do". A simple program might open a dataset, summarize the variables, create a codebook, and then do a frequency tabulation of the categorical variables. A program can include all the commands you use to label variables and values; it can recode variables, average variables, and define how you treat missing values. Such a program might be only a few lines long, but complicated programs can be thousands of lines long.

4.2 How Stata commands are constructed

Stata has many commands, and some of the commands covered in this book are

`list`	List values of variables
`summarize`	Summary statistics
`describe`	Describe data in memory or in file
`codebook`	Describe data contents
`tabulate`	Tables of frequencies
`generate`	Create or change contents of variable
`egen`	Extensions to generate
`correlate`	Correlations (covariances) of variables or coefficients
`ttest`	Mean-comparison tests
`anova`	Analysis of variance and covariance
`regress`	Linear regression
`logit`	Logistic regression, reporting coefficients
`factor`	Factor analysis
`alpha`	Compute interitem correlations (covariances) and Cronbach's alpha
`graph`	The graph command

Stata has a remarkably simple command structure. Stata commands are all lowercase. Virtually all Stata commands take the following form: `command varlist if/in, options`. The **`command`** is the name of the command, such as **`summarize`**, **`generate`**, and **`tabulate`**. The **`varlist`** is the list of variables used in the command. For many commands, listing no variables means that the command will be run on all variables. If we said **`summarize`**, Stata would summarize all variables in the dataset. If we said **`summarize age education`**, Stata would summarize just the participants' `age` and `education`. The variable list could include one variable or many variables. After the variable list come qualifiers on what will be included in the particular analysis. Suppose that we have a variable called `male`. A code of `1` means the participant is a male and a code of `0` means the participant is female, and we want to restrict the analysis to males. To do this, we would say `if male == 1`. Here we use two equal-signs, which is the Stata equivalent to the verb "is". So the command means "if male is coded with a value of `1`". Why the two equal-signs? The statement `male` = 1 literally means that the variable called `male` is a constant value of `1`, but males are coded as `1` and females are coded as `0` on this variable. Sometimes we want to run a command on a subset of observations, and we use the qualifier `in`. For example, we might have a command `summarize age education in 1/200`, which would summarize the first 200 observations.

Each command has a set of `options` that control what is done and how the results are presented. The options vary from command to command. One option for the command `summarize` is to obtain detailed results, summarizing the variables in more ways. If we wanted to do a detailed summary of scores on age and education for adult males, the command would be

```
. summarize age education if male == 1 & age > 17, detail
```

Although the command structure is fairly simple, it is absolutely rigid. This example used the ampersand (`&`), not the word "and". If we had entered the word "and", we would have received an error message. Here are more examples:

```
. summarize age education if sex == 0
. summarize age education if sex == 1 & age > 64
. summarize age sex if sex == 0 & age > 64 & education == 12
```

When you have missing values stored as `.` or `.a`, `.b`, etc., you need to be careful about using the `if` qualifier. Stata stores missing values internally as huge numbers that are bigger than any value in your dataset. If you had missing data coded as `.` or `.a` and entered the command `summarize age if age > 64`, you would include people who had missing values. The correct format would be

```
. summarize age if age > 64 & age < .
```

The `< .` qualifier at the end of the command is strange to read (less than dot) but necessary. Table 4.1 shows the relational operators available in Stata.

(Continued on next page)

Table 4.1. Relational operators used by Stata

Symbol	Meaning
`==`	is or is equal to
`!=` or `~=`	is not or is not equal to
`>`	is greater than
`>=`	is greater than or equal to
`<`	is less than
`<=`	is less than or equal to

The `in` qualifier specifies that you will perform the analysis on a subset of cases based on their order in the dataset. If we had 10,000 participants in a national survey and we wanted to list the values in the dataset for `age`, `education`, and `sex`, this would go on for screen after screen after screen, which would be a waste of time. We might want to list just the data on `age`, `education`, and `sex` for the first 20 observations by using `in 1/20`. The `1` is where you start (called the first case), the "`/`" is read as "to", and the `20` is the last case listed. Thus `in 1/20` tells Stata to do the command for the cases numbered from 1 to 20, or the first 20 cases. The full command is

```
. list age education sex in 1/20
```

Listing just a few cases is usually all you need to check for logical errors. Most Stata dialog boxes include an if/in tab for restricting data.

The final feature in a Stata command is a list of options. You must enter a comma before you enter the options. As you learn more about Stata, the options become increasingly important. If you do not list any options, Stata gives you what it considers basic results. Often, the basic results are all you will want. The options let you ask for special results or formatting. For example, in a graph, you might want to add a title. In frequency tabulation, you might want to include cases that have missing values. One of the best reasons for using dialog boxes is that you can discover options that can help you tailor your results to your personal taste. Dialog boxes either include an Option tab or have the options as boxes that you can check on the Main tab. The most common mistake a beginner makes when typing commands directly in the Command window is leaving the comma out before specifying the options.

Here are a few Stata commands and the results they produce. You can enter these commands in the Command window.

```
. use http://www.stata-press.com/data/agis2/firstsurvey_chapter4
. summarize

    Variable |       Obs        Mean    Std. Dev.       Min        Max
-------------+--------------------------------------------------------
          id |        20        10.5     5.91608          1         20
      gender |        20         1.5    .5129892          1          2
   education |        20       14.45    2.946452          8         20
      sch_st |        18    3.444444    1.149026          2          5
     sch_com |        20         3.5    1.395481          1          5
-------------+--------------------------------------------------------
      prison |        17    3.176471    1.550617          1          5
     conserv |        19    2.947368    1.544657          1          5
```

This `summarize` command does not include a variable list, so it assumes that all variables will be summarized. It has no `if/in` restrictions and no options, so it summarizes all the variables, giving us the number of observations with no missing values, the mean, the standard deviation, and the minimum and maximum values. The statistics for the `id` variable are not useful, but it is easier to get this for all variables than to list all the variables in a variable list, dropping only `id`.

We can add the option `detail` to our command to give more detailed information. Do this for just one variable.

```
. summarize education, detail

                      Years of education
-------------------------------------------------------------
      Percentiles      Smallest
 1%            8              8
 5%          9.5             11
10%         11.5             12       Obs                  20
25%           12             12       Sum of Wgt.          20

50%         14.5                      Mean              14.45
                        Largest       Std. Dev.      2.946452
75%         16.5             17
90%           18             18       Variance       8.681579
95%           19             18       Skewness      -.1636124
99%           20             20       Kurtosis       2.522208
```

As expected, this method gives us more information. The 50% value is the median, which is 14.5. We also get the values corresponding to other percentiles, the variance, a measure of skewness, and a measure of kurtosis (we will discuss skewness and kurtosis later).

Next we will use the `list` command. Here are four commands you can enter, one at a time, to get three different listings.

```
. list gender education prison in 1/5

     +-------------------------------------+
     | gender   educat~n            prison |
     |-------------------------------------|
  1. |  woman         15          too long |
  2. |    man         12   much too lenient|
  3. |    man         16       about right |
  4. |    man          8                 . |
  5. |  woman         12       about right |
     +-------------------------------------+

. list gender education prison in 1/5, nolabel

     +----------------------------+
     | gender   educat~n   prison |
     |----------------------------|
  1. |      2         15        4 |
  2. |      1         12        1 |
  3. |      1         16        3 |
  4. |      1          8        . |
  5. |      2         12        3 |
     +----------------------------+

. numlabel _all, add

. list gender education prison in 1/5

     +--------------------------------------------+
     |   gender   educat~n                 prison |
     |--------------------------------------------|
  1. | 2. woman         15            4. too long |
  2. |   1. man         12    1. much too lenient |
  3. |   1. man         16         3. about right |
  4. |   1. man          8                      . |
  5. | 2. woman         12         3. about right |
     +--------------------------------------------+
```

The first command shows the first five cases. Notice that the variable `education` appears at the top of its column as `educat~n`. We can use names with more than eight characters, but some Stata results will show only eight characters. Stata did this by keeping the first six characters and the last character and inserting the tilde (~) between them. Because we assigned value labels to `gender` and `prison`, the value labels are printed in the list. However, notice that the numerical values are omitted.

The second command adds the option `nolabel`, which gives us a listing with the numerical values we used for coding, but not the labels. The next command, `numlabel _all, add` adds the values to all variables (the `_all` tells Stata to apply this to all variables). If we wanted to turn this off later, we would enter `numlabel _all, remove`. Finally, the last listing gives us both the values and the labels for each variable.

4.3 Getting the command from the menu system

You may be asking yourself how you will ever learn to use all the options and qualifiers. One way is to read the Stata documentation, but it is often easier to use the menus and record the command in a do-file. Stata has a window called a Do-file Editor that is a simple text editor in which you can enter a series of commands. You can run all the commands in this file or just some of them. You can then edit, save, and open them at a later date. Saving these do-files means not only that you can replicate what you did and make any needed adjustments but also that you will develop templates you can draw on when you want to do a similar analysis. To open the Do-file Editor window, use Window ▷ Do-file Editor ▷ New Do-file. You can also open the Do-file Editor by clicking on the toolbar icon that looks like a notepad (figure 4.1); it is located below User.

Figure 4.1. The Do-file Editor icon

Stata 10 allows you to have multiple do-files open at once, but we will use only one do-file in this book. When you gain confidence using Stata, the ability to have several do-files open simultaneously is a useful feature. You might use one do-file primarily for data management purposes. You might use one for general analysis of the data. You might use a third for analyzing a subset of the data. You might even work on a couple of different projects in one session. The use of multiple do-files in a session is sort of like juggling balls, and some of us are better at that than others. For now, one Do-file Editor is all we need. When several do-files are open, clicking on the arrow to the right of the icon shows the name of each file. A blank do-file appears in figure 4.2.

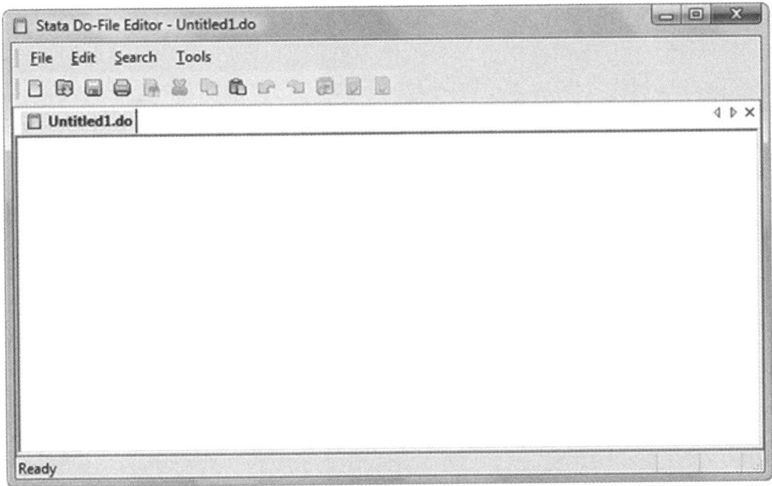

Figure 4.2. The Do-file Editor

When you click on another window, the Do-file Editor can be hidden by other windows. You can bring it to the front by clicking on it in the system toolbar or by using the Alt+Tab key combination to move through the windows until the one you want is on top. You can avoid having it disappear by arranging your desktop so it does not overlap with the Stata window. If you have two monitors, it is convenient to have the main Stata window on one monitor and the Do-file Editor on the other monitor.

The Do-file Editor lacks special features, such as underlining, special fonts, and graphic characters you get with a word processor such as MS Word or Apple's Pages. When you use a word processor, the program is adding features that you never realize are there because they are hidden from you. As a plain-text ASCII editor, the Do-file Editor includes only the letters and characters that are on your keyboard. Let's make a Do-file Editor program to summarize the variable `education` and then create a histogram of `education`.

First, we need to open the dataset `firstsurvey_chapter4.dta`. To do this, select File ▷ Open..., browse the directory that contains `firstsurvey_chapter4.dta`, and click on Open. This not only opens the dataset but also places the command in the Results window. Let's assume that the file is in `C:\data`. Here is the command the dialog box generates:

```
. use "C:\data\firstsurvey_chapter4.dta", clear
```

If you had it stored on a flash drive, you would see

```
. use "E:\firstsurvey_chapter4.dta", clear
```

The point is that Stata has written the actual command in the Results window, and we can copy this to the Do-file Editor. Then when we use this program in the future, we will be sure to use the same dataset. Here is how you copy it:

- First, highlight the command in the Results window. Do not include the `.` or the space in front of the command; highlight the command just as you would highlight a sentence in Word. Right-click on the highlighted text and select Copy Text from the menu. Switch to the Do-file Editor, click inside the blank window, and then right-click inside the window and select Paste from the menu.

- You can also right-click on the command in the Review window and select Send to Do-file Editor.

If you enter a path to the file, it helps to enter quotes around the path and filename. Notice that the comma appears after the quote rather than within the quote. Your English teacher would object, but this is the way Stata does it. The option `clear` is important to include. It clears any dataset that we might already have in memory. Stata does not want you to accidentally delete a dataset before you have a chance to save it, so Stata makes you explicitly clear the memory before you open a new dataset. If you already have a dataset open, you should save it before opening a new dataset.

Next let's do a summary statistical analysis; select Statistics ▷ Summaries, tables, and tests ▷ Summary and descriptive statistics ▷ Summary statistics. On the Main tab, we could list the variables to be summarized, but for this example we will leave it blank, meaning that we want to summarize all the variables. There is no Options tab with the `summarize` command because the options appear on the Main tab. Click on the radio button for *Display additional statistics*. Next click on the by/if/in tab; type `gender == 2` in the *If: (expression)* box; and check the box for *Use a range of observations* and enter the range from 1 to 15. If we had clicked on the box *Repeat command by groups*, we could have selected a variable from the pulldown menu that follows, and the `summarize` command would be done separately by each of these groups. For example, instead of using the `if` qualifier to restrict the `summarize` command to female (`gender == 2`), putting `gender` in the *Repeat command by groups* box would have given us separate analyses for female and for male.

Now let's see the features of the `summarize` dialog box. We have already discussed the role of the OK and Submit buttons at the bottom of the dialog box. To the left of them are three icons: ?, R, and Copy. The ? icon gives us a help screen explaining the various options. The explanations are brief, but there are examples at the bottom of the Viewer. The R icon clears the dialog box. Just to the right of the R icon is an icon that looks like two pages. If you click on this icon, the command is copied to a buffer. Figure 4.3 shows this Copy icon, which will copy the command to the clipboard rather than submit it to Stata for processing. Click on that once.

Figure 4.3. Copy dialog command to clipboard icon

Now switch to the Do-file Editor window, and paste the command where you want it. Put the cursor in the Do-file Editor window just below the command that opened our data, and press Ctrl+V to paste the command. This adds a second command to the do-file (see figure 4.4).

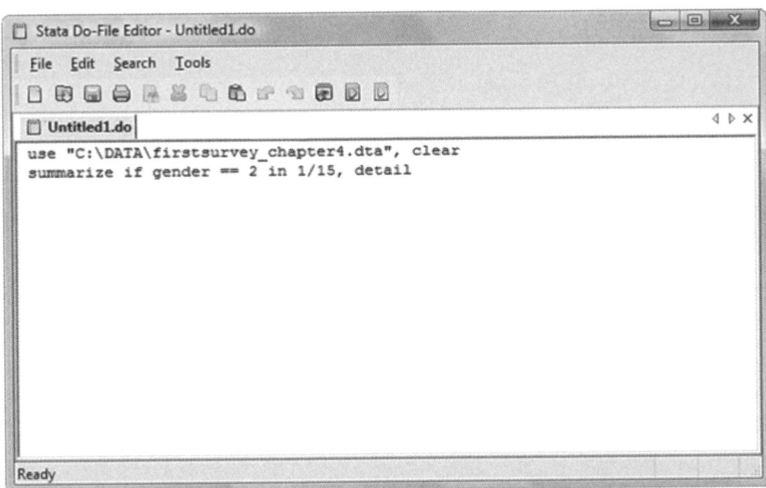

Figure 4.4. Commands in the Do-file Editor window

Now let's add a graph. We will create a pie chart of the prison variable by selecting Graphics ▷ Pie chart. On the Main tab, click on *Graph by categories*. Next type the variable prison in the box labeled *Category variable*. The variable prison contains the categories (much too lenient, too lenient, about right, too long, and much too long) to include in the pie chart. Next click on the Slices tab. In the middle of the dialog that opens is a section called *Labels*. Click on *Label properties (all)*. Here you want to select the *Label type*, which includes options that let us assign labels to the slices in the pie chart. The options are *None*, *Sum*, *Percent*, and *Name*. Select *Name* to label each piece of the pie with the name of the category it represents. You can vary the *Color*, *Size*, and *Orientation* of the labels as you judge appropriate. Now, go to the Options tab and check the box by *Exclude observations with missing values (casewise deletion)*. If there are any missing values, we will not want a piece of pie to represent them.

Next click on the Titles tab. Here we can enter a suitable title: Length of Prison Sentences. At the bottom of this tab is a place to enter a *Note*. Enter the filename of the dataset, firstsurvey_chapter4.dta.

At this point, we can click on the Submit button to make sure the graph looks okay. The pie chart looks pretty good on our computer screen, but if we are printing it in black and white, there will be a problem distinguishing between the slices of the pie. If we need a black and white version of the pie chart, we can go back to our dialog and select the Overall tab. Under *Scheme*, we can select a monochrome option. (It is fun to try a variety of these schemes.)

Once we like our pie chart, we want to record the command in our do-file. Click on the Copy icon on the bottom left of the dialog window, go to the Do-file Editor, and paste the command just below the summarize command. What happens? The Command window is not wide enough for us to see the entire line. We could leave it like this and scroll over to read the complete command, or we could break the command so it covers two or more lines in the Do-file Editor. This is a problem because Stata needs to know that both lines are part of the same command. One solution is to find a nice spot for a break, insert a space and three forward slashes, and press the Enter key to force the rest of the command to go to the next line. It is nice to indent the second line a couple of spaces so that we are reminded that it is part of the command for the graph. Here is what we get:

```
graph pie, over(prison) title(Length of Prison Sentences) ///
  note(firstsurvey_chapter4.dta) plabel(_all name) cw
```

Here is one example of why creating a do-file is useful. Let's assume that we have saved the do-file and, three weeks later, we decide that in addition to a pie chart for views on length of prison sentences, we want a pie chart on political conservatism. We could go back to our menu system and select all of the appropriate options. This could be a problem if we had adopted a special scheme and cannot remember its name or if we had picked some specialized options.

We could also work directly in the Do-file Editor. While in the Stata Do-file Editor, we could also copy the command we already have to do the prison graph and make the necessary changes. All we need to change is `over(prison)` to `over(conserv)` and `title(Length of Prison Sentences)` to `title(Political Conservatism)`.

We now have four commands in our do-file (`use`, `summarize`, and two `graph pie` commands). It is always good to add comments, which can go anywhere in the program. One way to make a comment is to start the line with an asterisk. Anything that comes after the asterisk will be treated as a comment and not as part of the program. Because we should keep a hard copy of the program, it would be good to include the filename and path for the do-file so you can find it later. It also makes sense to include a description of what the program does. This description is especially important in complex files. You might want to include other information, such as your name and the date you created the program.

Because Stata will have new versions of the program released that may be different, we should specify the version of Stata for which we wrote the program. If Stata 11 were to contain changes to the format so that the commands would not work, the program will know to use the commands from version 10, but only if we tell it that we wrote this program for version 10. StataCorp is remarkably committed to keeping older versions of their software functioning. Even though newer versions may change a few commands to add new capabilities, the program maintains the ability to execute commands from older versions.

Before we do anything with this program, save it under the name `program1.do` by selecting File ▷ Save As.... Our do-file should look like figure 4.5 at this point.

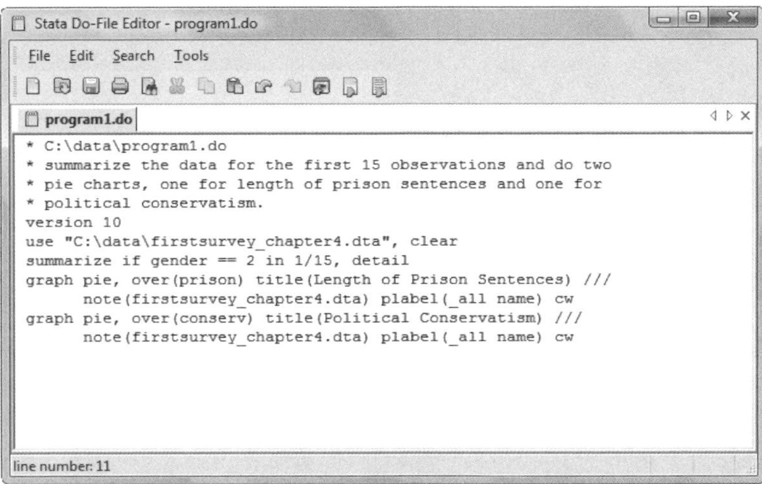

Figure 4.5. Do-file for summarizing data and creating two pie charts

We can run the whole program or just part of the program. If you hold the cursor over the second icon from the right in the Do-file Editor (see figure 4.6), a tiny box saying Run appears. Clicking on this icon runs the entire program. Try it. If you hold the cursor over the icon on the extreme right, a tiny box saying Do appears. To run just the graph for the belief about the length of prison terms, highlight that command. Hold the cursor over the icon on the extreme right; the tiny box should now say Do selected line. Click on this icon. If the data is not already open, you would need to highlight the lines for this graph and the lines above it that open the dataset. Try various combinations.

Figure 4.6. Row of icons in Do-file Editor

Most experienced Stata users use the Do-file Editor. It is a necessity if we want a compact record of our programs for when we need to run them again, revise them, or use them as a template for related applications. How Stata users use the Do-file Editor varies enormously, and you should decide for yourself how you want to use it. Some experienced users simply write the commands into the Editor without bothering with the dialog boxes. With some commands, this is the easiest way to do it because the commands are short and the syntax is simple to remember. Some use the dialog system, try the command to make sure it works, and then copy it to the Editor. Others will run the command in the Command window, and if they like the results, they will copy it from the Results window or the Review window (by right-clicking) and then paste it where they want it in the Do-file Editor. Most users use a mixture of methods. The important point is not how you do it but that you do it at all so you can have a permanent record of your work.

Stata do-files for this book

The web page for this book, http://www.stata-press.com/data/agis2/, has do-files for each chapter along with the datasets we have used. You can copy the do-files and datasets to your computer and reproduce the results in this book. These may also be useful as templates when you do your own work because they can be modified as you need. You will have to change the paths in these files to the paths where you have stored your data unless you have stored the data in Stata's default directory, `C:\Data`. You could also change the default directory. If you stored the data files in `C:\Data\agis2` you could enter the command `cd "C:\Data\agis2"` to make that the default directory. Then, to open a file for Stata to use, you would simply type `use` *filename*, where *filename* is the name of the data file you want to use.

4.4 Saving your results

Many people start using Stata to do their homework for a class in statistical methodology. In those cases, the datasets are often fully prepared, and you will not need to keep a record of how the data were created, how you labeled variables, how you recoded some items and dealt with missing values, or how you produced your results. In fact, many of the analyses will have short results. Here it may be simplest and best to save your results by highlighting the text that you want to save in the Results window and then right-click and select Copy Text from the menu (or use Ctrl+C). You can then paste it into your favorite word processor. It is a good idea to include the commands that are in the Results window, as these give you a record of what you did. Except when there is no data manipulation, commands like these are no substitute for a do-file that includes everything you did in preparing the data.

When you copy results from Stata's Results window to your word processor, the format may look like a hopeless mess because Stata output is formatted using a fixed-width font; when you paste your results, things will likely not line up properly. The simplest solution to this is to change the font and probably the font size, depending on your margins. I usually use the Courier or Courier New font at 9 point. Sometimes the lines may still wraparound, and you may need to widen the margins. Most Stata results will fit nicely if you have 1-inch margins and a 9- or 10-point Courier font. With most word processors, you change the font of the lines that contain the results by highlighting these lines and then selecting the Courier font and the 9-point font size. The rest of your document will then be in whatever font you like.

(*Continued on next page*)

Saving tabular output

If you are using the Windows version of Stata and have tabular output (say, the results of a `summarize` command), you may want to select just that portion of the text that appears as a table, right-click, and select Copy Table, Copy Table as HTML, or Copy as Picture from the menu. You can then paste this text into your word processor. The Copy Table as HTML option pastes it as a table like you would see on a web page. However, if you are working with a specific-style format, such as the APA requirements for tables, you will need to reenter the table in your word processor to meet those requirements. The Copy as Picture option works nicely as long as you do not need to format the output to a particular style. The Copy as Picture option takes a picture of the output you highlighted. If you paste this picture into a word processor, it will look just like it did in the Results window. You will then be able to resize it, but you will not be able to edit it.

The Mac version of Stata copies images to the Clipboard in the PDF format for the best possible output. If you are using the Mac version of Stata and are having a problem using the Copy as Picture option, open the General Preferences dialog and check the box by Copy Images to the Clipboard as PICT. Stata will now copy images to the Clipboard in the PICT format, which is supported by older Mac applications, such as Microsoft Office 2004.

However, the PICT format does not support rotated text and has been depreciated by Apple in favor of the PDF format. We encourage you to upgrade the software in which you are pasting Stata output to one that supports Clipboard data in the PDF format, such as Microsoft Office 2008. Once you upgrade, uncheck the box by Copy Images to the Clipboard as PICT in Stata so that Stata will copy images to the Clipboard in the PDF format, creating the best possible output.

As you progress in your class or as you start to use Stata in your own research, you will find that copying and pasting is really not up to the task of creating a record of your work. For more complex work, you will want to use log files, which we will now discuss.

4.5 Logging your command file

Stata can write a copy of everything that is sent to the Results window into a file. The file is usually called a *log file*. When you start logging, you can create a new file, or you can add on to the end of an existing log. You can temporarily suspend logging at any time and then restart it again. If you do not have a running Stata session, start one now, and let's take a look at output logs.

We can open a log by selecting File ▷ Log ▷ Begin..., which will bring up the file selector window. Navigate to the directory in which we want to keep our log, and then enter the name (or select it from the list). By default, Stata will save this log using a proprietary format called Stata Markup and Control Language (SMCL) that only Stata can read in a Viewer or Results window. The log will have a *filename*`.smcl` name, where `.smcl` is the extension. It looks nice in Stata but is limited because, although it is a text file, the SMCL tags will be displayed since the word processor or other text editors will not understand them.

The other format is called "log", and this is a simple text file that your word processor can read. We need to pick the option of having a log file rather than a SMCL file from the dialog box. At the bottom of the *Begin logging Stata output* dialog box, we can specify *Save as type*. From the pulldown menu, select *Log (*.log)* for the type. Like the Results window, it uses a fixed-width format, and if you insert this file into a word processor, you need to make sure that the font is fixed width (e.g., Courier) and the font size is small enough (e.g., 9 point). Make sure to select a location where we can find this log file. We can insert this log into an open file in our word processor, or we can open the log file itself in our word processor. Because the extension will be `.log`, when we open it into a word processor, we need to make sure that the word processor is not looking for just certain other extensions. For example, in MS Word, you would need to browse for the log file where the type is `*.*` rather than `*.doc`.

If you are using MS Word as your word processor, you have two options when working with a `*.log` file. You can open it as a new document in Word as long as you remember to change the file type from `doc` or `docx` to `*.*` or `show all`. If you do this, the Word document will have a fixed font. You may need to change the font size, say to 9 point, or make the margins wider. If you have an existing Word document, you need to use the Insert option rather than the Open option, and insert it where you want the text.

Open a new log file called `results.log`. Make sure to specify that it is a log file rather than a SMCL file. Then run a `summarize` command on the file. Now open the log by selecting File ▷ Log ▷ View..., which will show us the basic information about our file, where it is stored and the date and time it was created, and give the results of our summary in a Viewer. When we go back to our Command window, the Viewer will seem to disappear, but it will be on the taskbar at the bottom of our screen. We can click on it to open it again.

Run a few more commands, such as a tabulation and a graph. Now select the Viewer from the Windows toolbar, and the Viewer returns, but it goes only as far as the original `summarize` command. However, at the top of the Reviewer is a Refresh button. Clicking on this button will update the Viewer to include the `tabulate` and the `graph` commands. What happened to the graph? Unfortunately, the log does not include the graphic output in the log file.

Experienced users use log files a lot. For beginners, log files may not be necessary. Also, if you make a lot of mistakes and need to run each command several times before you get it just right, the log will have all the bad output that you do not want, along with the good output that you do want. You might want to pause the log file while you try out a command or program. Then when you have the command the way you want it, you can restart the log to minimize the bad output. To do this, select File ▷ Log ▷ Suspend or Resume, respectively.

A strength of using log files is that they provide a record of both your commands and your results. The commands precede each result, so you can immediately see what you did. The advantage of the program file from the Do-file Editor is that the commands are all together in a compact form, so it is easier to review the programming you did. A major limitation of the log file is that it does not save graphic output, such as the pie chart we did in this chapter. Graphs need to be saved by right-clicking while the pointer is on the graph and then choosing options, whether you want to save it (select Save Graph...) or copy it (select Copy Graph) so you can paste it into your word processor.

4.6 Summary

Working with Stata commands is useful. Let's go over what you have learned in this chapter:

- How to open a Do-file Editor and copy commands from the Results window
- How to use the dialog boxes in Stata to generate Stata commands and then copy these to the Do-file Editor
- How to string a series of commands together in the Do-file Editor and add comments to this file
- How to run an individual command and groups of commands from the Do-file Editor
- How to save your do-file and retrieve it for later analysis
- How to cut and paste between Stata and Word
- How to create and view a log file

This is a lot of new knowledge for you to absorb. Never feel bad if you need to review this material because you forgot a step along the way. Even as you gain experience with Stata, it will be useful to keep this chapter handy as a resource.

This chapter has focused on the mechanics of using the Do-file Editor, writing a simple program, and saving your results. As you go through the following chapters, you will learn how to write more complex programs and to do more complex data management.

The rest of the book will focus on performing graphic and statistical analysis, building on what we have done so far. Most people learning a statistics program want to learn how to do analyses rather than what we have done so far. Still, what we have done so far sets the groundwork for doing the analyses. Chapter 5 will go over graphic presentations and descriptive statistics.

4.7 Exercises

1. Open the `firstsurvey_chapter4.dta` file by selecting File ▷ Open.... Open a Do-file Editor, and copy the command that opened the dataset into the editor. Open the dialog box to summarize the dataset, run the **summarize** command for all the variables, and copy this command from the Results window to the Do-file Editor. Save this do-file as `4-1.do` in a place where you can find it.

2. Open the do-file you created in the first exercise, and add appropriate comments. Save the new do-file under the name `4-2.do`.

3. Open the file `4-2.do`. Put your cursor in the Do-file Editor just below the command that opened the dataset and above the command that summarized the variables (you will need to insert a new line to do this). Type the command **describe** in the Do-file Editor. Add a command at the bottom of the file that gives you the median score on education. Save the new do-file under the name `4-3.do`.

4. Open `4-3.do`, run the **describe** command and the command that gave you the median score on education, highlight the results, paste the results into your word processor, and change the font so that it looks nice.

5. Open `4-3.do`. Open a log file using the file type of log. Call the file `4results.log`. Run the entire `4-3.do` file, and exit Stata. Open a new session in your word processor, and open your log file into this session. Format it appropriately.

5 Descriptive statistics and graphs for one variable

5.1 Descriptive statistics and graphs

The most basic use of statistics is to provide descriptive statistics and graphs for individual variables. Advanced statistics and graph presentation can disentangle complex relationships between groups of variables, but for many purposes, simple descriptive statistics and graphs are exactly what is needed. Virtually every issue of a major city newspaper will have many descriptive statistics, and most issues will have one or more graphs. One article might report the percentage of teenagers who are smoking cigarettes. Another article might report the average value of new homes. Each spring, there will be one or more articles estimating the average salary new college graduates will earn.

If you are working in a position related to public health, economics, or any of the social sciences, you will be a regular consumer of descriptive statistics, and many of you will be producers of these statistics. A parole office may need a graph showing trends in different types of offenses. A public health agency may need to demonstrate the need for more programs focused on sexually transmitted diseases: How much of a problem are sexually transmitted diseases? Is the problem getting worse or better?

Our society depends more and more on descriptive statistics. Policy makers are reluctant to make decisions without knowing the appropriate descriptive statistics. Social programs need to justify themselves to survive, much less to grow, and descriptive statistics and graphs are critical parts of this justification. In this chapter, you will learn how to produce these statistics and graphs using Stata.

5.2 Where is the center of a distribution?

Descriptive statistics are used to describe distributions. Three measures of central tendency describe the middle of the distribution: mode, median, and mean. All three of these are called averages, but they can be different values. When you read a newspaper article, it may say that the average family in a community has two children (this is probably the median, but it could be the mode) or that the average income in the community is $55,218 (also probably the median, but it could be the mean). The article might say that the average SAT score at your university is 1,120 (probably the mean). It might say that the average person in a community has a high school diploma (this could be the mean, median, or mode). It is important to know when each central tendency measurement is appropriate.

The mode is the value or the category that occurs most often. We might say that the mode for political party in a parliament is the Labor Party. This would be the mode if there were more members of the parliament who were in the Labor Party than members of any other party. If we said the mode was 17 for age of high school seniors, this means that there are more high school seniors who are 17 than there are seniors at any other age. This would be a reasonable measure of central tendency because most high school seniors are 17 years old. The mode represents the average in the sense of being the most typical value or category. If there is not a category or value that characterizes a distribution clearly, the mode is not descriptive of the central tendency of a distribution. For example, you would not say the mode for height of an eighth-grade class because each adolescent might be a different height, and there is no single height that is typical of eight graders.

When you have unordered categorical variables such as gender, marital status, or race/ethnicity, the mode is the only measure of central tendency. Even here, the mode is helpful only if one category is much more common than the others. If 79% of the adults in a community are married, saying that the modal marital status is married is a fair description of the typical member of the community. However, if 52% of adults in a community are female and 48% are male, it does not make much sense to say that the modal gender is female because there are nearly as many men as there are women.

The median is the value or the category that divides a distribution into two parts. Half of the observations will have a higher value, and half will have a lower value than the median. The median can be applied to categories that are ordered (political liberalism, religiosity, job satisfaction) or to quantitative variables (age, education, income). If we said that the median household income of a community is $55,218, we mean that half the households in that community have an income more than $55,218 and half the households have an income less than $55,218. When we used the `summarize, detail` command in chapter 4, we saw that Stata refers to the median as the 50th percentile.

The median is not influenced by extreme cases. If Bill Gates moved to this community, his multibillion dollar income would not influence the median. He would simply be in the half of the distribution that made more than the median income. Because of this property, the median is sometimes used with quantitative variables that are skewed (a distribution is skewed if it trails off in one direction or the other). Income trails off at the high end because relatively few people have huge incomes.

The median is occasionally used with variables that are ordered categories. When there are relatively few ordered categories, there may not be a category that has exactly half the cases above and below it. You might ask people about their marital satisfaction and give them response options of (a) very dissatisfied, (b) somewhat dissatisfied, (c) neither satisfied nor dissatisfied, (d) somewhat satisfied, and (e) very satisfied. Because we usually code numbers rather than letters, we might code very dissatisfied as `1`, somewhat dissatisfied as `2`, neither satisfied nor dissatisfied as `3`, somewhat satisfied as `4`, and very satisfied as `5`. The median satisfaction for men might be in the category we coded with a `4`, somewhat satisfied. The median satisfaction for women might be `3`, neither satisfied nor dissatisfied.

More often, researchers compute the mean for variables like this, and the mean for men might be 4.21 compared with 3.74 for women. These values indicate that men are, on average, a little above the somewhat-satisfied level and the women are a little bit below the somewhat-satisfied level.

The mean is what lay people usually think of when they hear the word "average". It is the value every case would be, if every case had the same value. It is a fulcrum point that considers both the number of cases above and below it and how far they are above or below it. Although Bill Gates would scarcely change the median income of a community, his moving to a small town would raise the mean by a lot. Some people use M (recommended by the American Psychological Association) to represent the mean, and others use $\overline{X}$ (recommended by most statisticians). The formula for the mean is

$$\overline{X} = \frac{\Sigma X}{n}$$

In plain English, this says the mean is the sum of all the values, ΣX (pronounced sigma X or sum of X), divided by the number of observations, n. For example, if you had five college women who weighed 120, 110, 160, 140, and 210 pounds, respectively, the mean would be

$$\overline{X} = \frac{120 + 110 + 160 + 140 + 210}{5} = 148$$

From now on, we will use M instead of $\overline{X}$ to represent the mean.

What measure of central tendency should you use? This decision depends on the level of measurement you have, how your variable is distributed, and what you are trying to show (see table 5.1).

Table 5.1. Level of measurement and choice of average

Level of measurement	**Mode**	**Median**	**Mean**
Categorical, no order (nominal, e.g., gender)	Yes	No	No
Categorical, ordered (ordinal, e.g., social support)	Yes	Yes	Yes*
Quantitative (interval or ratio, e.g., age)	Yes	Yes	Yes

*Many researchers use the mean when there are several ordered categories.

- When you have categories with no order (gender, religion), you can use only the mode. The mode for religion in Saudi Arabia, for example, is "Muslim". Unordered categorical variables are called *nominal-level variables*.
- When you have ordered categories (religiosity, marital satisfaction), the median is often recommended. Such variables are often labeled as ordinal measures. You might read that the median religiosity response in Chicago is "somewhat religious". Ordered categories can be ordered along some dimension, such as low to high or negative to positive. When there are several categories, many researchers treat them as quantitative variables and use the mean. If religiosity has seven ordinal categories from `1` for not religious at all to `7` for extremely religious, you might use the mean by treating these numbers from 1 to 7 as if they are an interval-level measure. You might say that the mean is 3.4, for example.
- When you have quantitative data (meaningful numbers), you can use the mean, median, or mode. Quantitative data are often called *interval-level variables*. You will usually use the mean. If, however, the variable is extremely skewed, you would use the median.

Suppose that we wanted an average value for the number of children in households that have at least one child. The distribution is highly skewed in a positive direction because it trails off on the positive tail (figure 5.1).

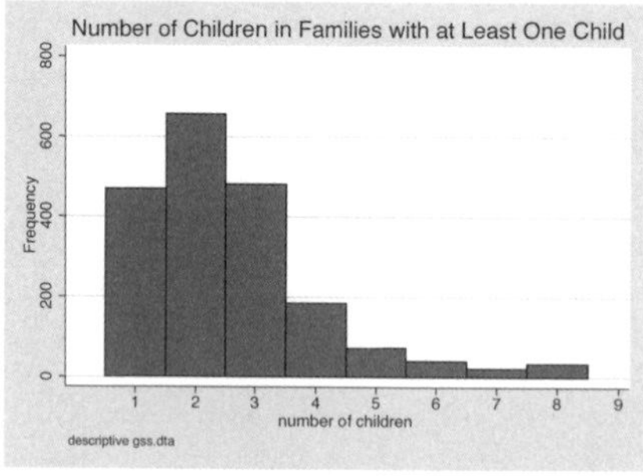

Figure 5.1. How many children do families have?

In this distribution, the mode is 2, the median (Mdn) is 2, and the mean (M) is 2.5. Notice how the small number of families with a lot of children drew the mean toward the tail but did not influence either the mode or the median. When a distribution is skewed, the mean will be bigger or smaller than the median, depending on the direction the distribution trails off.

There are two specialized averages you may want to use. The harmonic mean is useful when you want to average rates. Suppose you go on a 60 mile bicycle ride where the first half of the course is a major hill climb and the second half of the course is a major descent. You might average just 5 miles per hour for the first half and then average 25 miles per hour for the second half of the ride. If we call your rate a, the harmonic mean (H) is

$$\begin{aligned} H &= \frac{n}{\frac{1}{a_1} + \frac{1}{a_2} + \cdots + \frac{1}{a_k}} \\ &= \frac{2}{\frac{1}{5} + \frac{1}{25}} \\ &= 8.33 \end{aligned}$$

This harmonic mean of $H = 8.33$ miles per hour is a much better estimate of your average speed for the 60 mile ride than the arithmetic mean which would be $(5+25)/2 = 15$ miles per hour.

The geometric mean is useful when you have a growth process where the growth is at a constant rate. This happens with population size or annual income such as having your income grow at a rate of 3% per year. If you made 52,500 in 2000 and made 73,500 in 2010, what did you make in 2005? The arithmetic mean $(52{,}500 + 73{,}500)/2 = 63000$ exaggerates your income in 2005. The geometric mean (G) is

$$\begin{aligned} G &= \sqrt[n]{a_1 \cdot a_2 \cdots a_n} \\ &= \sqrt{52500 \cdot 73500} \\ &= 62118.84 \end{aligned}$$

where $G = \$62{,}118.84$ is a much better estimate of your 2005 income.

The Stata command `ameans` *varlist* computes the arithmetic mean, the geometric mean, and the harmonic mean.

5.3 How dispersed is the distribution?

Besides describing the central tendency or average value in a distribution, descriptive statistics describe the variability or dispersion of observations. Are they concentrated in the middle? Do they trail off in one direction? Are they widely dispersed? Some suburbs are extremely homogeneous with rows of houses that are similar in style and value. These communities are highly concentrated around the average on a range of variables (income, education, ethnic background). Other communities are heterogeneous, and although they may have the same average values as the first community, they differ by having a mix of people who range widely on income, education, and ethnic background. This means that to understand a distribution, we need to know how it is distributed as well as its average value.

When there are only a few values or categories, we can use a frequency distribution (tabulation) to describe the variable. This distribution shows each value or category and how many people have that value or fall into that category. Stata calls this a tabulation. We can also use graphs to describe the dispersion of a distribution. When there are only a few categories being shown, the most common graphs to use are pie charts and bar charts.

When a variable is quantitative, we will usually want one number to represent the dispersion in the same way that we use one number to represent the central tendency. The standard deviation (SD) is used, especially with variables that have many possible values. If we say that the mean SAT score at a college is 1,000 (M = 1,000) and the SD is 100 (SD = 100), this means that nearly all the students (about 95% of a normal distribution are within two SDs of the mean) had scores between 800 and 1,200.[1] This is the tall, but skinny, distribution in figure 5.2. If another school has the identical mean (M = 1,000) but has an SD of 200, then nearly all of the students had scores between 600 and 1,400. This dispersion is much greater at the second school than at the first. We can see this in a graph of the two schools (figure 5.2). The smaller the SD, the more homogeneous is the distribution. From the graph, you can see how students at the first school are much more clustered around the mean (M = 1,000) than are the students at the second school. The greater the SD, the more heterogeneous is the distribution.

1. The numbers used do not include the essay portion of the SAT.

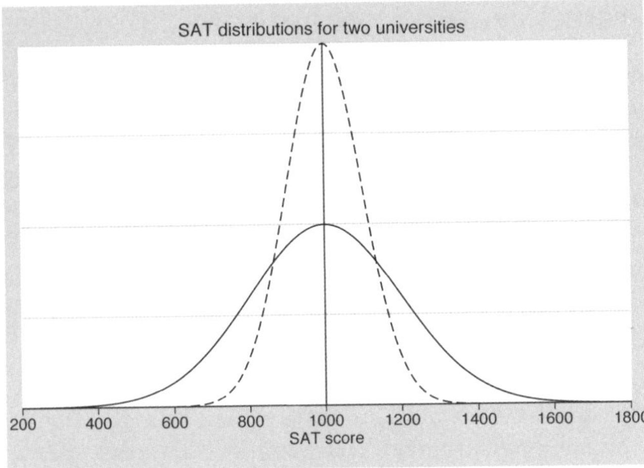

Figure 5.2. Distributions with same M = 1,000 but SDs = 100 or 200

Both distributions in figure 5.2 are normal and have identical means of 1,000 (M = 1000). The distribution that is tightly packed around the mean has an SD of 100 (SD = 100). The distribution that is more dispersed, the distribution that is low and wide in the figure, has an SD of 200 (SD = 200).

Skewness is known as the third moment of the distribution. A positive value indicates a positive skew, and we mentioned that income is often used as an example because there are relatively few people who have enormous incomes. A negative value indicates a negative skew. An example of a negatively skewed variable would be marital satisfaction. Surveys show that most married people are satisfied or very satisfied, few people are dissatisfied, and even fewer are very dissatisfied. Neither of the distributions in figure 5.2 is skewed. Both are symmetrical around their respective means.

Kurtosis is known as the fourth moment of the distribution. A distribution with high kurtosis tends to have a bigger peak value than a normal distribution. Correspondingly, a low kurtosis goes with a distribution that is too flat to be a normal distribution. The value of kurtosis for a normal distribution is 3. Some programs (both SAS and SPSS, for example) subtract 3 from the kurtosis to center it on zero, and some statistics books may use this approach, but Stata uses the correct formula. If you are reporting the kurtosis in a field where either SAS or SPSS are widely used programs, you need to know that for Stata a kurtosis of 3 indicates a normal distribution. In such fields, you might report the value of kurtosis minus 3 to be consistent with common practice in that field. When you do this, a kurtosis with an absolute value greater than 10 is problematic.

(Continued on next page)

5.4 Statistics and graphs—unordered categories

About all we can do to summarize a categorical variable that is unordered is to report the mode and show a frequency distribution or a graph (pie chart or bar chart). In this chapter, you will use a dataset, `descriptive_gss.dta`, that includes three categorical, unordered variables. `sex`, `marital`, and `polviews` are nominal-level variables. The variable `sex` is coded as `male` or `female`, `marital` is coded by marital status, and `polviews` is coded by political view. There is no order to `sex` in that being coded `male` or `female` does not make one higher or lower on `sex`. Similarly, there is no order to `marital` in that having a particular status (i.e., never married, married, separated, divorced, or widowed) does not make one higher or lower on marital status. These are just different statuses.

We can use the `tabulate` command to get frequency distributions for `sex`, `marital`, and `polviews`: the complete command is `tab1 sex marital polviews`. This is so simple that you probably want to enter it directly, but if you want to use the dialog box, select Statistics ▷ Summaries, tables, and tests ▷ Tables ▷ Multiple one-way tables; see figure 5.3. Be sure to select Multiple one-way tables rather than One-way tables.

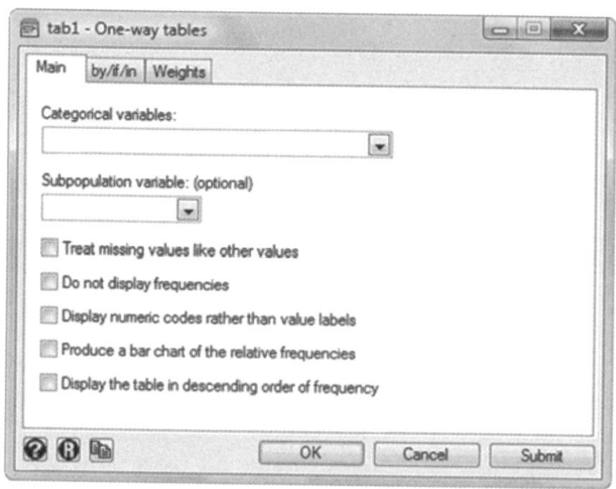

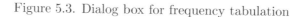

Figure 5.3. Dialog box for frequency tabulation

Tabulating a series of variables and including missing values

The command `tabulate` can be abbreviated in two ways. To do a tabulation of a variable, say, `educ`, the command is `tabulate educ`. To do a tabulation of a series of variables, you must change the command. Suppose that you want to do a tabulation on `educ`, `sex`, and `polviews`, and you want to do this using a single command. You would use `tab1 educ sex polviews`. Sometimes you might want to have the tabulation show missing values. To do this, add the option `missing`. To do a tabulation of the three variables in one command and to show the missing values, the command is `tab1 educ sex polviews, missing`.

Using the dialog box, we enter the variables `sex`, `marital`, and `polviews`. Clicking on OK produces the following result:

```
. tab1 sex marital polviews
-> tabulation of sex

respondents |
        sex |      Freq.     Percent        Cum.
------------+-----------------------------------
       male |      1,228       44.41       44.41
     female |      1,537       55.59      100.00
------------+-----------------------------------
      Total |      2,765      100.00
-> tabulation of marital

      marital |
       status |      Freq.     Percent        Cum.
--------------+-----------------------------------
      married |      1,269       45.90       45.90
      widowed |        247        8.93       54.83
     divorced |        445       16.09       70.92
    separated |         96        3.47       74.39
never married |        708       25.61      100.00
--------------+-----------------------------------
        Total |      2,765      100.00
-> tabulation of polviews

  think of self as |
        liberal or |
      conservative |      Freq.     Percent        Cum.
-------------------+-----------------------------------
  extremely liberal |         47        3.53        3.53
            liberal |        143       10.74       14.27
   slightly liberal |        159       11.95       26.22
           moderate |        522       39.22       65.44
slghtly conservative |        209       15.70       81.14
       conservative |        210       15.78       96.92
extrmly conservative |         41        3.08      100.00
-------------------+-----------------------------------
              Total |      1,331      100.00
```

The first tabulations for `sex` and `marital` tell us a lot. Some 55.6% of our sample of 2,765 adults are women, and 44.4% are men. We have 1,537 women and 1,228 men. For the `marital` variable, 45.9% (1,269) of adults are married. This is a clear mode because this marital status is so much more frequent than any of the other statuses. By contrast, the mode for `sex` is not as predominant a category.

A Stata user, Ben Jann, wrote a command called `fre`, which provides more details than `tab1`. To install the `fre` command, enter `findit fre` and follow the link. Instead of using `findit fre`, you could install it directly by running the command `ssc install fre`.

Enter the command `fre sex marital polviews`. This gives you the result we had before, but this is more useful when there are missing values. In this dataset, there are 1,434 people who were not asked—or at least did not report—their political views. The `fre` command provides the percentage of the total sample who selected each category

(18.88% picked moderate) and the percentage of valid (nonmissing) people who selected each category (39.22% picked moderate). We also get the cumulative percentages for the valid responses. Because many variables will have missing values, this is a nice command to use. A second advantage of using the `fre` command is that it shows the numerical value assigned to each category as well as the value label.

```
. fre sex marital polviews
sex — respondents sex
```

		Freq.	Percent	Valid	Cum.
Valid	1 male	1228	44.41	44.41	44.41
	2 female	1537	55.59	55.59	100.00
	Total	2765	100.00	100.00	

```
marital — marital status
```

		Freq.	Percent	Valid	Cum.
Valid	1 married	1269	45.90	45.90	45.90
	2 widowed	247	8.93	8.93	54.83
	3 divorced	445	16.09	16.09	70.92
	4 separated	96	3.47	3.47	74.39
	5 never married	708	25.61	25.61	100.00
	Total	2765	100.00	100.00	

```
polviews — think of self as liberal or conservative
```

		Freq.	Percent	Valid	Cum.
Valid	1 extremely liberal	47	1.70	3.53	3.53
	2 liberal	143	5.17	10.74	14.27
	3 slightly liberal	159	5.75	11.95	26.22
	4 moderate	522	18.88	39.22	65.44
	5 slghtly conservative	209	7.56	15.70	81.14
	6 conservative	210	7.59	15.78	96.92
	7 extrmly conservative	41	1.48	3.08	100.00
	Total	1331	48.14	100.00	
Missing	.	1434	51.86		
Total		2765	100.00		

Obtaining both numbers and value labels

> Before doing the tabulations, you might want to enter the command `numlabel _all, add`. After you enter this command, whenever you do the `tabulate` command, Stata reports both the numbers you use for coding the data (1, 2, 3, 4, and 5) and the value labels (married, widowed, divorced, separated, and never married). Later, if you do not want to include both of these, you can drop the numerical values by using the command `numlabel _all, remove`. This is left as an exercise. The tables with both numbers and value labels may not look great, so you may want two tables for each variable, with one showing the value labels but without the numeric codes and the other showing the numeric codes without the value labels. The default gives you the value labels. On the dialog box, there is an option to *Suppress displaying the value labels*. It is probably simpler to use Ben Jann's `fre` command if you have installed it. This command reports the values as well as the value labels in each tabulation.

In chapter 4, we created a pie chart. Here we will create a pie chart for marital status. Select Graphics ▷ Pie chart and look at the Main tab. If this dialog still has information entered from a previous pie chart, you should click on the R icon in the lower left of the view screen. Type `marital` as the *Category variable*. This uses the categories we want to show as pieces of the pie. Leave the *Variable: (optional)* box blank. Under the Titles tab, enter a nice title in the *Title* box and the name of the dataset we used as a *Note*. Under the Options tab, click on *Order by this variable* and type `marital`. Also check *Exclude observations with missing values (casewise deletion)* because we do not want these, if there are any, to appear as a separate piece of the pie. The dialog box for the Options tab is shown in figure 5.4.

(*Continued on next page*)

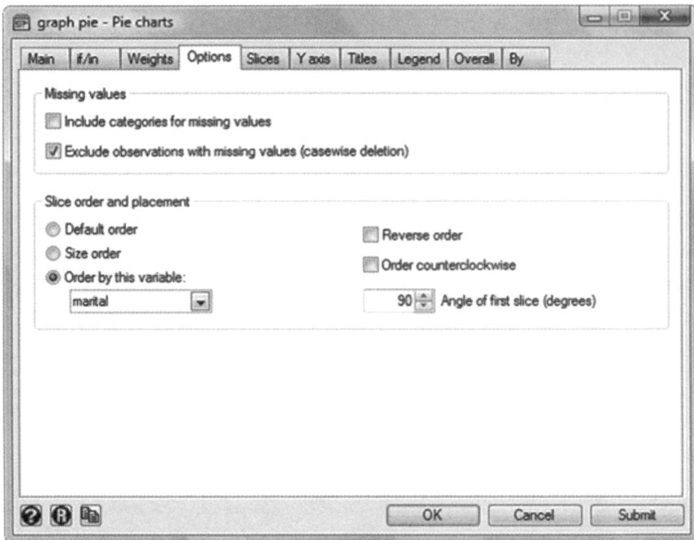

Figure 5.4. The Options tab for pie charts (by category)

The initial pie chart on the left in figure 5.5 provides a visual display of the distribution of marital statuses in the United States. The size of each piece of the pie is proportional to the percentage of the people in that status. This pie chart shows that the most common status of adults is married.

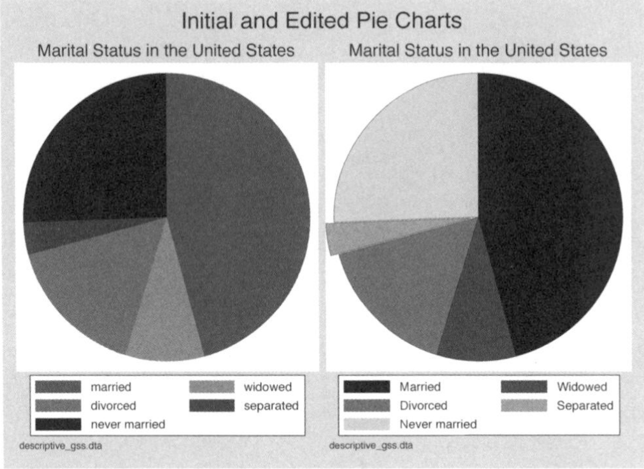

Figure 5.5. Pie charts of marital status in the United States

It is possible to make improvements on the default pie chart. The default pie chart is a bit hard to read because it assumes you want each slice a different color, and this will not work well when printing in black and white. Because many publications require black and white printing, we should edit the pie chart. From our dialog box for the pie chart, we could click on the Overall tab, and there we could select a monochrome scheme from the drop-down *Scheme* list. However, there are several other ways we can improve this pie chart, so let's open the Graph Editor.

We can open the Graph Editor by right-clicking on the pie chart, clicking on the icon above the graph that has a bar chart with a pencil or, in Windows, we can select File ▷ Start Graph Editor. This expands the window that has the graph and adds a panel on the side of the chart with things we might want to change. The Graph Editor is shown in figure 5.6. On the left is the pie chart we will edit, and on the right are the names of the parts of the pie chart.

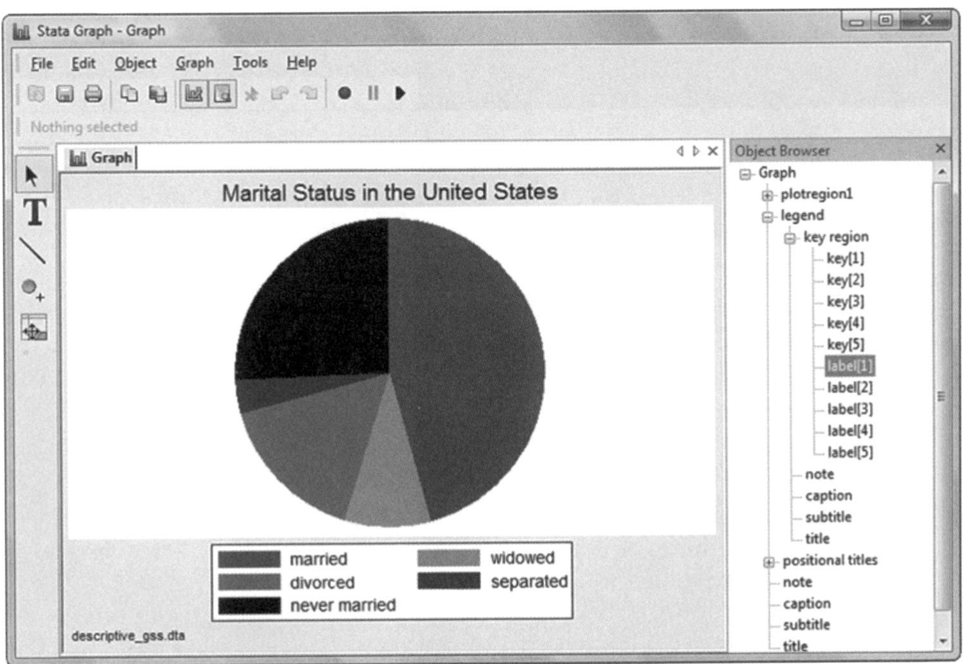

Figure 5.6. The Graph Editor

Notice that the labels in the legend of the initial pie chart are not capitalized. Click on the plus sign by *legend* and the plus sign by *key region*. Double-click on *label[1]* and change the *Text* from `married` to `Married`. Do the same for each of the other labels. We could also do this by double-clicking on the label `married`.

Next click on the plus sign by *plotregion1*. Then double-click on *pieslices[1]*. Here we will pick *Black* as the *Color* and *100%* as the *Fill intensity*. For *pieslices[2]*, pick *Black* and *70%*. For *pieslices[3]*, pick *Black* and *50%*. For *pieslices[4]*, pick *Black* and *30%*, check *Explode slice*, and make sure the *Distance* is *Medium*. Finally, for *pieslices[5]*, pick *Black* and *10%*. You can experiment with other options. The pie chart on the right in figure 5.5 is what we have created. The exploded slice, `Separated`, would be useful if you wanted to emphasize the size of this group.

We have only scratched the surface of what you can do with the powerful Graph Editor. For example, you can click somewhere on the figure, then click on the T to the left of the graph, and a dialog box opens so you can add text. You could then click on the \ (slash) just below the T, and draw a line from the text to the piece of the pie it describes. As an exercise, you might add the text "Less than half married" with a line pointing to the piece of the pie for married.

Within Stata's Graph Editor, you can record the changes you make to a graph by using the Graph Recorder. When the Graph Editor is open, there are symbols like you might see on a recorder or a video player (at the top right of the Editor). Clicking on the red circle starts the Recorder, and clicking on the pair of vertical bars pauses a recording. When you are done making your changes, click on the red circle again before you save the graph or exit the Graph Editor, and the Graph Recorder will prompt you to name the recording. Suppose that we call the recording `myscheme` and save it. The next time we do a similar graph and want to make the same changes, we can click on the arrow to the right of the pair of vertical bars, and it will give us a list of recordings we have saved. We can pick `myscheme` and the same changes will be applied to our current graph.

A bar chart is more attractive than a pie chart for many applications. Instead of selecting Bar Chart from the Graphics menu, select Histogram. Here we are creating a bar chart rather than a histogram, but this is the best way to produce a high-quality bar chart using Stata.

On the Main tab, type `marital` in the *Variable* box. Click on the button next to *Data are discrete*. In the section labeled *Y axis*, click on the button next to *Percent*. The trick to making this a bar chart is to click on the *Bar properties* in the lower left of the Main tab. This opens another dialog box where the default is to have no gap between the bars. Change this to a gap of `10`, which sets the gap between bars to 10 percent of the width of a bar. Click on Accept. If you switch to the Titles tab, you can enter a title, such as `Marital Status in the United States`. Next switch to the X axis tab and click on *Major tick/label properties*. This opens another dialog box where you select the Labels tab and check the box for *Use value labels*. Sometimes the value labels are too wide to fit under each bar. You may need to create new value labels that are shorter. If they are just a little bit too wide, you can change the angle. Click on *Angle* and select *45 degrees* from the drop-down menu. Click on Accept. Finally, switch back to the Main tab. In the lower right corner of the dialog box, click on *Add height labels to bars*. Because we are reporting percentages, this option will show the percentage in each marital status at the top of each bar. The Main tab is shown in figure 5.7.

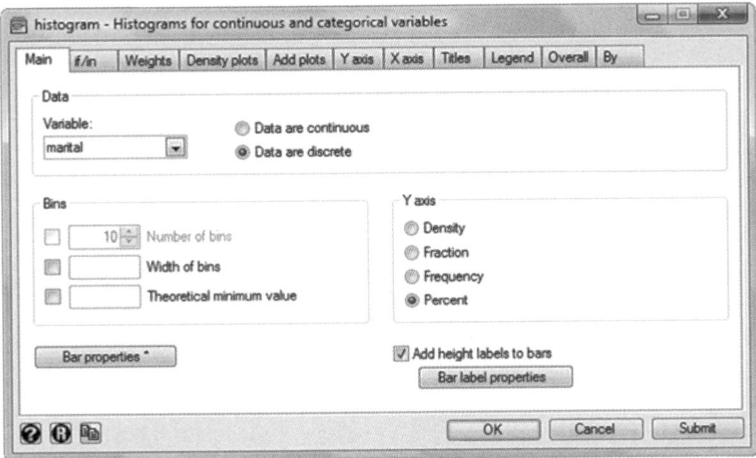

Figure 5.7. Using the `histogram` dialog box to make a bar chart

Figure 5.8 shows the resulting bar chart, which has the percentage in each status at the top of each bar. Married is the most common status, but never married is second. This dataset includes people who are 18 and older, and it is likely that many of those in the never-married status are between 18 and 30.

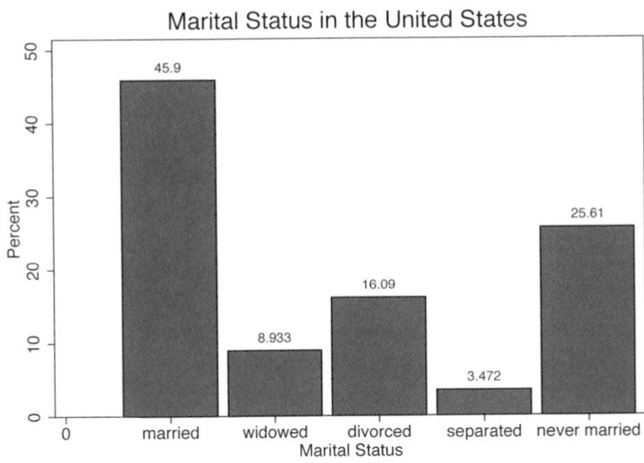

Figure 5.8. Bar chart of marital status of U.S. adults

Sometimes you may have a larger number of categories. When this happens, Stata's default will show only a limited number of value labels, so some of the bars will be unlabeled. If you want to label all of them, you need to go to the X axis tab and click on *Major tick/label properties* to open the dialog box we opened before. On the Rule tab, click on *Suggest # of ticks* and enter the number of bars in the box by *Ticks*. Another issue can arise if you have value labels that are no longer used. For the `polviews` variable, we could use seven ticks because there are seven categories for `polviews`. However, our missing category includes people who had originally been coded as `0 na` for not applicable. They were not asked this question for some reason. If you create a bar chart for `polviews`, you will need to change the value label or make a new set of value labels that do not have this option.

5.5 Statistics and graphs—ordered categories and variables

When our categories are ordered, we can use the median to measure the central tendency. When there are only a couple of categories, however, the median does not work well. Here is an example where there are several categories for the variable `polviews`, which asks people their political views on a seven-point scale from extremely liberal to extremely conservative. We might want to report the median or mean, and the SD. We have done a `summarize` and a `tabulate` already using the dialog system, so we will just enter the commands directly at this point: `tab1 polviews` and `summarize polviews, detail`. Before you run these commands, make sure that you run the `numlabel _all, add` command so both the numbers and the value labels are shown.

```
. numlabel _all, add
. tab1 polviews
-> tabulation of polviews
        think of self as |
  liberal or conservative |      Freq.     Percent        Cum.
--------------------------+-----------------------------------
     1. extremely liberal |         47        3.53        3.53
               2. liberal |        143       10.74       14.27
     3. slightly liberal  |        159       11.95       26.22
              4. moderate |        522       39.22       65.44
 5. slghtly conservative  |        209       15.70       81.14
          6. conservative |        210       15.78       96.92
 7. extrmly conservative  |         41        3.08      100.00
--------------------------+-----------------------------------
                    Total |      1,331      100.00
. summarize polviews, detail
          think of self as liberal or conservative
-------------------------------------------------------------
      Percentiles      Smallest
 1%            1              1
 5%            2              1
10%            2              1       Obs                1331
25%            3              1       Sum of Wgt.        1331
50%            4                      Mean           4.124718
                        Largest       Std. Dev.      1.385016
75%            5              7
90%            6              7       Variance       1.918268
95%            6              7       Skewness      -.1509408
99%            7              7       Kurtosis       2.693351
```

The frequency distribution produced by `tab1 polviews` is probably the most useful way to describe the distribution of an ordered categorical variable. We can see that it is fairly symmetrically distributed around the mode of moderate, with somewhat more people describing themselves as conservative rather than liberal.

Although the tabulation gives us a good description of the distribution, we often will not have the space in a report to show this level of detail. The median is provided by the `summarize` command, which shows that the 50th percentile occurs at the value of 4, so the Mdn is 4, corresponding to a political moderate. Even though these are ordinal categories, many researchers would report the mean. The mean assumes that the quantitative values, 1–7, are interval-level measures. However, the mean (M = 4.12) is usually a good measure of central tendency. The mean reflects the distribution somewhat more accurately here than does the median because the mean shows that the average response is a bit more to the conservative end than to the liberal. We know that the mean is a bit more conservative because the higher numeric score on `polviews` corresponds to more conservative views. You should be able to see this from reading the frequency distribution carefully. Although this variable is clearly ordinal, many researchers treat variables like this as if they were interval and rely on the mean as a measure of central tendency. If you are in doubt, it may be a good idea to report both the median and the mean.

Here is how you can create a histogram showing the distribution of political views. Type the command

```
. histogram polviews, discrete percent
> title(Political Views in the United States) subtitle(Adult Population)
> note(General Social Survey 2002) xtitle(Political Conservatism) scheme(s1mono)
```

Remember, this is the way Stata writes it in the Results window. You would actually need to use the `///`, preceded by a space at the end of each but the last line, so Stata will know the three lines are all parts of one command. You would write this in a Do-file Editor as

```
histogram polviews, discrete percent ///
  title(Political Views in the United States) subtitle(Adult Population) ///
  note(General Social Survey 2002) xtitle(Political Conservatism) scheme(s1mono)
```

Like most graph commands, this is an example of a complicated command that can easily be produced using the dialog box. Commands for making graphs can get complicated, so it is usually best to use the dialog box with graphs. Select Graphics ▷ Histogram.

We will not show the resulting dialogs. On the Main tab, select `polviews` from the *Variable* list, and click on the button by *Data are discrete*. Go to the X axis tab, and enter the title you want to appear on the x axis (the x axis is the horizontal axis of the graph). Go back to the Main tab and click on the radio button by *Percent*. Finally, go to the Titles tab, and enter the title, subtitle, and notes that you want to appear on the graph. The example graph appears in figure 5.9.

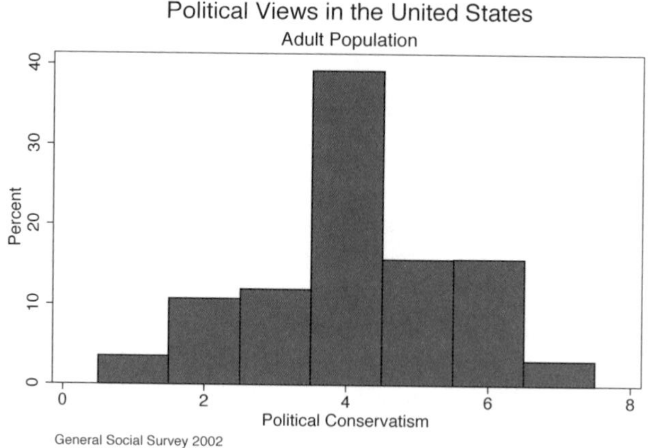

Figure 5.9. Histogram of political views of U.S. adults

This histogram is a nice combination for presenting the distribution. A reader can quickly get a good sense of the distribution. In 2002, moderate was the overwhelming choice of adults in the United States. Also, the bars on the right (conservative) are a bit higher than the bars on the left, which indicates a tendency for people to be conservative ($M = 4.12$, Mdn = 4, Mode = 4, SD = 1.39). Some researchers, when making a graph to show the distribution of an ordinal variable, are reluctant to have the bars for each value touch. Also some like to have the labels posted on the x axis, rather than the coded values.

5.6 Statistics and graphs—quantitative variables

We will study three variables: `age`, `educ`, and `wwwhr` (hours spent on the World Wide Web). Two types of useful graphs are the already familiar histogram and a new graph called the *box plot*. We will usually use the mean or median to measure the central tendency for quantitative variables. The SD is the most widely used measure of dispersion, but a statistic called the *interquartile range* is used by the box plots that are presented below.

Start with `wwwhr`, hours spent in the last week on the World Wide Web. These data were collected in 2002, and by now, the hours have probably increased a lot.

Computing descriptive statistics for quantitative variables is easy. Let's skip the dialog system and just enter the command:

```
. summarize wwwhr, detail
                       www hours per week
-------------------------------------------------------------
      Percentiles      Smallest
 1%            0              0
 5%            0              0
10%            0              0       Obs                1574
25%            1              0       Sum of Wgt.        1574

50%            3                      Mean           5.907878
                        Largest       Std. Dev.      8.866734
75%            7             60
90%           15             64       Variance       78.61897
95%           21            100       Skewness       3.997908
99%           40            112       Kurtosis       30.39248
```

This says that the average person spent a mean of 5.91 hours on the World Wide Web in the week before the survey was taken. The median is 3 hours. Because the mean is greater than the median when a distribution is positively skewed, we can assume that the distribution is positively skewed (tails off on the right side). This makes sense because the value for hours on the World Wide Web cannot be less than zero, but we all know a few people who spend many hours on the web. The SD is 8.87 hours, which tells us that the time on the web varies widely. For a normally distributed variable, about two-thirds of the cases are between the mean and one SD (−2.96 hours and 14.77 hours) and 95% will be between the mean and two SDs (−11.83 hours and 23.64 hours). Clearly, this does not make any sense because you cannot use the World Wide Web fewer than 0 hours a week. Still, this suggests that there is a lot of variation in how much time people spend on the web.

The skewness is 4.00, which means that the distribution has a positive skew (greater than zero), and the kurtosis is 30.39, which is huge compared with 3.0 for a normal distribution. Remember that a kurtosis greater than 10 is problematic; a kurtosis over 20 is very serious. This result suggests that there is a big clump of cases concentrated in one part of the distribution. Can you guess where this concentration was in 2002?

Stata can test for normality based on skewness and kurtosis. For most applications, this test is of limited utility. It is extremely sensitive to small departures from normality when you have a large sample, and it is insensitive to large departures when you have a small sample. The problem is that, when we do inferential statistics, the lack of normality is much more problematic with small samples (where the test lacks power) than it is with large samples (where the test usually finds a significant departure from normality, even for a small departure).

To run the test for normality based on skewness and kurtosis, we can use the dialog system by selecting Statistics ▷ Summaries, tables, and tests ▷ Distributional plots and tests ▷ Skewness and kurtosis normality test. Once the dialog box is open, enter the variable `wwwhr` and click on OK. Unlike the complex graph commands, with statistical tests it is often easier to simply enter the command (unless you cannot remember it). The command is simply `sktest wwwhr`.

```
. sktest wwwhr

                   Skewness/Kurtosis tests for Normality
                                                     ------- joint ------
    Variable | Pr(Skewness)   Pr(Kurtosis)  adj chi2(2)    Prob>chi2
-------------+-------------------------------------------------------
       wwwhr |      0.000          0.000             .         0.0000
```

These results show that based on skewness the probability that `wwwhr` is normal is 0.000 and that based on kurtosis the probability that `wwwhr` is normal is also 0.000. Anytime either probability is less than 0.05, we say that there is a statistically significant lack of normality. Testing for normality based on skewness and kurtosis jointly, Stata reports a probability of 0.000, which reaffirms our concern. It is best to report this as $\Pr < 0.001$ rather than as $\Pr = 0.000$. This test computes a statistic called chi-squared (χ^2), and it is so big that Stata cannot print it in the available space so instead inserts a ".".

When we are describing a lot of variables in a report, space constraints usually limit us to reporting the mean, median, and standard deviation. You can read these numbers along with the measure of skewness and kurtosis and have a reasonable notion of what each of the distributions looks like. However, it is possible to describe `wwwhr` nicely with a few graphs. First, we will do a histogram using the dialog system described previously, or for a basic histogram (shown in figure 5.6) we could enter

```
. histogram wwwhr, frequency
```

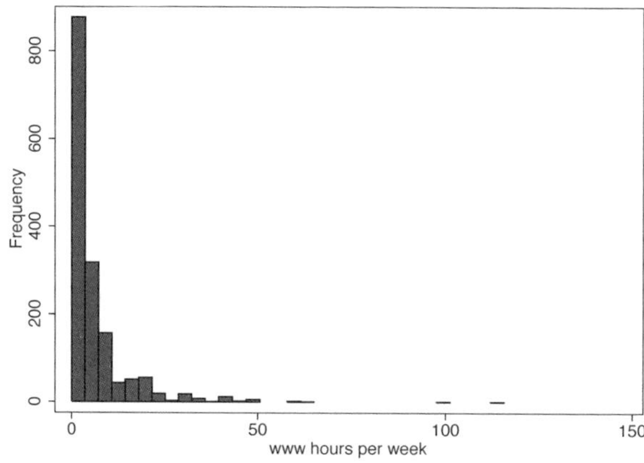

Figure 5.10. Histogram of time spent on the World Wide Web

This simple command does not include all the nice labeling features you can get using the dialog box, but it gives us a quick view of the distribution. This graph includes a few outliers (observations with extreme scores) who surf the World Wide Web more than 25 hours a week. Providing space in the histogram for the handful of people using the World Wide Web between 25 hours and 150 hours takes up most of the graph, and we do not get enough detail for the smaller number of hours that characterizes most of our web users.

We will get around this problem by doing a histogram for a subset of people who use the web fewer than 25 hours a week, and we will do a separate histogram for women and for men. You can get these using the dialog box by inserting the restriction `wwwhr < 25` in the *If: (expression)* box under the if/in tab and by clicking on *Draw subgraphs for unique values of variables* and then inserting the `sex` variable under the By tab. Here is the command we could enter directly:

```
. histogram wwwhr if wwwhr < 25, frequency by(sex)
```

Notice the `frequency by(sex)` part of the command appears after the comma. The new histograms appear in figure 5.11.

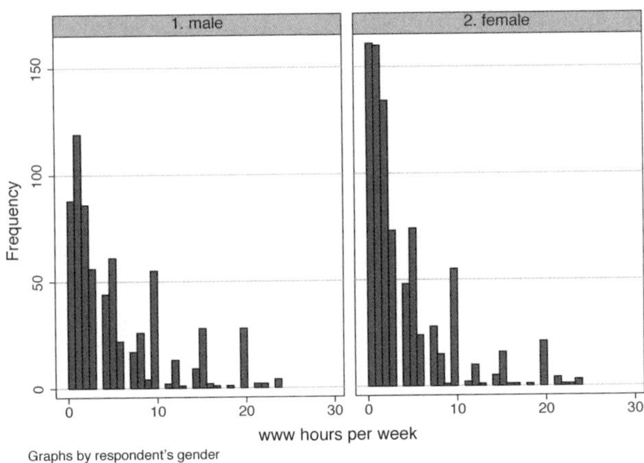

Figure 5.11. Histogram of time spent on the World Wide Web (fewer than 25 hours a week, by gender)

By using the dialog system, we could improve the format of figure 5.11 by adding titles, and we might want to report the results by using percentages rather than frequencies. We also could experiment with different widths of the bars. Still, figure 5.11 shows that the distribution is far from normal, as the measures of skewness and kurtosis suggested. By doing the histogram separately for women and men, we can see that, at the time these data were collected in 2002, far more women were in the lowest interval.

We could also open the Graph Editor. There we could click on the header for each of the histograms. We might want to replace `1.male` and `2.female` with `Men` and `Women`. You probably can think of additional changes that would make the graph nicer.

When you want to compare your distribution on a variable with how a normal distribution would be, you can click on an option for Stata to draw how a normal distribution would look right on top of this histogram. We will not show an illustration of this, but all you need to do is open the Density plots tab for the `histogram` dialog box, and check the box that says *Add normal-density plot*. This is left as an exercise.

There is another option for adding a kernel density plot on the dialog box. This is an estimate of the most likely population distribution for a continuous variable that would account for this sample distribution. This will smooth out some of the bars that are extremely high or low because of variation from one sample to the next.

To get the descriptive statistics for men and women separately (but not restricted to those using the web fewer than 25 hours a week), we need a new command:

```
. by sex, sort: summarize wwwhr
```

We can do this from the `summarize` dialog box by checking the *Repeat command by groups* and entering `sex` under the by/if/in tab. This command will sort the dataset by `sex` and then run the `summarize` separately for women and men.

Another way to obtain a statistical summary of the `wwwhr` variable is to use the `tabstat` command, which gives us a nicer display than what we obtained with the `summarize` command. Select Statistics ▷ Summaries, tables, and tests ▷ Tables ▷ Table of summary statistics (tabstat) to open the dialog box.

Under the Main tab, type `wwwhr` under *Variables*. Check the box next to *Group statistics by variable*, and type `sex`. Now pick the statistics we want Stata to summarize. Check the box in front of each row, and pick the statistic. The `tabstat` command gives us far more options than the `summarize` command. The dialog box in figure 5.12 shows that we asked for the mean, median, SD, interquartile range, skewness, kurtosis, and coefficient of variation.

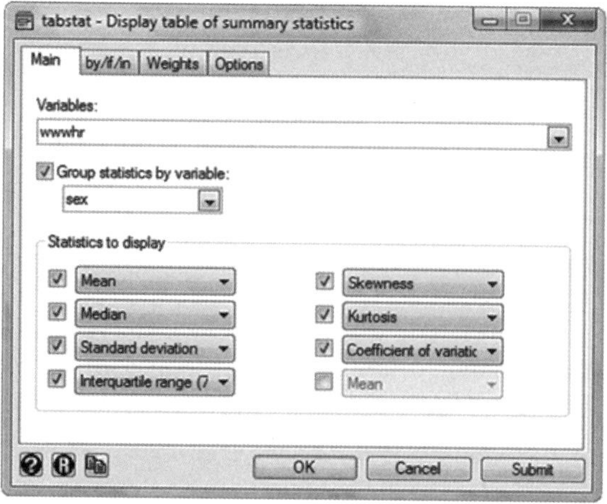

Figure 5.12. The Main tab for the `tabstat` dialog box

Under the Options tab, go to the box for *Use as columns* and select *Statistics*, which will greatly enhance the ease of reading the display. Next we could go to the by/if/in tab and enter `wwwhr < 25` under *If: (expression)*, but we will not do that here. Here is the resulting command:

```
. tabstat wwwhr, statistics(mean median sd skewness kurtosis cv iqr) by(sex)
> columns(statistics)
Summary for variables: wwwhr
     by categories of: sex (respondents sex)
      sex |      mean       p50        sd  skewness  kurtosis        cv       iqr
----------+----------------------------------------------------------------------
  1. male |  7.106892         4   9.98914  3.608189   25.2577  1.405557         9
2. female |  4.920046         2  7.688655  4.409274  36.78389   1.56272         4
----------+----------------------------------------------------------------------
    Total |  5.907878         3  8.866734  3.997908  30.39248  1.500832         6
---------------------------------------------------------------------------------
```

The table produced by the `tabstat` command summarizes the statistics we requested that it include, showing the statistics for males and females, and the total for males and females combined. Stata calls the median `p50` because the median represents the value corresponding to the 50th percentile. If you copied this table to a Word file, you might want to change the label to median to benefit readers who do not really know what the median is. If you highlight the `tabstat` output in the Results window and copy it as a picture to a Word document, you will not be able to make this change in Word. However, if you choose one of the other copy options, you will be able to make the change.

In addition to skewness and kurtosis, we selected two additional statistics we have not yet introduced. The coefficient of relative variation (CV) is simply the SD divided by the mean (i.e., CV = SD/M). This statistic is sometimes used to compare SDs for variables that are measured on different scales, such as income measured in dollars and education measured in years. The interquartile range is the difference between the value of the 75th percentile and the value of the 25th percentile. This range covers the middle 50% of the observations.

Men, on average, spent far more time using the World Wide Web in 2002 than did women. Because the means are bigger than the medians, we can assume that the distributions are positively skewed (as was evident in the histograms we did). Men are a bit more variable than women because their SD is somewhat greater. Both distributions are skewed and have heavy kurtosis. The CV is 1.41 for men and 1.56 for women. Women have slightly greater variance relative to their mean than men do (based on comparing the CV values), even though the actual SD is bigger for men. Finally, the interquartile range of 9 for men is more than double the interquartile range of 4 for women. Thus the middle 50% of men are more dispersed than the middle 50% of women. Comparing the CVs suggests the opposite finding to comparing interquartile ranges. Because the scale (hours of using the World Wide Web) is the same, we would not rely on the CV.

A horizontal or vertical box plot is an alternative way of showing the distribution of a quantitative variable such as `wwwhr`. Select Graphics ▷ Box plot. Here we will use four of the tabs: Main, Categories, if/in, and Titles. Under the Main tab, check the radio button by *Horizontal* to make the box plot horizontal, and enter the name of our variable, `wwwhr`. Under the Categories tab, check *Group 1* and enter the *grouping variable*, i.e., `sex`. This will create separate box plots for women and for men. We could have additional grouping variables, but these plots can get complicated. If we wanted a box plot that included both women and men, we would leave this tab blank. The Categories tab is similar to the by/if/in tab we used for the `tabstat` command. Under the if/in tab, we need to make a command so that the plots are shown only for those who spend fewer than 25 hours a week on the web. In the *If: (expression)* box, we type `wwwhr < 25`. Finally, under the Titles tab, enter the title and any subtitles or notes we want to appear on the chart. The command generated from the dialog box is

```
. graph hbox wwwhr if wwwhr < 25, over(sex)
> title(Hours Spent on the World Wide Web) subtitle(By Gender)
> note(descriptive_gss.dta)
```

and the resulting graph appears in figure 5.13.

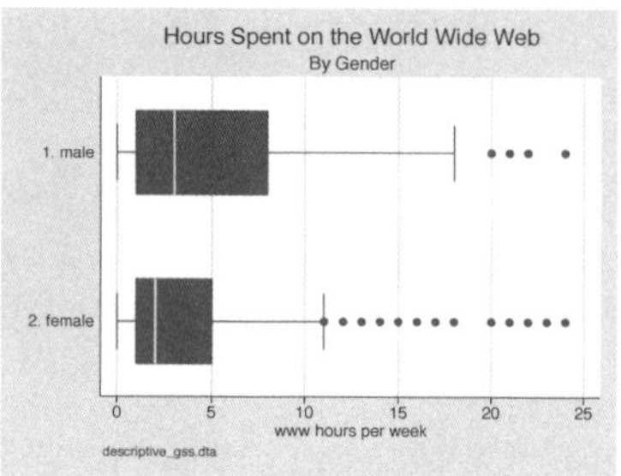

Figure 5.13. Box plot of time spent on the World Wide Web (fewer than 25 hours a week, by gender)

Histograms may be easier to explain to a lay audience than box plots. For a nontechnical group, histograms are usually a better choice. Many statisticians like box plots better because they show more information about the distribution. The white vertical line in the dark-gray boxes is the median. For women, you can see that the median is about 2 hours per week, and for men, about 4 hours per week.

The left and right sides of the dark-gray box are the 25th and 75th percentiles, respectively. Within this dark-gray box area are half of the people. This box is much wider for men than it is for women, showing how men are more variable than women. Lines extend from the edge of the dark-gray box 1.5 box lengths, or until they reach the largest or smallest cases. Beyond this, there are some dots representing outliers, or extreme values.

5.7 Summary

After four chapters about how to set up files and manage them, I hope you enjoyed getting to a substantive chapter showing you some of the output produced with Stata. We are just beginning to tap the power of Stata, but you can already summarize variables and create several types of attractive graphs. This chapter covered the following topics:

- How to compute measures of central tendency (averages), including the mean, median, and mode
- When to use the different measures of central tendency based on level of measurement, distribution characteristics, and your purposes
- How to describe the dispersion of a distribution by using
  - Statistics (standard deviations)
  - Tables (frequency distributions)
  - Graphs (pie charts, bar charts, histograms, and box plots)
- How to use Stata to give you these results for nominal-, ordinal-, and interval-level variables
- How to use Stata's Graph Editor

The graphs we have introduced in this chapter show just a few of the graph capabilities offered by Stata. We will cover a few more types of graphs in later chapters, but if you are interested in producing high-quality graphs, see *A Visual Guide to Stata Graphics, Second Edition* by Michael Mitchell (2008), which is available from Stata Press.

This is just the start of the useful output you can produce using Stata. Statistics books tend to get harder and harder as you move toward more complicated procedures. I cannot help that, but Stata is just the opposite. Managing data and doing graphs are the two hardest tasks for statistical programs because Stata is designed primarily to do statistical analysis. In the next chapter, we will examine how to use graphs and statistics when we are examining the relationship between two or more variables.

5.8 Exercises

1. Open `descriptive_gss.dta`, and do a detailed summary of the variable `hrs1` (hours worked last week). Also create a histogram of the variable. Interpret the mean and median. Looking at the histogram, explain why the skewness value is close to zero. What does the value of kurtosis tell us? Looking at the histogram, explain why the kurtosis is a positive value.

2. Open `descriptive_gss.dta`, and do a detailed summary of the variable `satjob7` (job satisfaction). Type the command `numlabel satjob7, add`, and then do a tabulation of `satjob7`. Interpret the mean and median values. Why would some researchers report the median? Why would other researchers report the mean?

3. Open `descriptive_gss.dta`, and do a tabulation of `deckids` (who makes decisions about how to bring up children). Do this using the by/if/in tab to select by `sex`. Create and interpret a bar chart by using the `histogram` dialog box. Why

would it make no sense to report the mean, median, or standard deviation for `deckids`? Use Stata's Graph Editor to make the bar chart look as nice as you can.

4. Open `descriptive_gss.dta`; do a tabulation of `strsswrk` (job is rarely stressful) and a detailed summary. Do this using the by/if/in tab to select by `sex`. Create and interpret a histogram, using the By tab to do this for males and females. In the Main tab, be sure to select the option Data are discrete. Carefully label your histogram to show value labels and the percentage in each response category. Each histogram should be similar to figure 5.9. Interpret the median and mean for men and women.

5. Open `descriptive_gss.dta`, and do a tabulation of `trustpeo`, `wantbest`, `advantge`, and `goodlife`. Use the `tabstat` command to produce a table that summarizes descriptive statistics for this set of variables by gender. Include the median, mean, standard deviation, and count for each variable. Interpret the means by using the variable labels you get with the tabulation command.

6. Open `descriptive_gss.dta`, and do a tabulation of `polviews`. Create a bar chart for this variable showing the percentage of people who are in each of the seven categories. Next create a chart that has labels at a 45-degree angle for each of the bars. Finally, change the chart by using the X axis tab's *Minor tick/label properties* and *Major tick/label properties* (using the *Custom* option) so the histogram does not have a null category and all the categories are labeled. Compare this final figure to figure 5.9.

7. Open `descriptive_gss.dta`. In figure 5.13, we created a box plot for the hours women and men spend watching the World Wide Web. First, create a similar box plot for `hrs1` (hours worked last week). Now add a second grouping variable, `marital`. Using this graph, give a detailed interpretation of how marital status and gender are related to hours a person works for pay.

6 Statistics and graphs for two categorical variables

6.1 Relationship between categorical variables

Chapter 5 focused on describing single variables. Even there, it was impossible to resist some comparisons, and we ended by examining the relationship between gender and hours per week spent using the web. Some research can stop by describing variables, one at a time. You do a survey for your agency and make up a table with the means and standard deviations for all the quantitative variables. You might include frequency distributions and bar charts for each key categorical variable. This is sometimes the extent of statistical information your reader will want. However, the more you work on your survey, the more you will start wondering about possible relationships.

- Do women who are drug dependent use different drugs from those used by drug-dependent men?
- Are women more likely to be liberal than men?

- Is there a relationship between religiosity and support for increased spending on public health?

You know you are "getting it" as a researcher when it is hard for you to look at a set of questions without wondering about possible relationships. Understanding these relationships is often crucial to policy decisions. If 70% of the nonmanagement employees at a retail chain are women, but only 20% of the management employees are women, there is a relationship between gender and management status that disadvantages women.

In this chapter, you will learn how to describe relationships between categorical variables. How do you define these relationships? What are some pitfalls that lead to misinterpretations? In this chapter, the statistical sophistication you will need increases, but there is one guiding principle to remember: the best statistics are the simplest statistics you can use—as long as they are not too simple to reflect the inherent complexity of what you are describing.

6.2 Cross-tabulation

Cross-tabulation is a technical term for a table that has rows representing one categorical variable and columns representing another. If you have one variable that depends on the other, you usually put the dependent variable as the column variable and the independent variable as the row variable. This layout is certainly not necessary, and several statistics books do just the opposite. That is, they put the dependent variable as the row variable and the independent variable as the column variable.

Let's start with a basic cross-tabulation of whether a person says abortion is okay for any reason and their gender. Say you decide that whether a person accepts abortion for any reason is more likely if the person is a woman because a woman has more at stake when she is pregnant than does her partner. Therefore, whether a person accepts abortion will be the dependent variable, and gender will be the independent variable. We will use the `gss2006_chapter6.dta` dataset that contains selected variables from the 2006 General Social Survey, and we will use the cross-tabulation command, `tabulate`, with two categorical variables, `sex` and `abany`. To open the dialog box for `tabulate`, select Statistics ▷ Summaries, tables, and tests ▷ Tables ▷ Two-way tables with measures of association. This dialog box is shown in figure 6.1.

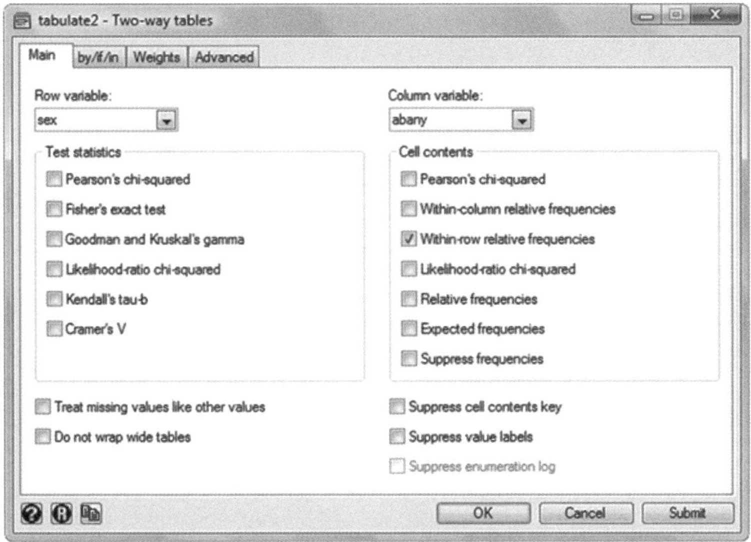

Figure 6.1. The Main tab for creating a cross-tabulation

Select `sex`, the independent variable, as the *Row variable* and `abany`, the dependent variable, as the *Column variable*. We are assuming that `abany` is the dependent variable that depends on `sex`. Also check the box on the right side under *Cell contents* for the *Within-row relative frequencies* option. This option tells Stata to compute the percentages so that each row adds up to 100%. Here are the resulting command and results:

```
. tabulate sex abany, row

+----------------+
| Key            |
|----------------|
|   frequency    |
| row percentage |
+----------------+

           |  ABORTION IF WOMAN
           | WANTS FOR ANY REASON
    Gender |       YES         NO |     Total
-----------+----------------------+----------
      MALE |       350        478 |       828 
           |     42.27      57.73 |    100.00 
-----------+----------------------+----------
    FEMALE |       434        677 |     1,111 
           |     39.06      60.94 |    100.00 
-----------+----------------------+----------
     Total |       784      1,155 |     1,939 
           |     40.43      59.57 |    100.00 
```

Independent and dependent variables

> Many beginning researchers get these terms confused. The easiest way to remember this is that the dependent variable "depends" on the independent variable. In this example, whether a person accepts abortion for any reason depends on whether the person is a man or a woman. By contrast, it would make no sense to say that whether a person is a man or a woman depends on whether they accept abortion for any reason.
>
> Many researchers call the dependent variable an "outcome" and the independent variable the "predictor". In this example, `sex` is the predictor because it predicts the outcome, `abany`.

The independent variable `sex` forms the rows with labels of `male` and `female`. The dependent variable, accepting abortion under any circumstance, appears as the columns labeled `yes` and `no`. The column on the far right gives us the total for each row. Notice that there are 828 males, 350 of whom find abortion for any reason to be acceptable, compared with 1,111 females, 434 of whom say abortion is acceptable for any reason. These frequencies are the top number in each cell of the table.

The frequencies at the top of each cell are hard to interpret because each row and each column have a different number of observations. One way to help interpret a table is to use the percentage, which takes into account the number of observations of each independent variable (predictor). The percentages appear just below the frequencies in each cell. Notice that the percentages add up to 100% for each row. Overall, 40.43% of the people said "yes", abortion is acceptable for any reason, and 59.57% said "no". However, men were relatively more likely (42.27%) than women (39.06%) to report that abortion is okay, regardless of the reason. We get these percentages because we told Stata to give us the *Within-row relative frequencies*, which in the command is the option `row`.

Thus men are more likely to report accepting abortion under any circumstance. We compute percentages on the rows of the independent variable and make comparisons up and down the columns of the dependent variable. Thus we say that 42.27% of the men compared with 39.06% of the women accept abortion under any circumstance. This is a small difference, but interestingly, it is in the opposite direction from what we expected.

6.3 Chi-squared test

The difference between women and men seems small, but could we have obtained this difference by chance? Or is the difference statistically significant? Remember, when you have a large sample like this one, a difference may be statistically significant even if it is small.

If we had just a handful of women and men in our sample, there would be a good chance of observing this much difference just by chance. With such a large sample, even a small difference like this might be statistically significant. We use a chi-squared (χ^2) statistic to test the likelihood that our results occurred by chance. If it is extremely unlikely to get this much difference between men and women in a sample of this size by chance, you can be confident that there was a real difference between women and men, but you still need to look at the percentages to decide whether the statistically significant difference is substantial enough to be important.

The chi-squared test compares the frequency in each cell with what you would expect the frequency to be by chance, if there were no relationship. The expected frequency for a cell depends on how many people are in the row and how many are in the column. For example, if we had asked a small high school group if they accept abortion for any reason, we might have only 10 males and 10 females. In that case, we would expect far fewer people in each cell than in this example, where we have 828 men and 1,111 women.

In the cross-tabulation, there were many options on the dialog box (see figure 6.1). To obtain the chi-squared statistic, check the box on the left side for *Pearson's chi-squared*. Also check the box for *Expected frequencies* that appears in the right column on the dialog box. The resulting table has three numbers in each cell. The first number in each cell is the frequency, the second number is the expected frequency if there were no relationship, and the third number is the percentage of the row total. We would not usually ask for the expected frequency, but now you know it is one of Stata's capabilities. The resulting command now has three options: `chi2 expected row`. Here is the command and the output it produces

```
. tabulate sex abany, chi2 expected row

  +--------------------+
  | Key                |
  |--------------------|
  |     frequency      |
  | expected frequency |
  |   row percentage   |
  +--------------------+

           |    ABORTION IF WOMAN
           |  WANTS FOR ANY REASON
    Gender |       YES         NO |     Total
-----------+----------------------+----------
      MALE |       350        478 |       828 
           |     334.8      493.2 |     828.0 
           |     42.27      57.73 |    100.00 
-----------+----------------------+----------
    FEMALE |       434        677 |     1,111 
           |     449.2      661.8 |   1,111.0 
           |     39.06      60.94 |    100.00 
-----------+----------------------+----------
     Total |       784      1,155 |     1,939 
           |     784.0    1,155.0 |   1,939.0 
           |     40.43      59.57 |    100.00 

          Pearson chi2(1) =   2.0254   Pr = 0.155
```

In the top left cell of the table, we can see that we have 350 men who accept abortion for any reason, but we would expect to have only 334.8 men here by chance. By contrast, we have 434 women who accept abortion for any reason, but we would expect to have 449.2. Thus we have $350 - 334.8 = 15.2$ more men accepting abortion than we would expect by chance and $434 - 449.2 = -15.2$ fewer women than we would expect. Stata uses a function of this information to compute chi-squared.

At the bottom of the table, Stata reports `Pearson chi2(1) = 2.0254` and `Pr = 0.155`, which would be written as $\chi^2(1, N = 1,939) = 2.0254$; p not significant. Here we have one degree of freedom. The sample size of $N = 1{,}939$ appears in the lower right part of the table. We usually round the chi-squared value to two decimal places, so 2.0254 becomes 2.03. Stata reports an estimate of the probability to three decimal places. We can report this, or we can use a convention found in most statistics books of reporting the probability as less than 0.05, less than 0.01, or less than 0.001. Since $p = 0.155$ is greater than .05, we say p not significant. What would happen if the probability were $p = 0.0004$? Stata would round this to $p = 0.000$. We would not report $p = 0.000$ but instead would report $p < 0.001$.

To summarize what we have done in this section, we can say that men are more likely to report accepting abortion for any reason than women are. In the sample of 1,939 people, 42.3% of the men say that they accept abortion for any reason compared with just 39.1% of the women. This relationship between gender and acceptance of abortion is not statistically significant.

6.3.1 Degrees of freedom

Because we assume that you have a statistics book explaining the formulas, we have not gone into detail. Stata will compute the chi-squared, the number of degrees of freedom, and the probability of getting your observed result by chance.

You can determine the number of degrees of freedom yourself. The degrees of freedom refers to how many pieces of independent information you have. In a two-by-two table, like the one we have been analyzing, the value of any given cell can be any number between 0 and the smaller of the number of observations in the row and the number of observations in the column. For example, the upper left cell (350) could be anything between 0 and 784. Let's use the observed value of 350 for the upper left cell. Now how many other cells are free to vary? By subtraction, you can determine that 434 people must be in the female/yes cell because $784 - 350 = 434$. Similarly, 478 observations must be in the male/no cell ($828 - 350 = 478$), and 677 observations must be in the female/no cell ($1{,}111 - 434 = 677$). Thus, with four cells, only one of these is free, and we can say that the table has 1 degree of freedom. We can generalize this to larger tables where degrees of freedom $= (R - 1)(C - 1)$, where R is the number of rows and C is the number of columns. If we had a three-by-three table instead of a two-by-two table, we would have $(3 - 1)(3 - 1) = 4$ degrees of freedom.

6.3.2 Probability tables

Many experienced Stata users have made their own commands that might be helpful to you. Philip Ender made a series of commands that display probability tables for various tests. The `findit` command finds user-contributed programs and lets you install them on your machine. Type the command `findit chitable`. This command takes you to the window shown in figure 6.2.

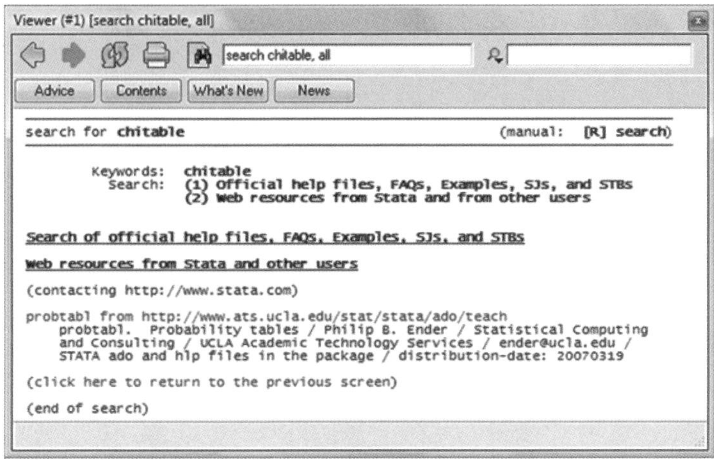

Figure 6.2. Results of `findit chitable`

From here, click on the blue web link

```
probtabl from http://www.ats.ucla.edu/stat/stata/ado/teach
```

This link takes you to another screen where you can click on the blue link labeled `click here to install`. Once you have done this, anytime you want to see a chi-squared table, you merely type the command `chitable`. (The installation may not work if you are using a server and do not have the permissions to update Stata.) This installation also gives you other probability tables that we will use elsewhere in this book, including t-test tables (`ttable`) and F-test tables (`ftable`). Simply entering `chitable` is a lot more convenient than having to look up a probability in a textbook. Try it now.

(*Continued on next page*)

```
. chitable

          Critical Values of Chi-square
  df      .50      .25      .10      .05     .025      .01     .001
   1     0.45     1.32     2.71     3.84     5.02     6.63    10.83
   2     1.39     2.77     4.61     5.99     7.38     9.21    13.82
   3     2.37     4.11     6.25     7.81     9.35    11.34    16.27
   4     3.36     5.39     7.78     9.49    11.14    13.28    18.47
   5     4.35     6.63     9.24    11.07    12.83    15.09    20.52
   6     5.35     7.84    10.64    12.59    14.45    16.81    22.46
   7     6.35     9.04    12.02    14.07    16.01    18.48    24.32
   8     7.34    10.22    13.36    15.51    17.53    20.09    26.12
   9     8.34    11.39    14.68    16.92    19.02    21.67    27.88
  10     9.34    12.55    15.99    18.31    20.48    23.21    29.59
  11    10.34    13.70    17.28    19.68    21.92    24.72    31.26
  12    11.34    14.85    18.55    21.03    23.34    26.22    32.91
  13    12.34    15.98    19.81    22.36    24.74    27.69    34.53
  14    13.34    17.12    21.06    23.68    26.12    29.14    36.12
  15    14.34    18.25    22.31    25.00    27.49    30.58    37.70
  16    15.34    19.37    23.54    26.30    28.85    32.00    39.25
  17    16.34    20.49    24.77    27.59    30.19    33.41    40.79
  18    17.34    21.60    25.99    28.87    31.53    34.81    42.31
  19    18.34    22.72    27.20    30.14    32.85    36.19    43.82
  20    19.34    23.83    28.41    31.41    34.17    37.57    45.31
  21    20.34    24.93    29.62    32.67    35.48    38.93    46.80
  22    21.34    26.04    30.81    33.92    36.78    40.29    48.27
  23    22.34    27.14    32.01    35.17    38.08    41.64    49.73
  24    23.34    28.24    33.20    36.42    39.36    42.98    51.18
  25    24.34    29.34    34.38    37.65    40.65    44.31    52.62
  26    25.34    30.43    35.56    38.89    41.92    45.64    54.05
  27    26.34    31.53    36.74    40.11    43.19    46.96    55.48
  28    27.34    32.62    37.92    41.34    44.46    48.28    56.89
  29    28.34    33.71    39.09    42.56    45.72    49.59    58.30
  30    29.34    34.80    40.26    43.77    46.98    50.89    59.70
  35    34.34    40.22    46.06    49.80    53.20    57.34    66.62
  40    39.34    45.62    51.81    55.76    59.34    63.69    73.40
  45    44.34    50.98    57.51    61.66    65.41    69.96    80.08
  50    49.33    56.33    63.17    67.50    71.42    76.15    86.66
  55    54.33    61.66    68.80    73.31    77.38    82.29    93.17
  60    59.33    66.98    74.40    79.08    83.30    88.38    99.61
  65    64.33    72.28    79.97    84.82    89.18    94.42   105.99
  70    69.33    77.58    85.53    90.53    95.02   100.43   112.32
  75    74.33    82.86    91.06    96.22   100.84   106.39   118.60
  80    79.33    88.13    96.58   101.88   106.63   112.33   124.84
  85    84.33    93.39   102.08   107.52   112.39   118.24   131.04
  90    89.33    98.65   107.57   113.15   118.14   124.12   137.21
  95    94.33   103.90   113.04   118.75   123.86   129.97   143.34
 100    99.33   109.14   118.50   124.34   129.56   135.81   149.45
```

In a `chitable`, the first row shows the significance levels. The first column shows the degrees of freedom. You can see that with 1 degree of freedom, you need a chi-squared value of 3.84 to be significant at the 0.05 level. If you had 4 degrees of freedom, you would need a chi-squared of 9.49 to be significant at the 0.05 level. You might try the other commands for the other tables. If you have ever had to search for one of these tables in a textbook, you will appreciate these commands.

Reporting chi-squared results

How do we report the significance level of chi-squared? How do we report that chi-squared varies somewhat from one field to another? A safe way to report the significance is as $p < 0.05$, $p < 0.01$, or $p < 0.001$. Suppose we had a chi-squared of 10.05 with 1 degree of freedom and $p = 0.002$. This is less than 0.01 but not less than 0.001. We would say that $\chi^2(1) = 10.05$, $p < 0.01$. Notice that we put the degrees of freedom in parentheses. Some disciplines would also like you to list the sample size, e.g., $\chi^2(1, N = 435) = 10.05$, $p < 0.01$. Still other disciplines would rather you report the probability value, e.g., $p = 0.002$.

This last approach has the advantage of providing more detailed information. For example, one result might have a $p = 0.052$ and another might have a $p = 0.451$. The first of these is almost statistically significant and is an unlikely result if the null hypothesis of no relationship is correct. The second of these is just about what you would expect to get by chance flipping a coin. Saying that both of these have the same classification of p not significant conceals the clear difference between these two results.

6.4 Percentages and measures of association

We have already discussed the use of percentages. These are often the easiest and best way to describe a relationship between two variables. In our last example, the percentage of men who said abortion was okay for any reason was slightly greater than, although not statistically significantly greater than, the percentage of women. Percentages often tell us what we want to know. There are other ways of describing an association. These are called measures of association, and they try to summarize the strength of a relationship with a number.

The value of chi-squared depends on two things. First, the stronger the association between the variables, the bigger chi-squared will be. Second, because we have more confidence in our results when we have larger samples, then the more cases we have, the bigger chi-squared will be. In fact, for a given relationship expressed in percentages, chi-squared is a function of sample size. If you had the same relationship as in our example but, instead of having 1,939 observations, you had 19,390 (10 times as many), chi-squared would be 20.254, also 10 times as big. There would still be one degree of freedom, but here the results would be statistically significant, $p < 0.001$. When you have large samples, researchers sometimes misinterpret a statistically significant chi-squared value as indicating a strong relationship. With a large sample, even a weak relationship can be statistically significant. This makes sense because with a large sample we have the power to detect even small effects.

One way to minimize this potential misinterpretation is to divide chi-squared by the maximum value it could be for a table of a particular shape and number of observations.

This is simple in the case of two-by-two tables, such as the one we are using. The maximum value of chi-squared for a two-by-two table is the sample size, N. Thus, in our example, if the relationship were as strong as possible, chi-squared would be 1,939. Our chi-squared of 2.03 is pretty tiny in comparison. The coefficient ϕ (phi) is defined as the square root of the quantity chi-squared divided by N.

$$\phi = \sqrt{\frac{\chi^2}{N}}$$

For larger tables, such as three-by-three, three-by-four, and so on, we call the coefficient Cramér's V, but it is still the square root of the quantity chi-squared divided by its maximum value. The maximum value of chi-squared for a table with R rows and C columns is $N \times \text{Min}(R-1, C-1)$. For example, in a four-by-three table (this means four rows and three columns) with an $N = 1939$, the maximum value would be $N \times \text{Min}(4-1, 3-1) = 1939 \times 2 = 3878$.

Because both V and ϕ are the square root of chi-squared divided by its maximum possible value and because ϕ can be thought of as a special case of V, Stata simply has an option to compute Cramér's V. However, if you have a two-by-two table, you should call it ϕ to avoid confusion. On the dialog box for doing the cross-tabulation, simply check *Cramer's V* under the list of *Test statistics*. This results in the command `tabulate sex abany, chi2 row V`. If you type this command directly, it is important to remember to capitalize the `V`. This is a rare example where you must use an uppercase letter.

In the example, you would say $\phi = 0.03$. A value of V or ϕ of less than 0.20 is generally considered a weak relationship. Values between 0.20 and 0.49 are considered moderate, and values more than 0.50 are considered strong. If we had the same distribution based on percentages, the ϕ would be the same whether we had 1,939 observations, 194 observations, or 19,390 observations indicating that the relationship is weak. This is important because, with the largest of these examples, the chi-squared would tell us the relationship was statistically significant. In that case, we would want to report that the relationship was statistically significant but substantively weak.

Odds ratios for two-by-two tables

Odds ratios are useful when the dependent variable has just two categories. We define odds ratios by using the independent variable. What are the odds that a man will say that abortion is okay for any reason? We know that 350 men say this and 478 do not, so the odds are $350/478 \approx 0.73$. In other words, there are 73 men agreeing with the statement for every 100 men who disagree. Thus the odds of a man supporting abortion for any reason are 0.73 times less than his not supporting it for any reason.

What about women? The odds for a woman supporting abortion for any reason are $434/677 \approx 0.64$. Because the ratio is less than 1.0, fewer women say yes to the question than say no to it. There are 64 women agreeing with the statement for every 100 women who disagree. The odds of a man saying yes, abortion is okay for any reason, 0.73, are slightly greater than the odds of a woman saying yes, 0.63.

Next compute the odds ratio by calculating the ratio of the two odds. Thus we say the odds ratio is (approximately) $0.73/0.64 = 1.14$. The odds of a man answering yes are 1.14 times as great as the odds of a woman answering yes. We could say men have 14% greater odds of supporting abortion for any reason than do women $[(1.14 - 1) \times 100]$. If the odds ratio had been less than 1.00, say, 0.80, then we would say the men had 20% lower odds $[(1.00 - 0.80) \times 100]$. When the odds ratio is greater than 1, we subtract 1 from the odds ratio; when it is less than 1, we subtract the odds ratio from 1.

Stata has a collection of commands, such as `tabodds`, for epidemiological analyses that use odds ratios. Other procedures, such as logistic regression, also use odds ratios. Here we introduce the idea because these other commands are beyond the scope of this chapter.

6.5 Ordered categorical variables

The example we have covered involves two unordered categorical variables (nominal level). Sometimes the categorical variables have an underlying order. For example, we might be interested in the relationship between health (`health`) and happiness (`happy`). Are people who are healthier also happier? When the question is asked this way, `health` is the independent variable because being happy is said to depend on your health.

`health` → `happy`

Another researcher might reverse this premise and argue that the happier a person is then the more likely they are to rate everything, including their own health, as better than people who are unhappy with their life. If this is your argument, then happiness is the independent variable and health depends on how happy you are.

$$\texttt{health} \leftarrow \texttt{happy}$$

A third researcher may simply say the two variables are related without claiming the direction of the relationship. We say that happiness and health are reciprocally related. In other words, the happier you are, the more positive you will report your health to be, and the healthier you are, the happier you will report being.

$$\texttt{health} \leftrightarrow \texttt{happy}$$

The example above uses a double-headed arrow, meaning that happiness leads to better perceived health and better perceived health leads to happiness. Because both variables depend on each other, there is no single variable that we can call independent or dependent. This probably makes sense. We have all known people who are happy and see the world through rose-colored lenses, where they rate nearly everything positively. They are likely to rate their health as positive. We have also known people for whom their health has a big influence on their happiness. If they have a health condition that varies, then when they are feeling relatively well they will report being happy, and when they are feeling bad they will report being unhappy.

If there is no clear independent or dependent variable, it is usually best to make the row variable the one with the most categories. Running the command `codebook happy health, compact` shows that `health` has four categories (`excellent`, `good`, `fair`, and `poor`) and `happy` has just three categories (`very happy`, `pretty happy`, and `not too happy`). So we will make `health` the row variable and `happy` the column variable.

Select Statistics ▷ Summaries, tables, and tests ▷ Tables ▷ Two-way tables with measures of association. The resulting dialog box is the same as the one in figure 6.1. We need to enter the names of the variables and the options we want. Enter `health` for the row variable and `happy` for the column variable. We pick both *Within-column relative frequencies* and *Within-row relative frequencies*. Also check *Pearson's chi-squared*, *Goodman and Kruskal's gamma*, *Kendall's tau-b*, and *Cramer's V* to obtain these statistics. Here are the command and results that are produced:

```
. tabulate health happy, chi2 column gamma row taub V

+-------------------+
| Key               |
|-------------------|
|     frequency     |
|  row percentage   |
| column percentage |
+-------------------+

 CONDITION  |         GENERAL HAPPINESS
 OF HEALTH  | VERY HAPP  PRETTY HA  NOT TOO H |     Total
------------+---------------------------------+----------
  EXCELLENT |       271        247         33 |       551 
            |     49.18      44.83       5.99 |    100.00 
            |     42.74      22.50      12.50 |     27.61 
------------+---------------------------------+----------
       GOOD |       261        567        103 |       931 
            |     28.03      60.90      11.06 |    100.00 
            |     41.17      51.64      39.02 |     46.64 
------------+---------------------------------+----------
       FAIR |        82        231         92 |       405 
            |     20.25      57.04      22.72 |    100.00 
            |     12.93      21.04      34.85 |     20.29 
------------+---------------------------------+----------
       POOR |        20         53         36 |       109 
            |     18.35      48.62      33.03 |    100.00 
            |      3.15       4.83      13.64 |      5.46 
------------+---------------------------------+----------
      Total |       634      1,098        264 |     1,996 
            |     31.76      55.01      13.23 |    100.00 
            |    100.00     100.00     100.00 |    100.00 
          Pearson chi2(6) = 182.1737   Pr = 0.000
                 Cramér's V =   0.2136
                      gamma =   0.3917  ASE = 0.030
            Kendall's tau-b =   0.2492  ASE = 0.020
```

Notice that Stata makes some compromises in making this table. If a value label is too big to fit, Stata simply truncates the label. You might need to do a codebook on your variables to make sure you have the labels correct. In this table, the value label "very happy" appears as "very happ", and "not too happy" appears as "not too h". If you were preparing this table for publication, you would want to edit it so that it has proper labels.

Just like with unordered categories, we use chi-squared to test the significance of the relationship. The relationship between perceived health and happiness is statistically significant: $\chi^2(6, N = 1996) = 182.17$, $p < 0.001$.

The percentages are also useful. Here we will pick one variable, `health`, arbitrarily as the independent variable. The box just above the table tells us that the row percentages are the second number in each cell. Only 18.35% of those with poor health said they are very happy, compared with 49.18% of those in excellent health. Similarly, only 5.99% of those in excellent health said they were not too happy, but 33.03% of those in poor health said they were not too happy.

If you decide to treat happiness as the independent variable, you would say that 42.74% of those who were very happy reported being in excellent health, compared with just 12.50% of those who were not too happy. Analyzing the percentages provides far richer information about the relationship than a measure of association can provide. However, researchers often want the simplicity of having a number that summarizes the strength of the association, and this is exactly why we have measures of association.

Cramér's V can be used as a measure of association, but it does not use the ordered nature of the variables. Both gamma (γ) and tau-b (τ_b) are measures of association for ordinal data. Both of these involve the notion of concordance. If one person is happier than another, we would expect that the person will report being in better health. We call this "concordance". If a person has worse health, we expect this person to be less happy. This is what we mean when we say that health and happiness are positively related. Gamma and tau-b differ in how they treat people who are tied on one or the other variable, but both measures are bigger when there is a predominance of concordant pairs. Because of the way it is computed, gamma tends to be bigger than tau-b, and tau-b is closer to what you would get if you treated the variables as interval level and computed a correlation coefficient. Values of tau-b less than 0.2 signify a weak relationship. Values between 0.2 and 0.49 indicate a moderate relationship. Values of 0.5 and higher indicate a strong relationship. Stata tells us that tau-b is 0.25, so we can say the relationship is moderate. After we study the percentages, they seem consistent with this judgment provided by tau-b.

The chi-squared test is an appropriate test for the significance of Cramér's V. Stata does not provide a significance level of gamma or tau-b, but it does provide an asymptotic standard error (ASE) for each. The asymptotic standard error for tau-b is ASE $= 0.020$. If you divide tau-b by this estimated standard error, you get the z test value. For our example, $z = 0.249/0.020 = 12.45$. A $z \leq 1.96$ is significant at the $p < 0.05$ level, $z \leq 2.60$ is significant at the 0.01 level, and $z \leq 3.32$ is significant at the .001 level. Considering these levels, we can say that our tau-b $= 0.25$, $z = 12.45$, $p < 0.001$, meaning that there is a moderate relationship and that it is statistically significant. A note of caution: because these are asymptotic standard errors, they are only good estimates when you have a large sample.

6.6 Interactive tables

If you are reading a report, you might find a cross-tabulation where the authors did not compute percentages, did not compute chi-squared, or did not compute any measures of association. Here is an example of a table showing the cross-tabulation of `sex` and `abany` (abortion is OK for any reason). We have been using data from the 2006 General Social Survey, and this table is from the 2002 General Social Survey. You might find this table and want to study its contents to see if there is a statistically significant relationship and how strong the association is. The author may have presented this table to show that men are less supportive of abortion than women, and you want to make sure the author has interpreted the table correctly.

```
. tabulate sex abany

           |    abortion if woman
respondent |   wants for any reason
       sex |       yes         no |     Total
-----------+----------------------+----------
      male |       215        269 |       484
    female |       172        244 |       416
-----------+----------------------+----------
     Total |       387        513 |       900
```

If you do not have access to the author's actual data, you need to enter the table into Stata. You can enter the raw numbers in the cells of the table, and Stata will compute percentages, chi-squared, and measures of association.

Select Statistics ▷ Summaries, tables, and tests ▷ Tables ▷ Table calculator to open a dialog box in which you enter the frequencies in each cell. The format for doing this is rigid. Enter each row of cells, separating the rows with a backslash, \. Enter only the cell values and not the totals in the margins of the table. This is illustrated in figure 6.3. You need to check the table Stata analyzes to make sure that you entered it correctly. Check the boxes to request chi-squared and within-row relative frequencies. Because this is not ordinal data, the only measure of association you request is Cramér's V.

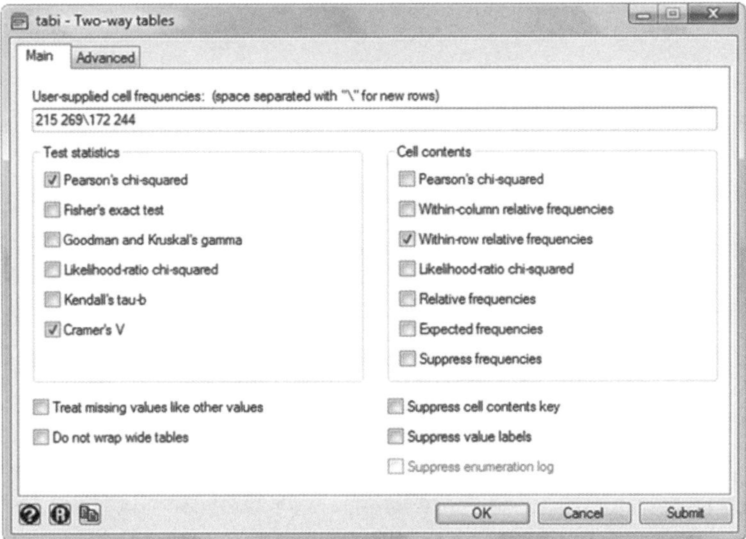

Figure 6.3. Entering data for a table

If you click on Submit in this dialog box, you obtain the same results as if you had entered all the data in a big dataset.

```
. tabi 215 269\172 244, chi2 row V

  +-------------------+
  | Key               |
  |-------------------|
  |     frequency     |
  |  row percentage   |
  +-------------------+

           |          col
       row |         1          2 |     Total
-----------+----------------------+----------
         1 |       215        269 |       484 
           |     44.42      55.58 |    100.00 
-----------+----------------------+----------
         2 |       172        244 |       416 
           |     41.35      58.65 |    100.00 
-----------+----------------------+----------
     Total |       387        513 |       900 
           |     43.00      57.00 |    100.00 

          Pearson chi2(1) =   0.8633   Pr = 0.353
               Cramér's V =   0.0310
```

These results indicate that the author overstated his findings. The results are not statistically significant; $p = .353$ and Cramér's V is .03. A slightly higher percentage of men in the 2002 sample said that abortion was acceptable for any reason (44.42%) compared with women (41.35%), but this difference is not statistically significant.

6.7 Tables—linking categorical and quantitative variables

Many research questions can be answered by tables that mix a categorical variable and a quantitative variable. We may want to compare income for people from different racial groups. A study may need to compare the length of incarceration for men and for women sentenced for the same crime. A drug company may want to compare the time it takes for its drug to take effect with the time it takes another drug.

Here we will use `hrs1`, hours a person worked last week, as our quantitative variable and compare men with women. Our hypothesis is that men spend more time working for pay each week than do women. We will use a formal test of significance in the next chapter, but here we will just show a table of the relationship. Select Statistics ▷ Summaries, tables, and tests ▷ Tables ▷ Table of summary statistics (table) to get to the dialog box in figure 6.4.

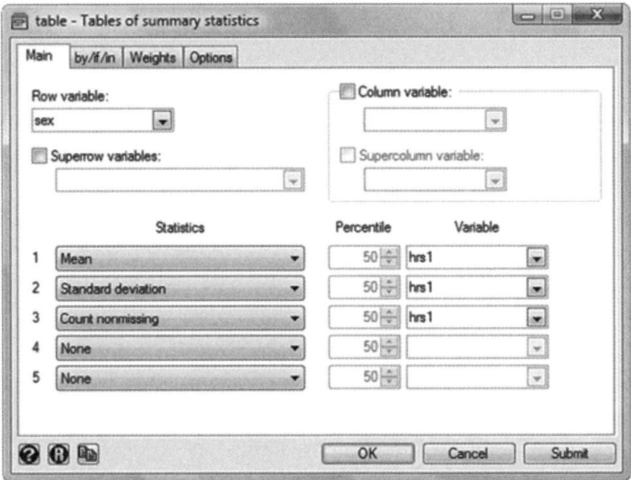

Figure 6.4. Summarizing a quantitative variable by categories of a categorical variable

The dialog box is getting pretty complicated. Look closely at where we enter the variable names and statistics we want Stata to compute. The *Row variable* is the categorical variable, `sex`. Under *Statistics*, select the following from the menus: *Mean*, *Standard deviation*, and *Count nonmissing*. On the far right, there are places to enter variables. We want `hrs1` for the mean, standard deviation, and count. Click on Submit so that we can return to this dialog box at a later time. The resulting command is

```
. table sex, contents(mean hrs1 sd hrs1 count hrs1)
```

Gender	mean(hrs1)	sd(hrs1)	N(hrs1)
MALE	44.8767	14.27598	1,387
FEMALE	39.2034	13.60452	1,352

This table shows that the week before the survey men spent on average over 5 more hours working for pay: 44.88 hours for men versus 39.20 for women.

Now suppose that you want to see the effect of marital status and gender on hours spent working for pay. Go back to the dialog box, and add `marital` as the *Column variable*. Go to the Options tab and select *Add row totals*. The command for this is

```
. table sex marital, contents(mean hrs1 sd hrs1 count hrs1) row
```

We will not show the results here.

These summary tables are useful, but to communicate to a lay audience, a graph often works best. We can create a bar chart showing the mean number of hours worked by gender and by marital status. We created one type of bar chart in chapter 5, where we used the dialog box for a histogram to create the bar chart. Now we are doing a bar chart for a summary statistic, mean hours worked, over a pair of categorical variables, `sex` and `marital`. To do this, access the dialog box for a bar chart by selecting Graphics ▷ Bar chart. Under the Main tab, make sure that the first statistic is checked, and, by default, this is the mean. To the right of this, under *Variables*, type `hrs1` to make the graph show the mean hours worked. Next click on the Categories tab. Check *Group 1* and type `sex` as the *Grouping variable*. Check *Group 2* and type `marital` as the *Grouping variable*. There is a button labeled *Properties* to the right of where you entered `sex`. Click on this button and change the *Angle* to *45 degrees*.

It would be nice to have the actual mean values for hours worked at the top of each bar, so switch to the Bars tab, and check *Label with bar height*. If we stop here, the numerical labels will have too many decimal places to look nice. We can fix these by making the numerical labels have a fixed format. Click on the *Properties* button in the section labeled *Bar labels*. In the box labeled *Format*, we replace the default value with `%9.1f`. The *Bar label properties* dialog box appears in figure 6.5.

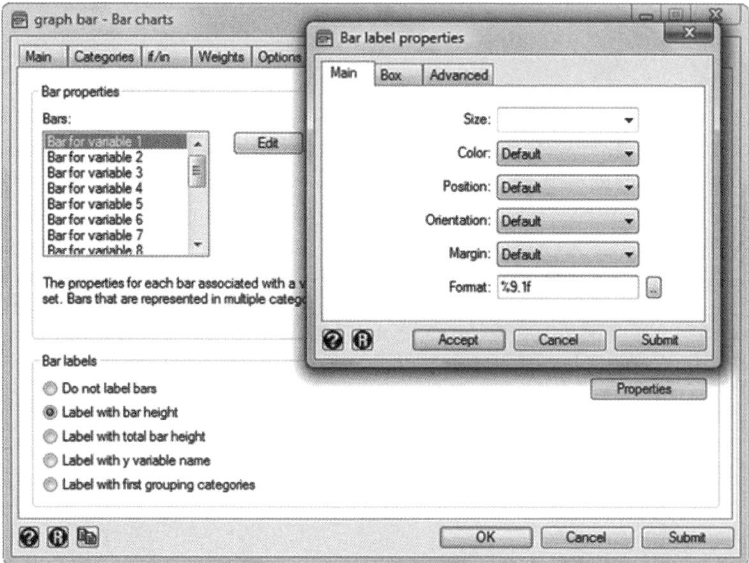

Figure 6.5. Labeling the bars

This produces a nice bar chart, but there are some things we can improve. The marital status labels are too wide and run into each other. Also, the label of the hours worked for pay could be better. Open the Graph Editor and change the font of the labels of marital status and gender to small. Then add the title `Hours Worked Last Week` and the subtitle `By Sex and Marital Status`. Your final bar chart appears in figure 6.6.

```
. graph bar (mean) hrs1, over(sex, label(angle(forty_five))) over(marital)
> blabel(bar, format(%9.1f)) title(Hours Worked Last Week)
> subtitle(By Sex and Marital Status) scheme(s1mono)
```

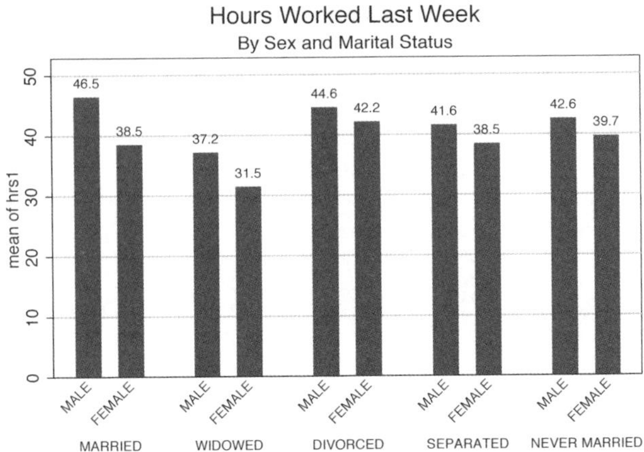

Figure 6.6. Bar graph summarizing a quantitative variable by categories of a categorical variable

6.8 Summary

Cross-tabulations are extremely useful ways of presenting data, whether the data are categorical or a combination of categorical and quantitative variables. This chapter has covered a lot of material. We have learned

- How to examine the relationship between categorical variables
- How to develop cross-tabulations
- How to perform a chi-squared test of significance
- How to use percentages and compute appropriate measures of association for unordered and ordered cross-tabulations
- What the odds ratio is
- How to use an interactive table calculator
- How to extend tables to link categorical variables with quantitative variables
- How a bar graph can show how means on a quantitative variable differ across groups on a categorical variable

Stata has many capabilities for dealing with categorical variables that we have not covered. Still, think about what you have learned in this chapter. Policies are made and changed because of highly skilled presentations. Imagine your ability to make a presentation to a research group or to a policy group. Now you can "show them the numbers" and do so effectively. Much of what you have learned would have been impossible without statistical software. The next chapter will continue to show how Stata helps us analyze the relationship between pairs of variables. It will focus on quantitative outcome variables and categorical predictors and move beyond the point biserial correlation.

6.9 Exercises

1. Open `gss2006_chapter6.dta`, and do a `codebook` on `pornlaw`. Do a cross-tabulation of this with `sex`. Which variable is the independent variable? Which variable will you put on the row, and how will you do the percentages? If you were preparing a document, how would you change the labels of the response options for `pornlaw`?

2. Based on the first exercise, what is the chi-squared value? How would you report the chi-squared and the level of significance? What is the value of Cramér's V, and how strong of an association does this show? Finally, interpret the percentages to answer the question of how much women and men differ in their attitude about legalizing pornography.

3. Open `gss2006_chapter6.dta`, and do a cross-tabulation of `pres00` (whom you voted for in 2000) and `pres04` (whom you voted for in 2004). From the dialog box, check the option to include missing values. Do a `codebook` on both variables, and then use the by/if/in tab in the `tabulate` dialog box to repeat the table just for those who voted for Gore or Bush in 2000 and for Kerry or Bush in 2004. Treat the 2000 vote as the independent variable. Is there a significant relationship between how people voted in 2000 and 2004? Interpret the percentages and phi, as well as the statistical significance.

4. Open `gss2006_chapter6.dta`, and do a cross-tabulation of `polviews` and `premarsx`. Treating `polviews` as the independent variable, compute percentages on the rows. Because these are ordinal variables, compute gamma and tau-b. Is there a significant relationship between political views and conservatism? Interpret the relationship using gamma and tau-b. Interpret the relationship using the percentages.

5. In the last exercise, Stata reports that there are 18 degrees of freedom. How does it get this number?

6. Open `gss2002_chapter6.dta`. Do a table showing the mean hours worked in the last week (`hrs1`) for each level of political views (`polviews`). In your table, include the standard deviation for hours and the frequency of observations. What do the means suggest about people who are extreme in their views (in either direction)?

7. Based on the last exercise, create a bar chart showing the relationship between hours worked in the last week and political views.

8. You are given the following dataset:

RESPONDENT S SEX	FEELINGS ABOUT PORNOGRAPHY LAWS ILLEGAL T	ILLEGAL U	LEGAL	Total
MALE	234	568	38	840
FEMALE	541	557	29	1,127
Total	775	1,125	67	1,967

Calculate the chi-squared test of significance, percentages, and Cramér's V using the `tabi` command. Interpret Cramér's V.

7 Tests for one or two means

7.1 Introduction to tests for one or two means

Imagine that you are in the public relations department for a small liberal arts college. The mean score on the SAT at your college is $M = 1180$.[1] Suppose that the mean for all small liberal arts colleges in the United States is $\mu = 1080$. We use M to refer to the mean of a sample and μ, pronounced 'mu', to refer to the mean of a population. For our purposes, we are treating the students at your college as the sample and all students at small liberal arts colleges in the United States as the population. Say that our supervisor asked if we could demonstrate that our college is more selective than its peer institutions. Could this difference, our $M = 1180$, and the mean for all colleges, $\mu = 1080$, occur just by chance?

1. The numbers used do not include the essay portion of the SAT.

Suppose that last year, our Boys and Girls Club had 30% of the children drop out of programs before the programs were completed. We have a new program that we believe does a better job of minimizing the dropout rate. However, 25% of the children still drop out before the new program is completed. Because a smaller percentage of children drop out, this is better, but is this enough difference to be statistically significant?

These two examples illustrate a one-sample test where we have a single sample (students at our college or the children in the new Boys and Girls Club program), and we want to compare our sample mean with a population value (the mean SAT score for all liberal arts colleges or mean dropout rate for all programs at our Boys and Girls Club).

At other times, we may have two groups or samples. For example, we have a program that is designed to improve reading readiness among preschool children. We might have 100 children in our preschool and randomly assign half of them to this new program and the other half to a traditional approach. Those in the new program are in the treatment group (because they get the new program), and those in the old program are in the control group (because they did not get the new program). If the new program is effective, the children in the treatment group will score higher on a reading-readiness scale than the children in the control group. Let's say that, on a 0–100 point scale, the 50 children randomly assigned to the control group have a mean of $M = 71.3$ and a standard deviation of $\text{SD} = 10.4$. Those in the treatment group, who were exposed to the new program, have $M = 82.5$ and $\text{SD} = 9.9$. It sounds like the program helped. The children in the treatment group have a mean that is more than one standard deviation higher than that of the students in the control group. This sounds like a big improvement, but is this statistically significant? We have a problem for a two-group or two-sample t test.

We have a 20-item scale that measures the importance of voting. Each item is scored from 1, signifying that voting is not important, to 5, signifying that voting is very important. Thus the possible range for the total score on our 20-item scale is from 20 (all items answered with a 1, indicating voting is not important) to 100 (all items answered with a 5, signifying that voting is very important). We hypothesize that minority groups will have a lower mean than whites. Can we justify this hypothesis? Perhaps the lower mean results because minority groups have experienced a history of their votes not counting or the perception that their vote does not matter in the final outcome. To do an experiment with a treatment and control group, we would need to randomly assign people to minority status or majority status. Because we cannot do this, we can use a random sample of 200 people from the community. If this sample is random for the community, it is a random sample for any subgroup. Let's say that 70 minority community members are in our sample and they have a mean importance of voting score of $M = 68.9$. The 130 white community members in our sample have a mean importance of voting score of $M = 83.5$. This is what we hypothesized would happen. A two-sample t test will tell us if this difference is statistically significant.

Sometimes researchers use a two-sample t test even when they do not have random assignment to a treatment or control group or when they do not have random sampling.

You might want to know if a highly developed recreation program at a full-care retirement center leads to lower depression among center members. Your community might have two full-care retirement centers, one with a highly developed recreation program and the other without such a program. We could do a two-sample test to see whether the residents in the center with the highly developed recreation program had a lower mean score on depression.

Can you see the problems with this? First, without randomization to the two centers, we do not know whether the recreation program makes the difference or if people who are more prone to be depressed selected the center that does not offer much recreation. A second problem is that the programs being compared may be different in many respects other than our focus. A center that can afford a highly developed recreation program may have, e.g., better food, more spacious accommodations, and better trained staff. Even if we have randomization, we want the groups to differ only on the issue we are testing. More-advanced procedures, such as analysis of covariance and multiple regression, can help control for other differences between the groups being compared, but a two-sample t test is limited in this regard without randomization or random sampling.

Random sample and randomization

It is easy to confuse random sample and randomization. A random sample refers to how we select the people in our study. Once we have our sample, randomization is sometimes used to randomly assign our sample to groups. Some examples may illustrate:

Random sample without randomization. We obtain a list of all housing units that have a water connection from our city's utility department. We randomly sample 500 of these houses.

Randomization without a random sample. We solicit volunteers for an experiment. After 100 people volunteer, we randomly assign 50 of them to the treatment group and 50 to the control group. We randomized the assignment, but we did not start with a random sample. Because we did not start with a random sample, we could have a **sample selection bias** problem because volunteers are often much more motivated to do well than the rest of the population.

Randomization with a random sample. Obtaining a list of all students enrolled in our university, we randomly sample 100 of these students. Next we randomly assign 50 of them to the treatment group and 50 to the control group.

7.2 Randomization

How can we select the people we want to randomly assign to the treatment group and the control group? There are two types of random selection. One is done with replacement, and one is done without replacement. When we sample with replacement, each observation has the same chance of being selected. However, the same observation may be selected more than once. Suppose that you have 100 people and want to do random sampling with replacement. You randomly pick one case, say, Mary. She has a 1/100 probability of being selected. If you put her back in the pool (replacement) before selecting your second case, whoever you select will have the same 1/100 chance of being selected. However, you may sample Mary again doing it this way.

Many statistical tests assume sampling with replacement, but in practice, we rarely do this. We usually sample without replacement. This means that the first person would have a 1/100 chance of being selected, but the second person would have a 1/99 chance of being selected. This method violates the assumption that each observation has the same chance of being selected. Sampling without replacement is the most practical way to do it. If we have 100 people and want 50 of them in the control group and 50 in the treatment group, we can sample 50 people without replacement (this gives us 50 different people) and put them in the control group. Then we put the other 50 people in the treatment group. If we had sampled with replacement, the 50 people we sampled might actually be only 45 people, with a few of them selected twice.

Sampling without replacement is a simple task in Stata. If you have $N = 20$ observations, you would number these from 1 to 20. Then you would enter the command `sample 10, count` to select 10 observations randomly but without replacement. These 10 people would go into your treatment group, and the other 10 would go into your control group. Here is the set of commands:

```
clear
set obs 20
gen id = _n
list
set seed 220
sample 10, count
list
```

The first command clears any data we have in memory. You should save anything you were doing before entering this command. The second command, `set obs 20`, tells Stata that you want to have 20 observations in your dataset. The next command, `gen id = _n`, will generate a new variable called `id`, which will be numbered from 1 to the number of observations in your dataset, in this case, 20. You can see this in the first listing below. We then use the command `set seed 220`, where the number 220 is arbitrary. This is not necessary, but by doing this, we will be able to reproduce our result. The random process will start at the same place when you set the seed. Finally, we take the sample using the command `sample 10, count` and show a listing of the 10 cases that Stata selected at random without replacement. These cases would go into the control group, and the other 10 would then go into the treatment group.

Thus observations numbered 8, 19, 9, 13, 18, 12, 6, 11, 17, and 3 go into the control group, and the other 10 observations go into the treatment group. By doing this, any differences between the groups that are not due to the treatment are random differences.

```
. clear

. set obs 20
obs was 0, now 20

. gen id = _n

. list

     +----+
     | id |
     |----|
  1. |  1 |
  2. |  2 |
  3. |  3 |
  4. |  4 |
  5. |  5 |
     |----|
  6. |  6 |
  7. |  7 |
  8. |  8 |
  9. |  9 |
 10. | 10 |
     |----|
 11. | 11 |
 12. | 12 |
 13. | 13 |
 14. | 14 |
 15. | 15 |
     |----|
 16. | 16 |
 17. | 17 |
 18. | 18 |
 19. | 19 |
 20. | 20 |
     +----+

. set seed 220

. sample 10, count
(10 observations deleted)

. list

     +----+
     | id |
     |----|
  1. |  8 |
  2. | 19 |
  3. |  9 |
  4. | 13 |
  5. | 18 |
     |----|
  6. | 12 |
  7. |  6 |
  8. | 11 |
  9. | 17 |
 10. |  3 |
     +----+
```

7.3 Random sampling

The process of selecting a random sample is similar. Suppose you had 20,000 students in your university. Assume each student has a unique identification number. You would have a Stata dataset with these numbers. Assume too that you need a sample of $N = 500$ students. You would simply set the seed at some arbitrary value, 3 with the command `set seed 3`, and then run the command `sample 500, count`. You could then list these ID numbers and this would be your sample.

If we have a target sample of 500, we can expect that several of our initial list of $N = 500$ either will not be located or will refuse. Therefore, it would be good to draw a second sample of say, $N = 300$, and go down this list as we need replacements.

7.4 Hypotheses

Before we can run a z test or a t test, we need to have two hypotheses: a null hypothesis (H_0) and an alternative hypothesis (H_A). We will mention these briefly because they are covered in all statistics texts.

Although one-tailed tests are appropriate if we can categorically exclude negative findings (results in the opposite direction of our hypothesis), we will rarely be this confident in the direction of the results. Most statisticians report two-tailed tests routinely and make note that the direction of results is what they expected. A two-tailed test is always more conservative than a one-tailed test, so the tendency to rely on two-tailed tests can be viewed as a conservative approach to statistical significance.

Stata reports both one- and two-tailed significance for all the tests covered in this chapter. For more-advanced procedures, such as those based on regression, Stata reports only two-tailed significance. It is easy to convert two-tailed significance to one-tailed significance. If something has a probability of .1 using a two-tailed test, its probability would be .05 using a one-tailed test—we simply cut the two-tailed probability in half.

7.5 One-sample test of a proportion

Let's use items from the 2002 General Social Survey dataset, `gss2002_chapter7.dta`, to test how adults feel about school prayer. An existing variable, `prayer`, is coded `1` if the person favors school prayer and `2` if the person does not. We must recode the variable `prayer` into a new variable, `schpray`, so that a person has a score of `1` if they support school prayer and a score of `0` if they do not. We can use Stata's `recode` command to accomplish this. We will do a cross-tabulation as well to make sure we did not make a mistake:

```
. recode prayer (1 = 1 Approve) (2 = 0 Disapprove), gen(schpray)
. tabulate schpray prayer, missing
```

Stata will not let you accidentally write over an existing variable, so if you try to create a variable that already exists, Stata will issue an error message. When we give `schpray` a score of `1` for those adults who favor school prayer and a score of `0` for those who oppose it, Stata can compute and test proportions correctly. If we ran a proportions test on the variable `prayer`, we would get an error message saying that `prayer` is not a 0/1 variable.

Although we can speculate that most people oppose school prayer, i.e., have a score of `0` on `schpray`, let's take a conservative track and use a two-tailed hypothesis:

Alternative hypothesis H_A: $p \neq 0.5$

Null hypothesis H_0: $p = 0.5$

The null hypothesis, $p = 0.5$, uses a value of 0.5 because this represents what the proportion would be if there were no preference for or against school prayer.

Open the dialog box by selecting Statistics ▷ Summaries, tables, and tests ▷ Classical tests of hypotheses ▷ One-sample proportion test. Enter the variable `schpray` and the hypothesized proportion for the null hypothesis, `0.5`, and click on OK. The resulting command and results are

```
. prtest schpray == .5
One-sample test of proportion                  schpray: Number of obs =       859
------------------------------------------------------------------------------
    Variable |       Mean   Std. Err.                      [95% Conf. Interval]
-------------+----------------------------------------------------------------
     schpray |   .3969732   .0166937                       .3642542    .4296923
------------------------------------------------------------------------------
    p = proportion(schpray)                                       z =  -6.0392
Ho: p = 0.5

     Ha: p < 0.5                 Ha: p != 0.5                   Ha: p > 0.5
 Pr(Z < z) = 0.0000         Pr(|Z| > |z|) = 0.0000          Pr(Z > z) = 1.0000
```

Let's go over these results and see what we need. All the results in this chapter use this general format for output, so it is worth the effort to review it closely. The first line repeats the Stata command. The second line indicates that this is a one-sample test of a proportion, the variable is `schpray`, and we have $N = 859$ observations. The table gives a mean, $M = 0.397$, and standard error, 0.017. Because we recoded `schpray` as a dummy variable, where a `1` signifies that the person supports school prayer and a `0` signifies that the person does not, the mean has a special interpretation. The mean is simply the proportion of people coded `1`. (This is true only of 0/1 variables and would not work if we had used the original variable, `prayer`.) Thus 0.397, or 39.7%, of the people in the survey say that they support school prayer. The 95% confidence interval tells us that we can be 95% confident that the interval 0.364 to 0.430 includes adults who support school prayer. Using percentages, we could say that we are 95% confident that the interval of 36.4% to 43.0% contains the true percentage of adults who support school prayer.

Below the table, we can see the null hypothesis that the proportion is exactly 0.5, i.e., H_0: $p = 0.5$. Also just below the table, but to the far right, is the computed z test, $z = -6.039$.

Below the null hypothesis are three results that depend on how we stated the alternative hypothesis. Here we used a two-tailed alternative hypothesis, and this result appears in the middle group. Stata's `Ha: p != 0.5` is equivalent to stating the alternative hypothesis, H_A: $p \neq 0.5$. Because Stata results are plain text, Stata uses `!=` in place of $\neq$. Below the alternative hypothesis is the probability that the null hypothesis is true: `Pr(|Z| > |z|) = 0.0000`. The $p = 0.0000$ does not mean that there is no probability that the null hypothesis is true, just that the probability is all zeros to four decimal places. If the $p = 0.00002$, this is rounded to $p = 0.0000$ in Stata output. A $p = 0.0000$ is usually reported as $p < 0.001$. This finding is highly statistically significant and allows us to reject the null hypothesis. Fewer than 1 time in 1,000 would we obtain these results by chance if the null hypothesis were true.

We would write this as follows: 39.7% of the sample support school prayer. With a $z = -6.039$, we can reject the null hypothesis that $p = 0.5$ at the $p < 0.001$ level. The 39.7% is in the direction expected, namely, that fewer than half of the adult population supports school prayer.

On the left side is a one-tailed alternative hypothesis that $p < 0.5$. If another researcher thought that only a minority supported school prayer and could rule out the possibility that the support was in the opposite direction, then the researcher would have this as a one-tailed alternative hypothesis, and it would be statistically significant with a $p < 0.001$.

On the right side is a one-tailed alternative hypothesis that $p > 0.5$. This means the researcher thought that most adults favored school prayer, and the researcher (incorrectly) ruled out the possibility that the support could be the other way. The results show that we cannot reject the null hypothesis. When we observe only 39.7% of our sample supporting school prayer, we clearly have not received a significant result that most adults do support school prayer.

Distinguishing between two *p*-values

In testing proportions, we have two different p-values. Do not let this confuse you. One of these is the proportion of the sample coded `1`. In this example, this is the proportion that supports school prayer, 0.397 or 39.7%.

The second p-value is the level of significance. In this example, $p < 0.001$ refers to the probability that we would obtain this result by chance if the null hypothesis were true. Because we would get our result fewer than 1 time in 1,000 by chance, $p < 0.001$, we can reject the null hypothesis. Many researchers will reject a null hypothesis whenever the probability of the results is less than 0.05, i.e., $p < 0.05$.

Proportions and percentages

Often we report proportions as percentages because many readers are more comfortable trying to understand percentages than proportions. Stata, however, requires us to use proportions. As you may recall, the conversion is simple. We divide a percentage by 100 to get a proportion, so 78.9% corresponds to a proportion of 0.789. Similarly, we multiply a proportion by 100 to get a percentage, so a proportion of 0.258 corresponds to a percentage of 25.8%.

7.6 Two-sample test of a proportion

Sometimes a researcher wants to compare a proportion across two samples. For example, you might have an experiment testing a new drug (`wide.dta`). You randomly assign 40 study participants so that 20 are in a treatment group receiving the drug and 20 are in a control group receiving a sugar pill. You record whether the person is cured by assigning a 1 to those who were cured and a 0 to those who were not. Here are the data:

```
. list

     +-----------------+
     | treat   control |
     |-----------------|
  1. |     1         1 |
  2. |     0         0 |
  3. |     1         0 |
  4. |     1         0 |
  5. |     1         0 |
     |-----------------|
  6. |     1         1 |
  7. |     1         1 |
  8. |     0         0 |
  9. |     1         0 |
 10. |     1         0 |
     |-----------------|
 11. |     1         0 |
 12. |     1         1 |
 13. |     1         1 |
 14. |     0         1 |
 15. |     1         1 |
     |-----------------|
 16. |     1         0 |
 17. |     0         0 |
 18. |     1         0 |
 19. |     0         0 |
 20. |     1         0 |
     +-----------------+
```

In the treatment group, we have 15 of the 20 people, or 0.75, cured; i.e., they have a score of `1`. In the control group, just 7 of the 20 people, or 0.35, are cured. Before proceeding, we need a null and an alternative hypothesis. The null hypothesis is that the two groups have the same proportion cured. The alternative hypothesis is that the proportions are unequal and, therefore, the difference between them will not equal zero: $p_{(\text{treat})} - p_{(\text{control})} \neq 0$. Your statistics book may state the null hypothesis as $p_{(\text{treat})} \neq p_{(\text{control})}$. The two ways of stating the null hypothesis are equivalent. They are both two-tailed tests because we are saying the proportions cured in the two groups are not equal. We could argue for a one-tailed test that the proportion in the treatment group is higher, but this means that we need to rule out the possibility that it could be lower. The null hypothesis is that the two proportions are equal; hence, there is no difference between them: $p_{(\text{treat})} - p_{(\text{control})} = 0$.

Alternative hypothesis H_A: $p_{(\text{treat})} - p_{(\text{control})} \neq 0$

Null hypothesis H_0: $p_{(\text{treat})} - p_{(\text{control})} = 0$

Notice that these are independent samples and that the data for the two groups are entered as two variables. To open the dialog box for this test, select Statistics ▷ Summaries, tables, and tests ▷ Classical tests of hypotheses ▷ Two-sample proportion test. This is a fairly simple dialog box. Type `treat` in the box for the *First variable*, and type `control` in the box for the *Second variable*. That is all there is to it, and our results are

```
. prtest treat == control
Two-sample test of proportion                  treat: Number of obs =        20
                                             control: Number of obs =        20

    Variable |       Mean   Std. Err.       z    P>|z|     [95% Conf. Interval]
-------------+----------------------------------------------------------------
       treat |        .75   .0968246                        .5602273    .9397727
     control |        .35   .1066536                        .1409627    .5590373
-------------+----------------------------------------------------------------
        diff |         .4   .1440486                        .1176699    .6823301
             |  under Ho:   .1573213     2.54   0.011
------------------------------------------------------------------------------
        diff = prop(treat) - prop(control)                        z =   2.5426
    Ho: diff = 0

    Ha: diff < 0                 Ha: diff != 0                 Ha: diff > 0
 Pr(Z < z) = 0.9945         Pr(|Z| < |z|) = 0.0110          Pr(Z > z) = 0.0055
```

These results have a layout that is similar to that of the one-sample proportion test. The difference is that we now have two groups, so we get statistical information for each group. Under `Mean` is the proportion of `1` codes in each group. We have 0.75 (75%) of the treatment group coded as cured, compared with just 0.35 (35%) of the control group. The difference between the treatment group mean and the control group mean is 0.40; i.e., $0.75 - 0.35 = 0.40$. This appears in the table as the variable `diff` with the mean of 0.4.

Directly below the table is the null hypothesis that the difference in the proportion cured in the two groups is 0, and to the right is the computed z test, `z = 2.5426`. Below this are the three hypotheses we might have selected. Using a two-tailed approach, we can say that $z = 2.54$, $p < 0.05$. If we had a one-tailed test that the treatment group proportion was greater than the control group proportion, our z would still be 2.54, but our p would be $p < 0.01$. This one-tailed test has a p-value, 0.0055, which is exactly one-half of the two-tailed p-value. If someone had a hypothesis that the treatment group success would have a lower proportion than the control group, they would have the results on the far left. Here the results would not be significant because the $p = .9945$ is far greater than the required $p = 0.05$.

This difference-of-proportions test requires data to be entered in what is called a wide format. Each group (treatment and control) is treated as a variable with the scores on the outcome variable coded under each group, as illustrated in the listing that appeared above. When dealing with survey data, it is common to use what is called a long format in which one variable is a grouping variable of whether someone is in the treatment group, coded `1`, or control group, coded `0`. The second variable is the score on the dependent variable, which is also a binary variable coded as `1` if the person is cured and `0` if the person is not cured. This appears in the following long-format listing (`long.dta`).

```
. list

     +--------------+
     | group   cure |
     |--------------|
  1. |     1      1 |
  2. |     1      0 |
  3. |     1      1 |
  4. |     1      1 |
  5. |     1      1 |
     |--------------|
  6. |     1      1 |
  7. |     1      1 |
  8. |     1      0 |
  9. |     1      1 |
 10. |     1      1 |
     |--------------|
 11. |     1      1 |
 12. |     1      1 |
 13. |     1      1 |
 14. |     1      0 |
 15. |     1      1 |
     |--------------|
  (output omitted)
     |--------------|
 31. |     0      0 |
 32. |     0      1 |
 33. |     0      1 |
 34. |     0      1 |
 35. |     0      1 |
     |--------------|
 36. |     0      0 |
 37. |     0      0 |
 38. |     0      0 |
 39. |     0      0 |
 40. |     0      0 |
     +--------------+
```

When your data are entered this way, you need to use a different test for the difference of proportions. Select Statistics ▷ Summaries, tables, and tests ▷ Classical tests of hypotheses ▷ Two-group proportion test to open the appropriate dialog box. Type `cure` under *Variable name* (the dependent variable) and `group` under *Group variable name* (the independent variable). Click on OK to obtain the following results:

```
. prtest cure, by(group)
Two-sample test of proportion                    0: Number of obs =        20
                                                 1: Number of obs =        20
------------------------------------------------------------------------------
    Variable |       Mean   Std. Err.      z    P>|z|     [95% Conf. Interval]
-------------+----------------------------------------------------------------
           0 |        .35   .1066536                      .1409627    .5590373
           1 |        .75   .0968246                      .5602273    .9397727
-------------+----------------------------------------------------------------
        diff |        -.4   .1440486                     -.6823301   -.1176699
             |  under Ho:   .1573213   -2.54   0.011
------------------------------------------------------------------------------
        diff = prop(0) - prop(1)                                  z =  -2.5426
    Ho: diff = 0

    Ha: diff < 0                 Ha: diff != 0                 Ha: diff > 0
 Pr(Z < z) = 0.0055         Pr(|Z| < |z|) = 0.0110          Pr(Z > z) = 0.9945
```

Here we have two means: one for the group coded `0` of 0.35 and one for the group coded `1` of 0.75. These are, of course, the means for our control group and our treatment group, respectively. Below this is a row labeled `diff`, the difference in proportions, which has a value of -0.40 because the mean of the group coded as `1` is subtracted from the mean of the group coded as `0`. Be careful interpreting the sign of this difference. Here a negative value means that the treatment group is higher than the control group: 0.75 versus 0.35. The z test for this difference is $z = -2.54$. This is the same absolute value that we had with the test for the wide format, but the sign is reversed. Be careful interpreting this sign just like with interpreting the difference. It happens to be negative because we are subtracting the group coded as `1` from the group coded as `0`.

We can interpret these results, including the negative sign on the z test, as follows. The control group has a mean of 0.35 (35% of participants were cured), and the treatment group has a mean of 0.75 (75% of the participants were cured). The $z = -2.54$, $p < 0.05$ indicates that the control group had a significantly lower success rate than the treatment group. Pay close attention to the way the difference of proportions was computed so that you interpret the sign of the z test correctly.

This long form is widely used in surveys. For example, in the General Social Survey 2002 data (`gss2002_chapter7.dta`), there is an item, `abany`, asking if abortion is okay anytime. The response option is binary, namely, `yes` or `no`. If we wanted to see if more women said yes than did men, we could use a difference-of-proportions test. The grouping variable would be `sex`, and the dependent variable would be `abany`. Because the dependent variable needs to be coded with 0s and 1s, we first need to generate a new variable that is coded this way for the abortion item. The independent variable, `sex`, does not need to be coded with 0s and 1s, but it must be binary. Try this.

7.7 One-sample test of means

You can do z tests for one-sample tests of means, or you can do t tests. We will cover only the use of t tests because these are by far the most widely used. A z test is appropriate when you know the population variance (which it is highly unusual to know). Unless you have a small sample, both tests yield similar results.

Several decades ago, social theorists thought that the workweek for full-time employees would get shorter and shorter. This has happened in many countries but does not seem to be happening in the United States. A full-time workweek is defined as 40 hours. Among those who say they work full time, do they work more or less than 40 hours? The General Social Survey dataset (`gss2002_chapter7.dta`) has a variable called `hrs1`, which represents the reported hours in the workweeks of the survey participants. The null hypothesis is that the mean will be 40 hours. The alternative hypothesis is that the mean will not be 40 hours. Using a two-tailed hypothesis, you can say

Alternative hypothesis H_A: $\mu_{(\text{hours})} \neq 40$

Null hypothesis H_0: $\mu_{(\text{hours})} = 40$

Notice that we are using the Greek letter, μ, to refer to the population mean in both the alternative and the null hypotheses. To open the dialog box for this test, select Statistics ▷ Summaries, tables, and tests ▷ Classical tests of hypotheses ▷ One-sample mean-comparison test. In the resulting dialog box, enter `hrs1` (outcome variable) under the *Variable name* and `40` (hypothesized mean for the null hypothesis) under the *Hypothesized mean*.

Because we are interested in how many hours full-time employees work, we should eliminate part-time workers. To do this, click on the by/if/in tab, and type `wrkstat == 1` as the *If: (expression)*. The variable `wrkstat` is coded as `1` if a person works full time (you can run a `codebook wrkstat` to see the frequency distribution). The resulting by/if/in tab is shown in figure 7.1.

(Continued on next page)

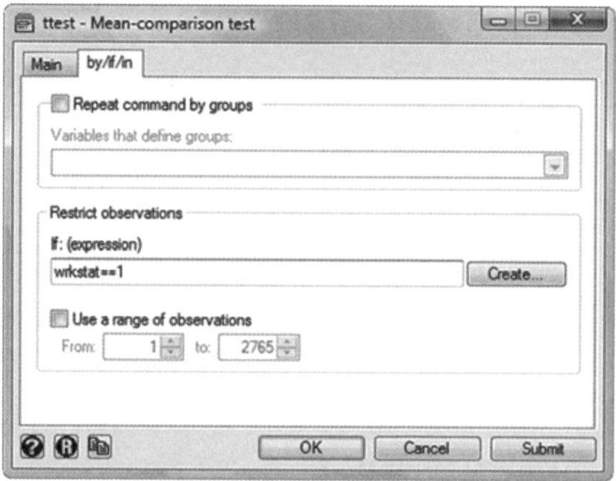

Figure 7.1. Restrict observations to those who score "1" on `wrkstat`

Clicking on OK produces the following command and results:

```
. ttest hrs1 == 40 if wrkstat == 1

One-sample t test

------------------------------------------------------------------------------
Variable |     Obs        Mean    Std. Err.   Std. Dev.   [95% Conf. Interval]
---------+--------------------------------------------------------------------
    hrs1 |    1419    45.97111    .3156328    11.88977    45.35195    46.59026
------------------------------------------------------------------------------
    mean = mean(hrs1)                                             t =  18.9179
Ho: mean = 40                                    degrees of freedom =     1418

    Ha: mean < 40               Ha: mean != 40                Ha: mean > 40
 Pr(T < t) = 1.0000         Pr(|T| > |t|) = 0.0000          Pr(T > t) = 0.0000
```

From these results, we can see that the mean for the variable `hrs1` is $M = 45.97$ hours, and the standard deviation is $\text{SD} = 11.89$ hours. Because we are using a two-tailed test, we can look at the right column under the table to see that, for $t = 18.92$, $p < 0.001$. Thus full-time workers work significantly more than the traditional 40-hour work week. In fact, they work about 6 hours more than this standard on average.

When we report a t test, we need to also report degrees of freedom. Look closely at the results above. The number of observations is 1,419, but the degrees of freedom is 1,418 (found just below the $t = 18.92$). For a one-sample t test, the degrees of freedom is always $N - 1$. If we report the degrees of freedom as 1,418, a reader will also know that the sample size is one more than this, or 1,419.

Different specialty areas report one-sample t tests slightly differently. The format I recommend is $t(1418) = 18.918$, $p < 0.001$. This is read as saying that t with 1,418 degrees of freedom is 18.918, p is less than 0.001. The results show that the average full-time employee works significantly more than the standard 40-hour week.

Degrees of freedom

> The idea of degrees of freedom was discussed in chapter 6 on cross-tabulations. For t tests, the degrees of freedom is a little different. Your statistics book will cover this in detail, but here we will present just a simple illustration. Suppose that we have four cases with $M = 10$. The first three of the four cases have scores of 8, 12, and 12. Is the fourth case free? Because the mean of the four cases is 10, the sum of the four cases must be 40. Because $40 - 8 - 12 - 12 = 8$, the fourth case must be 8. It is not free. In this sense, we can say that a one-sample test of means has $N - 1$ degrees of freedom.

Those who like to use confidence intervals can see these in the results. Stata gives us a 95% confidence interval of 45.4 hours to 46.6 hours. The confidence interval is more informative than the t test. We are 95% confident that the interval of 45.4 to 46.6 hours contains the mean number of hours full-time employees report working in a week. Because the value specified in the null hypothesis, $\mu = 40$, is not included in this interval, we know the results are statistically significant. The confidence interval also focuses our attention on the number of hours and helps us understand the substantive significance (importance), as well as the statistical significance.

7.8 Two-sample test of group means

A researcher says that men make more money than women because men work more hours a week. The argument is that a lot of women work part-time jobs, and these neither pay as well nor offer opportunities for advancement. What happens if we consider only people who say they work full time? Do men still make more than women when both the men and women are working full time?

The General Social Survey 2002 dataset (`gss2002_chapter7.dta`) has a question asking the respondent's income, `rincom98`. Like many surveys, it does not report the actual income but reports it in categories (e.g., under \$1,000, \$1,000 to 2,999). Run a `tabulate` command to see the coding the surveyors used. For some reason, they have not defined a score coded as 24 as a "missing value", but this is what a code of 24 represents. Even with highly respected national datasets like the General Social Survey, you need to check for coding errors. A code of 24 was assigned to people who did not report their income. Many researchers have used `rincom98` as it is coded (I hope after defining a code of 24 as a missing value). However, this coding is problematic because the intervals are not equal. The first interval, under \$1,000, is \$1,000 wide, but the second interval, \$1,000 to \$2,999, is virtually \$2,000 wide. Some intervals are \$10,000 wide.

Before we can compare the means for women and men, we need to recode the `rincom98` variable. We could recode by using the dialog box as described earlier in this book, but instead let's type the commands. You could enter the following commands in the Command window, one by one. A much better approach would be to enter them using a do-file. Remember that we should add some comments at the top of the do-file that include the name of the file and its purpose. For this example, you must use Stata/SE because the number of columns exceeds the 20-column limit of Stata/IC.

```
* recode income.do        (sample do-file)
* This is a short program that recodes income. It does a
* tabulation to see how income is coded (tab rincom98). People
* given a value of 24 are recoded as missing (mvdecode rincom98,
* mv(24)). We generate a new variable called inc that is equal to the
* old variable, rincom98. We recode each interval with its
* midpoint. We do a cross-tabulation of the new and old income
* variables as a check.
tabulate rincom98, missing
mvdecode rincom98, mv(24)
gen inc = rincom98
replace inc = 500   if rincom98 == 1
replace inc = 2500  if rincom98 == 2
replace inc = 3500  if rincom98 == 3
replace inc = 4500  if rincom98 == 4
replace inc = 5500  if rincom98 == 5
replace inc = 6500  if rincom98 == 6
replace inc = 7500  if rincom98 == 7
replace inc = 9000  if rincom98 == 8
replace inc = 11250 if rincom98 == 9
replace inc = 13250 if rincom98 == 10
replace inc = 16250 if rincom98 == 11
replace inc = 18750 if rincom98 == 12
replace inc = 21250 if rincom98 == 13
replace inc = 23750 if rincom98 == 14
replace inc = 27500 if rincom98 == 15
replace inc = 32500 if rincom98 == 16
replace inc = 37500 if rincom98 == 17
replace inc = 45000 if rincom98 == 18
replace inc = 55000 if rincom98 == 19
replace inc = 67500 if rincom98 == 20
replace inc = 82500 if rincom98 == 21
replace inc = 100000 if rincom98 == 22
replace inc = 110000 if rincom98 == 23
tabulate inc rincom98, missing
```

There are several lines summarizing what the do-file program will do. The first line after the comments runs a tabulation (`tabulate`), including missing values. The results help us understand how the variable was coded. The second line makes the code of `24` into a missing value so that anybody who has a score on `rincome98` of `24` is defined as having missing values (`mvdecode`). The third line generates (`gen`) a new variable called `inc` that is equal to the old variable `rincom98`. Following this are a series of commands to replace (`replace`) each interval with the value of its midpoint. Thus a code of `8` for `income98` is given a value of `9000` on `inc`. The final command does a cross-tabulation (`tabulate`) of the two variables to check for coding errors. Economists and demographers may not be happy with this coding system. Those who have a `rincom98` code of `1` may include people who lost a fortune, so substituting a value of `500` may not be ideal. The commands make no adjustment for these possible negative incomes.

Those who have a code of 23 include people who make \$110,000 but also may include people who may make \$1,000,000 or more. We hope that there are relatively few such cases at either end of the distribution, so the values we use here may be correct.

Now that we have income measured in dollars, we are ready to compare the income of women and men who work full time with a two-sample t test. Open the dialog box by selecting Statistics ▷ Summaries, tables, and tests ▷ Classical tests of hypotheses ▷ Two-group mean-comparison test; this dialog box is shown in figure 7.2.

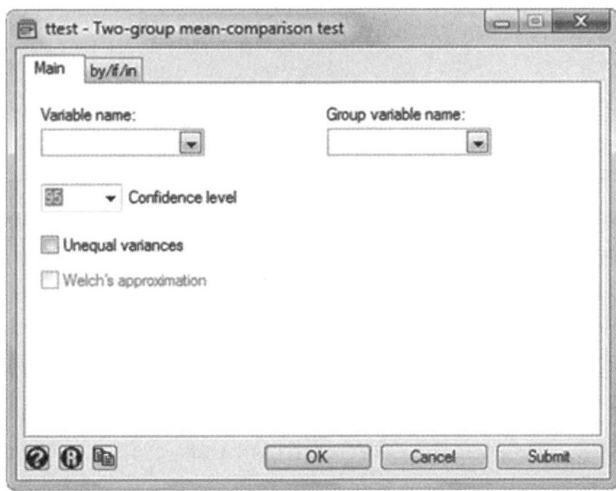

Figure 7.2. Two-group mean-comparison test dialog box

In this dialog box, type `inc` (outcome variable) as the *Variable name*. Then type `sex` as the *Group variable name*. Statistics books discuss assumptions for doing this t test, and one of these is that the variances are equal. If we believed that the variances were unequal, we could click on the *Unequal variances* box, and Stata would automatically adjust everything accordingly.

We can click on Submit at this point, and we will find a huge difference between women and men with men making much more than women on average. However, remember that we wanted to include only people who work full time. To implement this restriction, click on the by/if/in tab and enter this restriction. In the *Restrict observations* section, type `wrkstat == 1` (remember to use the double equal-signs) in the *If: (expression)* box. Here are the results:

(*Continued on next page*)

```
. ttest inc if wrkstat == 1, by(sex)
Two-sample t test with equal variances

------------------------------------------------------------------------------
   Group |     Obs        Mean    Std. Err.   Std. Dev.   [95% Conf. Interval]
---------+--------------------------------------------------------------------
    male |     671    44567.81    1054.665     27319.7    42496.96    46638.66
  female |     589    33081.07    895.9353    21743.74    31321.45    34840.69
---------+--------------------------------------------------------------------
combined |    1260    39198.21    718.7267    25512.27    37788.18    40608.25
---------+--------------------------------------------------------------------
    diff |            11486.74    1404.217                8731.874    14241.61
------------------------------------------------------------------------------
    diff = mean(male) - mean(female)                              t =   8.1802
Ho: diff = 0                                     degrees of freedom =     1258

    Ha: diff < 0                 Ha: diff != 0                 Ha: diff > 0
 Pr(T < t) = 1.0000         Pr(|T| > |t|) = 0.0000          Pr(T > t) = 0.0000
```

The layout of these results is similar to what we had for the one-sample t test. Using a two-tailed hypothesis (right column, below the main table), we see that the `t = 8.1802` has a $p < 0.001$ ($p = 0.0000$). The $N = 671$ men who work full time have an average income of just under \$44,568, compared with just over \$33,081 for the $N = 589$ women who are employed full time. Notice that the degrees of freedom, 1,258, are two fewer than the total number of observations, 1,260, because we used two means and lost one degree of freedom for each of them. For a two-sample t test, we will always have $N - 2$ degrees of freedom. A good layout for reporting a two-sample t test is $t(1258) = 8.18$, $p < 0.001$.

Is this result substantively significant? This question is pretty easy to answer because we all understand income. Men make about \$11,487 more than women on average, and this is true even though both our men and our women are working full time. It is sometimes helpful to compare the difference of means with the standard deviations. The \$11,487 difference of means is roughly one-half of a standard deviation if we average the two standard deviations as a guide. A difference of less than 0.2 standard deviations is considered a weak effect, a difference of 0.2 to 0.49 is considered a moderate effect, and a difference of 0.5 or more is considered a strong effect. If your statistics book covers the delta statistic, δ, you can get a precise number, but here we have just eyeballed the difference.

So far, we have been using the group comparison t test, which assumes that the data are in the long format. The dependent variable `income` is coded with the income of each participant. This is compared across groups by `sex`, which is coded `1` for each participant who is a man and `2` for each participant who is a woman. If the data were arranged in the wide format, we would still have two variables, but they would be different. One variable would be the income for each man and would have 671 entries. The other variable would be the income for each woman and would have 589 entries. This would look like the wide format shown previously for comparing proportions, except that the variables would be called `maleinc` and `femaleinc`.

When our data are in a wide format, we use a different dialog box, which is accessed by selecting Statistics ▷ Summaries, tables, and tests ▷ Classical tests of hypotheses ▷

Two-sample mean-comparison test. Here we would simply enter the names of the two variables, `maleinc` and `femaleinc`. The resulting command would be `ttest maleinc == femaleinc, unpaired`. We cannot illustrate this process here because the data are in the long format. If you are interested, you can use Stata's `reshape` command to convert between formats; type `help reshape`.

Effect size

There are two measures of effect size that are sometimes used to measure the strength of the difference between means. These are R^2 and Cohen's delta (δ). At the time this book was written, neither of these is directly computed by Stata, but they are easy to compute using Stata's built-in calculator. The formula is $R^2 = t^2/(t^2 + \text{df})$. Using the results of the two-sample t test comparing income of women and men, $R^2 = 8.1802^2/(8.1802^2 + 1258)$. We can compute this using a hand calculator or the Stata command

```
. display "r-squared = " 8.1802^2/(8.1802^2 + 1258)
r-squared = .05050561
```

The square root of this value is sometimes called the point biserial correlation. Values of 0.01 to 0.09 are a small effect, 0.10 to 0.25 are a medium effect, and over 0.25 are a large effect. If you use the Stata calculator with a negative t, it is important to insert parentheses correctly so Stata does not see the negative sign as making the t^2 a negative value. If we had a $t = -4.0$ with 100 degrees of freedom, the Stata command would be

```
. display "r-squared = " (-4.0)^2/((-4.0)^2 + 100)
```

or you could simply use the absolute value of t when doing the calculations.

Cohen's δ measures how much of a standard deviation separates the two groups.

$$\text{Cohen's } \delta = \frac{\text{mean difference}}{\text{pooled standard deviation}}$$

$$s_p = \sqrt{\frac{(N_1 - 1)s_1^2 + (N_2 - 1)s_2^2}{\text{df}}}$$

Here this would be

```
. display "Cohen's d: = " (44567.81 - 33081.07) / sqrt((670*(27319.7)^2 +
> 588*(21743.74)^2) / 1258)
Cohen's d: = .46187975
```

A Cohen's δ of 0.01 to 0.19 is a small effect, 0.20 to 0.49 is a medium effect, and over 0.50 is a large effect.

7.8.1 Testing for unequal variances

In the dialog box for the group comparison t test, we did not click on the option that would adjust for unequal variances. If the variances or standard deviations are similar in the two groups, there is no reason to make this adjustment. However, when the variances or standard deviations differ sharply, we may want to test to see if the difference is significant. Before doing the t test, some researchers will do a test of significance on whether variances are different. This test is problematic because the test of equal variances will show small differences in the variances to be statistically significant in large samples (where the lack of equal variance is less of a problem), but the test will not show large differences to be significant in small samples (where unequal variance is a bigger problem). To open the dialog box, select Statistics ▷ Summaries, tables, and tests ▷ Classical tests of hypotheses ▷ Two-group variance-comparison test. Here we enter our dependent variable `inc` and the grouping variable `sex` just like we did for the t test. We also should use the by/if/in tab to make sure to restrict the sample to those who work full time, `wrkstat == 1`. The command and results are

```
. sdtest inc if wrkstat == 1, by(sex)
Variance ratio test
------------------------------------------------------------------------------
   Group |     Obs        Mean    Std. Err.   Std. Dev.   [95% Conf. Interval]
---------+--------------------------------------------------------------------
    male |     671    44567.81    1054.665     27319.7    42496.96    46638.66
  female |     589    33081.07    895.9353    21743.74    31321.45    34840.69
---------+--------------------------------------------------------------------
combined |    1260    39198.21    718.7267    25512.27    37788.18    40608.25
------------------------------------------------------------------------------
    ratio = sd(male) / sd(female)                                 f =   1.5786
Ho: ratio = 1                                  degrees of freedom = 670, 588

    Ha: ratio < 1               Ha: ratio != 1                 Ha: ratio > 1
  Pr(F < f) = 1.0000         2*Pr(F > f) = 0.0000           Pr(F > f) = 0.0000
```

This test produces an F test that is used to test for a significant difference in the variances (although the output reports only standard deviations). The F test is simply the ratio of the variances of the two groups. Just below the table, we can see on the left side that the null hypothesis is `Ho: ratio = 1` and on the right side the `f = 1.5786` along with two degrees of freedom, `degrees of freedom = 670, 588`. The first number, 670, is the degrees of freedom for males ($N_1 - 1$), and the second number is the degrees of freedom for females ($N_2 - 1$). The variance for males is 1.58 times as great as the variance for females. In the middle bottom of the output, we can see the two-tailed alternative hypothesis, `Ha: ratio != 1`, and a reported probability of 0.0000. Because the F test has two numbers for degrees of freedom, we would report this as $F(670, 588) = 1.58, p < 0.001$.

Although the variances are significantly different, remember that this test is sensitive to large sample sizes, and in such cases, the assumption is less serious. If the groups have roughly similar sample sizes and if the standard deviations are not dramatically different, most researchers choose to ignore this test. If you want to adjust for unequal variances, however, look back at the dialog box for the two-sample t test. All you need to do is check the box for *Unequal variances*. When you get the results, the degrees of freedom are no longer $N - 2$ but are based on a complex formula that Stata will compute for you.

7.9 Repeated-measures t test

The repeated-measures t test goes by many names. Some statistics books call it a repeated-measures t test, some call it a dependent-sample t test, and some call it a paired-sample t test. A few examples of when it is used may help. A repeated-measures t test is used when one group of people is measured at two points. An example would be an experiment in which you measured everybody's weight at the start of the experiment and then measured their weight a second time at the end. Did they have a significant loss of weight? As a second example, we may want to know if it helps to retake the GRE. How much better do students perform when they repeat the GRE? We have one group of students, but we measure them twice—the first time they take the GRE and again the second time. The idea is that we measure the group on a variable, something happens, and then we measure them a second time. A control group is not needed in this approach because the participants serve as their own control.

An alternative use of the repeated-measures t test is to think of the group as being related people, such as parents. Husbands and wives would be paired. An example would be to compare the time spent on household chores by wives and their husbands. Here each husband and each wife would have a score on time spent for the wife and a score on time spent for the husband.

Who spends more time on chores—wives or husbands? Here we do not measure time spent on chores twice for the same person, but we measure it twice for each related pair of people. The way data are organized involves having two scores for each case. With paired data, the case consists of two related people (see `chores.dta`). Here is what five cases (couples) might look like:

```
. list

     +----------------+
     | husband   wife |
     |----------------|
  1. |      11     31 |
  2. |      10     40 |
  3. |      21     44 |
  4. |      15     36 |
  5. |      12     29 |
     +----------------+
```

We will illustrate the related-sample t test with data from the General Social Survey 2002, gss2002_chapter7.dta. Each participant was asked how much education his or her father had completed. This survey is not like the example of mothers and fathers because we have two measures for each participant. We know how much education each person reports for his or her mother and for his or her father. Do these respondents report more education for their mother or for their father? Here we will use a two-tailed test with a null hypothesis that there is no difference on average and an alternative hypothesis that there is a difference:

Alternative hypothesis: H_A: $\mu_{\text{diff}} \neq 0$

Null hypothesis: H_0: $\mu_{\text{diff}} = 0$

Remembering that we use M for a sample mean and μ for a population mean, we express these hypotheses as mean differences in the population (μ_{diff}). To open the dialog box, select Statistics ▷ Summaries, tables, and tests ▷ Classical tests of hypotheses ▷ Mean-comparison test, paired data. Once we open the dialog box, enter the two variables. It does not really matter which we enter as the *First variable* and which as the *Second variable*, but we need to remember this order because changing the order will reverse the signs on all the differences. Type paeduc as the *First variable* and maeduc as the *Second variable*. Remember that different statistics books will call this test the repeated-measures t test, paired-sample t test, or dependent-sample t test. All these tests are the same.

Here are the command and results:

```
. ttest paeduc == maeduc

Paired t test
------------------------------------------------------------------------------
Variable |     Obs        Mean    Std. Err.   Std. Dev.   [95% Conf. Interval]
---------+--------------------------------------------------------------------
  paeduc |    1903    11.42091    .0917118    4.000779    11.24105    11.60078
  maeduc |    1903    11.58276    .0781763    3.410314    11.42944    11.73608
---------+--------------------------------------------------------------------
    diff |    1903   -.1618497    .0729315    3.181518   -.3048838   -.0188156
------------------------------------------------------------------------------
     mean(diff) = mean(paeduc - maeduc)                           t =  -2.2192
 Ho: mean(diff) = 0                              degrees of freedom =     1902

 Ha: mean(diff) < 0           Ha: mean(diff) != 0           Ha: mean(diff) > 0
 Pr(T < t) = 0.0133         Pr(|T| > |t|) = 0.0266          Pr(T > t) = 0.9867
```

Mothers have more education on average than fathers. The mean for mothers is $M = 11.58$ years, compared with $M = 11.42$ years for fathers. Before going further, we need to ask if this difference (father's education − mother's education $= -0.16$) is important. Because we entered father's education as the first variable and mother's education as the second variable, the negative difference indicates that the mothers had slightly more education than the fathers. However, 0.16 years of education does not sound like an important difference. This would amount to $0.16 \times 12 = 1.92$ months of education, which does not sound like an important difference, even though it is statistically significant.

Just below the table, on the left, is the null hypothesis that there is no difference, `Ho: mean(diff) = 0`, and on the right is the t test, `t = -2.2192`, and its degrees of freedom, 1,902. The middle column under the main table has the alternative hypothesis, namely, that the mean difference is not 0. The $t = -2.22$ has a $p < 0.05$; thus, the difference is statistically significant.

This example illustrates a problem with focusing too much on the tests of significance. With a large sample, almost any difference will be statistically significant; it still may not be important. We have seen that this difference represents fewer than 2 months of education ($M_{\text{diff}} = 1.92$ months) so it might be fair to say that we have a statistically significant difference that is not substantively significant.

7.10 Power analysis

Most researchers report the significance level but do not report the power of a test. The power of a test is how likely it is to reject the null hypothesis when the null hypothesis is really wrong. If men make more money than women, failing to reject the null hypothesis that there is no difference would be a serious error. To avoid such an error, it is good to have a power of at least 0.80. This means that, 80% of the times we do our test of significance, we will be able to reject a null hypothesis that is really wrong.

Working with power is tricky because we will rarely know the true situation. Instead, we can propose a true difference of means that is the minimum that we would find substantively significant. Once we do this, we can assess the power for a particular size sample, or we can assess how big a sample we would need to obtain a particular power, say, 0.90.

Ideally, you make an estimate of the power of a test before you go to the expense and effort of gathering your data. Collecting data is often expensive and sometimes exposes participants to risks. Funding agencies do not want to fund you to collect more data than you need to have adequate power to test your hypotheses. If you have a power of 0.90 with just 500 observations, it would be inefficient to fund you to collect data on 1,000 observations. In our example of comparing the mean income of women and men who are employed full time, we might want to know how much power we had to detect a certain difference with a given sample size. We might also want to know how big a sample we needed to detect a certain difference with a given power.

In either case, we need to estimate a standard deviation and mean for each group before we do the test. How could we do this? We might have seen published data from a previous study reporting that the mean income in the United States was $40,000 and the standard deviation was $25,000. We also need to decide how much of a difference is interesting. Some researchers use Cohen's delta (δ) for this purpose. A δ is how much of a standard deviation separates the two groups, measured as a portion of a standard deviation. A delta of 0.2 would be one-fifth of a standard deviation. A delta of 0.5 would be one-half of a standard deviation. Usually a $\delta < 0.2$ is considered a weak relationship, a δ of 0.2 to 0.49 is considered a moderate relationship, and a $\delta \geq 0.5$ is considered a strong relationship.

Let's say that a difference needs to have $\delta = 0.2$ or more to be important to us. How much is this? This is the lower end of a moderate relationship. Remember that delta is the portion of a standard deviation that separates the groups, so a delta of 0.2 means a difference of $\delta(\sigma)$ or $0.2 \times \$25,000 = \$5,000$. Using this information, we need to estimate what the mean for women and for men would have to be to be this far apart. Given that the U.S. mean was \$40,000, we might say that the mean for women is \$37,500 and the mean for men is \$42,500. These are only estimates we make before collecting data to estimate the power or sample size we need. Instead of using δ, you might be able to say that some actual difference is the minimum difference you would find interesting. One researcher may say that she or he is interested only in a difference of at least \$1,000, and another researcher might be interested only in a difference of \$10,000.

To open the dialog box for doing power analysis, select Statistics ▷ Summaries, tables, and tests ▷ Classical tests of hypotheses ▷ Sample size and power determination. In the dialog box, check two-sample comparison of means and enter the estimated mean and standard deviation for each group. It does not matter which group is the women and which is the men. Type `37500` under *Mean one* and `42500` under *Mean two*. Type `25000` under *Std. deviation one* and `25000` under *Std. deviation two*. Moving to the Options tab, click on *Compute sample size*, and type `0.05` for *Significance level (alpha)* and `0.90` for *Power of the test*. The resulting command and output are

```
. sampsi 37500 42500, sd1(25000) sd2(25000) alpha(0.05) power(0.90)
Estimated sample size for two-sample comparison of means
Test Ho: m1 = m2, where m1 is the mean in population 1
                    and m2 is the mean in population 2
Assumptions:
         alpha =   0.0500  (two-sided)
         power =   0.9000
            m1 =    37500
            m2 =    42500
           sd1 =    25000
           sd2 =    25000
         n2/n1 =     1.00
Estimated required sample sizes:
            n1 =      526
            n2 =      526
```

This output says that, for an alpha of 0.05 with a power of 0.90, we need 526 women and 526 men to detect a difference between \$37,500 and \$42,500 if both groups have a standard deviation of \$25,000. Sometimes a power of 0.80 is considered adequate. If we are satisfied with a power of 0.80, we can redo the command, changing only the *Power of the test* on the Options tab from `0.90` to `0.80`. You can verify that we would need only 393 women and 393 men for this level of power.

What if we knew that we could sample only 50 women and 50 men? What would our power be? On the Options tab, click on *Compute power*, and keep *Significance level (alpha)* at 0.05. The box for *Power of the test* is grayed out because you are not fixing power but estimating it. Under *Sample-based calculations* put 50 in the box for *Sample one size*. Then check the box so that you can type 50 for *Sample two size*. Here are the results:

```
. sampsi 37500 42500, sd1(25000) sd2(25000) alpha(0.05) n1(50) n2(50)
Estimated power for two-sample comparison of means
Test Ho: m1 = m2, where m1 is the mean in population 1
                    and m2 is the mean in population 2
Assumptions:
         alpha =   0.0500  (two-sided)
            m1 =    37500
            m2 =    42500
           sd1 =    25000
           sd2 =    25000
sample size n1 =       50
            n2 =       50
         n2/n1 =     1.00
Estimated power:
         power =   0.1701
```

This result shows us that, to detect a difference of \$5,000 between means with just 50 women and 50 men, our power is only 0.17. What does this mean? It means that we have only a $p = .17$ chance of having a statistically significant result with this sample size when the actual difference in the population is \$5,000. You would not want to go through all the trouble of doing the research with this probability of finding a statistically significant result, even when there really is this much difference in salary between women and men.

Can you have too big of a sample? You can verify that with a sample of 1,000 women and 1,000 men, the power would be 0.994. There would be no reason to have a bigger sample than that. You would just be wasting time and money.

Although we do not have space to cover them, this dialog box offers other examples for doing power analysis either for a sample mean or for proportions. At the lower left of the dialog box is a question mark. As with the dialog box for any Stata command, by clicking on the question mark, you can get help that shows examples of things you can do with the command.

Power analysis may seem like a lot of work, and you may be wondering why someone would do this. There are several reasons why it is worth the effort. Funding agencies want to support research that will have sufficient power to detect an important difference. If there are too few participants in a study, the study has little chance of being successful. If there are too many participants, the agency is spending more money than it needs to spend. In studies where participants are at some risk, this is especially important. You do not want to put more people at risk than necessary.

Power analysis software

> Although Stata offers power analysis for a variety of statistical tests, there are many other software packages that provide more elaborate power analyses and do this for a greater variety of statistical tests. One program that is both easy to use and covers a wide variety of basic statistical tests is called `G*Power 3`. This program is free and has versions for both PCs and Macs. You can obtain it at http://www.psycho.uni-duesseldorf.de/abteilungen/aap/gpower3/.

7.11 Nonparametric alternatives

The examples in this chapter for t tests have assumed that the outcome or dependent variable is measured on at least the interval level. They also assumed that both groups have equal variances in their populations and are normally distributed. The tests we have covered are remarkably robust against violations of these assumptions, but sometimes we want to use tests that make less challenging assumptions. Nonparametric alternatives to conventional t tests are one way to do this. These tests have some limitations of their own, including being somewhat less powerful than the t test. Here are examples of such tests you can do using Stata.

7.11.1 Mann–Whitney two-sample rank-sum test

If we assume that we have ordinal measurement rather than interval measurement, we can do the rank-sum test. We can test the hypothesis that two independent samples (i.e., unmatched data) are from a population distribution by using the Mann–Whitney two-sample rank-sum test.

This test involves a simple idea. It combines the two groups into one group and ranks the participants from top to bottom. The highest score gets a rank of N (8,871 in this example), and the lowest score gets a rank of 1. If one group has higher scores than the other group, they should have a predominance of higher ranks. If girls report that more of their friends smoke, they should have more of the top ranks, and boys should have more of the lower ranks. If girls and boys have the same distribution, then both of them should have about the same number of high ranks and low ranks. In other words, the sum of the ranks girls have should be about the same as the sum of the ranks boys have.

The test computes the sum of ranks for each group and compares this with what we would expect by chance. Using data from the National Longitudinal Survey of Youth 1997 (`nlsy97_chapter7.dta`), we can compare the answers girls and boys give on how many of their friends smoke. They were asked what percentage of their friends smoke, and they answered in broad categories that were coded `1`–`5`. Treating these categories as ordinal, we can do a rank-sum test. We will not show the dialog box for this, but the command and results are

```
. ranksum psmoke97, by(gender97)
Two-sample Wilcoxon rank-sum (Mann-Whitney) test
      gender97 |      obs    rank sum    expected
---------------+---------------------------------
             1 |     4540    19130393    20139440
             2 |     4331    20221363    19212316
---------------+---------------------------------
      combined |     8871    39351756    39351756
unadjusted variance    1.454e+10
adjustment for ties   -7.360e+08
                      ----------
adjusted variance      1.380e+10
Ho: psmoke97(gender97==1) = psmoke97(gender97==2)
              z =  -8.589
    Prob > |z| =   0.0000
```

This output shows that there is a significant difference of `z = -8.589`. We should also report the median and means so that the direction of this difference is clear. The rank test compares the entire distribution rather than a particular parameter, such as the mean or median.

7.11.2 Nonparametric alternative: median test

Sometimes you will want to compare the medians of two groups rather than the means. Whereas the Mann–Whitney rank-sum test compares entire distributions, the median test performs a nonparametric K-sample test on the equality of medians. It tests the null hypothesis that the K samples were drawn from populations with the same median. In the case of two samples ($K = 2$), a chi-squared statistic is calculated with and without a continuity correction.

Ideally, there would be no ties, and a median could be identified that had exactly 50% of the observations above it and 50% below it. With this example, the dependent variable is on a 1-to-5 scale, and there are many ties; for example, hundreds of students have a score of 2. This means that there is not an equal number of observations that are above and below the median for either group. We will not illustrate the dialog box for this command, but the command and results are

(Continued on next page)

```
. median psmoke97, by(gender97) exact medianties(split)
Median test

  Greater
  than the      youth gender 1997
    median          1           2    |      Total
-----------+-------------------------+-----------
        no        2,904       2,470  |      5,374
       yes        1,636       1,861  |      3,497
-----------+-------------------------+-----------
     Total        4,540       4,331  |      8,871
          Pearson chi2(1) =  44.6269   Pr = 0.000
           Fisher's exact =                 0.000
   1-sided Fisher's exact =                 0.000
  Continuity corrected:
          Pearson chi2(1) =  44.3371   Pr = 0.000
```

The median test makes sense for comparing skewed variables, such as income. We would report the example as $\chi^2(1) = 44.63$; $p < .001$. If we set up a hypothesis, we would reject the null hypothesis, H_0: $\text{Med}_1 = \text{Med}_2$, in favor of the alternative hypothesis, H_A: $\text{Med}_1 \neq \text{Med}_2$.

7.12 Summary

The tests in this chapter are among the most frequently used statistical tests. As you read this chapter, you may have thought of applications in which you can use a t test or a z test for testing individual means and proportions or for testing differences. These are extremely useful tests for people who work in an applied setting and need to make decisions based on statistics. It is becoming a requirement to demonstrate that any new program has advantages over older programs; it is not enough to say that you believe that the new program is better or that you can "see" the difference. You need to find appropriate outcomes, such as reading readiness, retention rate, loss of weight, increased skill, or participant satisfaction. Then you need to show that the new program has a statistically significant influence on the outcomes you select. If you are designing an exercise program for older adults, there is not much reason for implementing your program unless you can demonstrate that it has a good retention rate, that participants are satisfied with the program, and that behavioral outcome goals are met.

In this chapter, we have discussed statistical significance and substantive significance, both of which are extremely important. If something is not statistically significant, we do not have confidence that there is a real effect or difference. What we observed in our data may be something that could have happened just by chance. There are two major reasons why a result may not be statistically significant. The first is that the result represents a small effect where there is little substantive difference between the groups. The second is that we have designed our study with too few observations to show significance, even when the actual difference is important. Therefore, we have introduced the basic Stata commands for estimating the sample size needed to do a study with sufficient power to show that an important result is statistically significant.

We have touched only the surface of what Stata can do with power analysis, and we encourage you to check the Stata *Reference* manuals for more ways of estimating power and sample size requirements.

When we have statistical significance, we are confident that what we observed in our data represents a real effect that should not be attributed to chance. It is still essential to evaluate how substantively significant the result is. With a large sample, we may find a statistically significant difference of means when the actual difference is small and substantively insignificant. Finding a significant result begs the question of how important the result is. A statistically significant result may or may not be large enough to be important.

Finally, we introduced nonparametric alternatives to z tests and t tests. These alternatives usually have less power but may be more easily justified in the assumptions they make.

Sometimes we have more than two groups, and the t test is no longer adequate. Chapter 9 discusses analysis of variance (ANOVA), which is an extension of the t test that allows us to work with more than two groups. Before we do analysis of variance, however, we will cover bivariate correlation and regression in chapter 8.

7.13 Exercises

When doing these exercises, it is important that you create a do-file for each assignment. You might name the do-file for the first exercise `c7_1.do` and put these in the directory where you are keeping the Stata programs related to this book. This approach will be useful in the future if you need to redo one of these examples or want to do a similar task. For example, you could use the do-file for the second exercise anytime you needed to do randomization of participants in a study.

1. According to the registrar's office at your university, 52% of the students are women. You do a web-based survey of attitudes toward student fees supporting the sports program. You have 20 respondents, 14 of whom are men and 6 of whom are women. Is there a gender bias in your sample? To answer this, create a new dataset that has one variable, namely, `sex`. Enter a value of `1` for your first 14 observations (for your 14 males) and a value of `0` for your last 6 observations (for your 6 females). Your data will have one column and 20 rows. Then do a one-sample z test against the null hypothesis that $p = 0.52$. Explain your null hypothesis. Interpret your results.

2. You have 30 volunteers to participate in an exercise program. You want to randomly assign 15 of them to the control group and 15 to the treatment group. You list them by name—the order being arbitrary—and assign numbers from 1 to 30. What are the Stata commands you would use to do the random assignment (randomization without replacement)? Show how you would do this.

3. We showed you how to do a random sample without replacement. To do a random sample with replacement, you use the `bsample` command. Repeat the last exercise, but use the command `bsample 15`. Then do a tabulation to see if any observations were selected more than once.

4. Use the approach you used in the first exercise, but this time to draw a random sample of 10 students from a large class of 200 students. You use the class roster in which students each have a number from 1 to 200. Show the commands you use, and set the seed at 953. List the numbers for the 10 students you select.

5. Open up the `nlsy97_chapter7.dta` dataset. A friend says that Hispanic families have more children than other ethnic groups. Use the variables `hh18_97` (number of children under age 18) and `ethnic97` (`0` being non-Hispanic and `1` being Hispanic) to test whether this is true or not. Are the means different? If the result is statistically significant, how substantively significant is the difference?

6. Use the same variables as the previous exercise. Use `summarize`, `detail`, and `tabulate` to check the distribution of `hh18_97`. Do this separately for Hispanics and non-Hispanics. What are the medians of each group? Run a median test. Is the difference significant? How can you reconcile this with the medians you computed? (Think about ties and the distributions.)

7. You want to compare Democrats and Republicans on their mean score on abortion rights. From earlier uses of the scale, you know that the mean is somewhere around 50 and the standard deviation is about 15. Select an alpha level and the minimum difference that you would find important. Justify your minimum difference (this is pretty subjective, but you might think in terms of a proportion of a standard deviation difference). How many cases do you need to have 80% power? How many cases do you need to have 90% power? How many do you need to have 99% power?

8. A friend believes that women are more likely to feel abortion is okay under any circumstances than men because women have more at stake in a decision about whether to have an abortion. Use the General Social Survey 2002 dataset (`gss2002_chapter7.dta`), and test whether there is a significant difference between women and men (`sex`) on whether abortion is acceptable in any case (`abany`).

8 Bivariate regression and correlation

8.1 Introduction to bivariate correlation and regression

Bivariate correlation and regression are used to examine the relationship between two variables. They are usually used with quantitative variables that we assume are measured on the interval level. Some social science statistics books present correlation first and then regression, whereas others do just the opposite. You need to understand both of these, but which comes first is sort of like asking whether the chicken or the egg came first. In this chapter, we will use the following order:

1. Construct a scattergram. This is a graphic representation of the relationship between two variables.

2. Superimpose a regression line. This is a straight line that best describes the linear form of the relationship between the two variables.

3. Estimate and interpret a correlation. This tells us the strength of the relationship.

4. Estimate and interpret the regression coefficients. This tells us the functional form of the relationship.

5. Estimate and interpret Spearman's rho as a rank-order correlation for ordinal data.

6. Estimate and interpret alpha. This uses correlation to assess the reliability of a measure.

7. Estimate and interpret kappa as a measure of agreement for categorical data.

In chapter 10, we will expand this process to include multiple correlation and multiple regression. If you have a good understanding of bivariate correlation and regression, you will be ready for chapter 10.

8.2 Scattergrams

Suppose that we are interested in the relationship between an adult man's education and the education his father completed. We believe that the advantages of education are transmitted from one generation to the next. Fathers who have limited education will have sons who have restricted opportunities, and hence they too will have limited education. On the other hand, fathers who have high education will offer their sons opportunities to get more education. When examining a relationship like this, we know that it is not going to be "perfect". We all know adult men who have far more education than their own fathers, and we may know some who have far less.

To understand the relationship between these two variables, let's use the dataset `gss2006_chapter8.dta` and create a graph. Because this is a large dataset, a scattergram will have so many dots that it will not make much sense. Suppose that we have 20 father–son dyads that have a father with a score of 15 and a son with a score of 15, but just one father–son dyad that has a father with a score of 20 and a son with a score of 10. The 20 dyads and the single dyad would each appear as one dot on the graph, and this would create a visual distortion of the relationship. (We will discuss ways to work around this later.) To manage this problem, limit the scattergram to a random sample of 100 observations by using the command `sample 100, count`.

How can you get the same result each time?

When you take a random sample of your data for a scattergram or for any other purpose, you may want to make sure you get the same sample if you repeat the process later. Random sampling is done by first generating a `seed` that instructs Stata on where to begin the random process. If you have a table of random numbers, this is equivalent to picking where you will start in the table. By setting the seed at a specific value, you can replicate your results. The easiest way to select your subsample and make sure that you get the same subsample if you repeat the process is with this pair of commands:

```
. set seed 123
. sample 100, count
```

Here we picked a seed of 123. You can pick any number, and when you repeat the command using the same number, you will get the same random sample.

Because a father's education comes before his son's, the father's education will be the predictor (the X or independent variable). The son's education will be the outcome (the Y or dependent variable). By convention, a scattergram places the predictor on the horizontal axis and the outcome on the vertical axis.

The dialog system is helpful with scattergrams because there are so many options that it would be hard to remember all of them. Open the dialog box by selecting Graphics ▷ Twoway graph (scatter, line, etc.). The resulting dialog box is shown in figure 8.1.

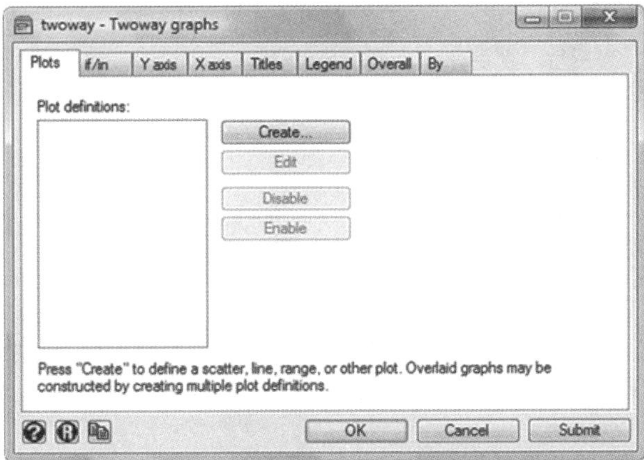

Figure 8.1. Dialog box for a scattergram

Under the Plots tab, click on *Create....* We will use the *Basic plots* and make sure *Scatter* is highlighted. Type `educ` (Y variable) and `paeduc` (X variable), and click on Accept. This returns you to the main Plots tab. Here you need to make sure that *Plot 1* is highlighted, and you can add features to the graph by using the other tabs. Click on the if/in tab and type `sex==1` in the *If: (expression)* box, because we want to limit this graph to sons, and `sex==1` is the code for males. You can click on the Y axis tab and, under *Title*, type `Son's education`. Click on the X axis tab and, under *Title*, type `Father's education`. Using the Titles tab, you can give the graph a title of `Scattergram relating father's education to his son's education` and add a note at the bottom of the graph, `Data source: GSS 2006; random sample of N = 100`. If you click on Submit, you will probably decide the title is too wide for the graph. To make it fit better, you can click on *Properties* (to the right of the title) in the dialog box and change the *Size* to *Medium*.

Click on Submit. This gives a nice graph, but the y-axis title might be too close to the numbers on the y axis. If you want to move that title away from the numbers, then go back to the Y axis tab and click on *Properties*, which is to the right of the title. This gives you several options. Going to the Advanced tab allows you to change the *Inner gap*. Try putting a 5 here. Click on Submit and see how you like the graph.

These dialog boxes are complex, but remember that you are not going to hurt Stata or your data by trying different options. If you do not understand what an option does, just try it and see what happens. You can always return to the default value for the option. Here is the resulting Stata command. (Notice that in addition to what we have done with the dialog box, I turned the legend off in the Legend tab. Also the command is listed as you would see it in a do-file with the three forward slashes (///) used as a line break.) The resulting graph is shown as figure 8.2.

```
twoway (scatter educ paeduc) if sex==1, ytitle(Son's education) ///
    yscale(titlegap(5)) xtitle(Father's education) ///
    title(Scattergram relating father's education to his son's education, ///
    size(medium)) note(Data source: GSS 2006; random sample of N = 100) ///
    legend(off)
```

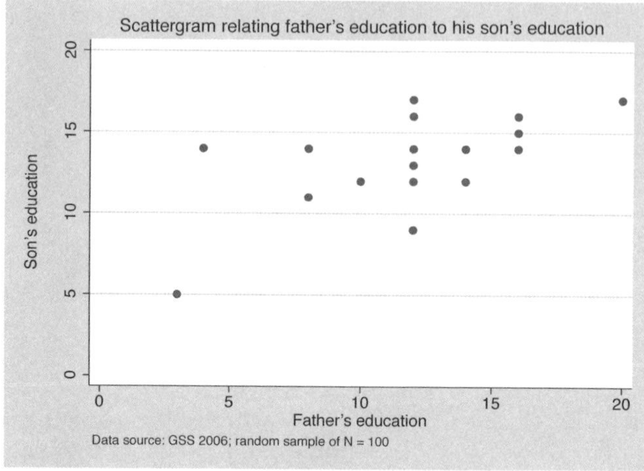

Figure 8.2. Scattergram of participant's education on father's education

How do we interpret the resulting scattergram? There is a general pattern that the higher the father's education, the higher his son's education, but there are many exceptions. The data are for the United States. Some countries have a virtual caste system, where there is little opportunity for upward mobility and little risk of downward mobility. In those countries, the scattergram would fall close to a straight line where the son's education was determined largely by his father's education. In liberal democracies, such as the United States, the relationship is only moderately strong, and there is a lot of room for intergenerational mobility.

Can you think of relationships for which a scattergram would be useful? A gerontologist might want to examine the relationship between a score on the frequency of exercise and the number of accidental falls an older person has had in the last year. Unlike the results in figure 8.2, the older people who exercise more often might have fewer falls. The scatter would start in the upper left and go down if those who exercise more regularly are less prone to accidental falls. An educational psychologist might examine the relationship between parental support and academic performance. All you need is a pair of variables that are interval level or that you are willing to treat this way. What happens to the relationship between self-confidence among college students and how long they have been in college? Is the initial level of self-confidence high? Does it drop to a low point during the freshman year, only to gradually increase during the sophomore, junior, and senior years? This is an empirical question whose answer could be illustrated with a scattergram.

Rarely does a scattergram help when you have a very large sample, say, 500 or more observations. However, there is a clever approach in Stata that tries to get around this problem and that works for reasonably sized datasets: the `sunflower` command. Sunflowers are used to represent clusters of observations. Sunflowers vary in the number of "petals" they have. The more petals there are, the more observations are at that particular coordinate point. Even the `sunflower` command is of limited value with really large samples. If you want to try the `sunflower` command, you can enter **`help sunflower`** to see the various options. When you are in the Viewer window with the help file for the `sunflower` command open, notice that in the top right corner it says `dialog: sunflower`. Click on `sunflower` and it takes you directly to the `sunflower` dialog box. If you have trouble finding a dialog box, but you know the name of the command, this is a useful way to open the dialog box.

A more common approach is to add "jitter" to the markers (dots) where there are several observations with the same score on both the X variable and the Y variable. To do this, return to the `twoway` dialog box and click on the Plots tab. Make sure that *Plot 1* is highlighted, and click on *Edit*. Next you will click on *Marker properties* and click on the Advanced tab. You can check the box for *Add jitter to markers* and type 6 for the *Noise factor*. You might want to set the seed at some number, here 222. The resulting graph (figure 8.3) is a more realistic representation of our 100 observations.

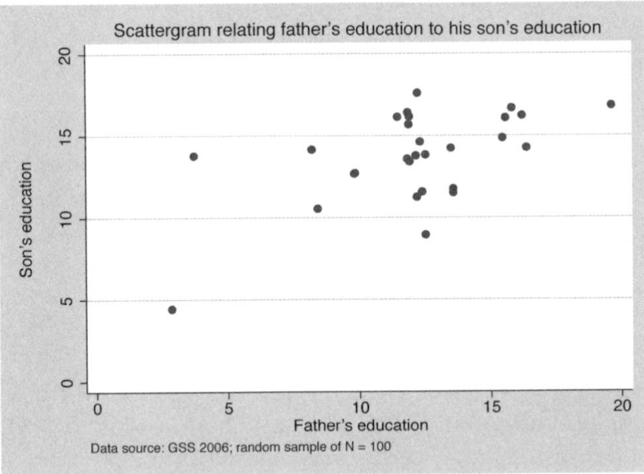

Figure 8.3. Scattergram of participant's education on father's education with "jitter"

The command that results from the `twoway` dialog box and produces the above graph is

```
twoway (scatter educ paeduc, jitter(6) jitterseed(222)) if sex==1,
    ytitle(Son's education) yscale(titlegap(5)) xtitle(Father's education)
    title(Scattergram relating father's education to his son's education,
    size(medium)) note(Data source: GSS 2006; random sample of N = 100)
    legend(off)
```

Predictors and outcomes

Different statistics books and different substantive areas vary in the terms they use for predictors and outcomes. We need to be comfortable with all these different terms. A scattergram is a great place to really understand the terminology because we must have one variable on the horizontal x axis and another variable on the vertical y axis. Sometimes it helps to think of the X variable as a cause and the Y variable as an effect. This works if we can make the statement "X is a cause of Y". It makes sense to say that the father's education is a cause of his son's education rather than the other way around. We say "a" cause rather than "the" cause because many variables will contribute to how much education a son will achieve.

People who are uncomfortable with cause–effect terminology often call the X variable the independent variable and the Y variable the dependent variable. Which variable is dependent? In the scattergram, we would say that the son's education is dependent on his father's education. We would not say that the father's education depends on his son's education. The dependent variable is the Y variable, and the other variable is the X variable or the independent variable—because it does not depend. Father's education does not depend on his son's education.

Other people prefer to think of an outcome variable and a predictor. We do not need to know what causes something, but we can discover what predicts it. The son's education is an outcome, and the father's education is the predictor. Predictors may or may not meet a philosopher's definition of a cause. Couples who fight a lot before they get married are more likely to fight after they are married, so we would say that conflict prior to marriage predicts conflict after marriage. The premarriage conflict is the predictor, and the postmarriage conflict is the outcome.

Sometimes none of these terms make a lot of sense. Wives who have high marital satisfaction more often have husbands who have high marital satisfaction. Both variables depend on each other and simultaneously influence each other. Sometimes we refer to these as reciprocal relationships. In such a case, which variable is the X variable and which is the Y variable is arbitrary.

8.3 Plotting the regression line

Later in this chapter, we will learn how to do a regression analysis. For now, we will simply introduce the concept and show how to plot it on the scattergram. The regression line shows the form of the relationship between the two variables. Ordinarily, we assume that the relationship is linear. For example, what is the relationship between income and education? To answer this question, we need to know how much income you could expect to make if you had no education (intercept or constant) and then how much more income you can expect for each additional year of education (slope). Suppose that you could expect to make $10,000 per year, even with no formal education, and for each additional year of education you could expect to earn another $3,000. You could write this as an equation

$$\text{Estimated income} = 10000 + 3000(\texttt{educ})$$

where 10000 is the estimated income if you had no education, i.e., `educ` = 0. This is called the intercept or constant. The 3000 is how much more you could expect for each additional year of `educ`, and it is called the slope. Thus a person who has 12 years of education would have an estimated income of $10000 + 3000(12) = 46000$ or $46,000.

A symbolic way of writing a regression equation is

$$\widehat{Y} = a + b(X)$$

where a is the intercept or constant, and b is the slope. The $\widehat{Y}$ (pronounced Y-hat) is the dependent variable (`income`), and the circumflex means that it is estimated (many statistics texts use a "^" over the Y for an estimated value). The X is the independent variable (`educ`). Let's use the example relating educational attainment of fathers and sons to see how the regression line appears. We will not get the values of a and b until later, but we will get a graph with the line drawn for us.

Return to your Plots tab, or if you closed the dialog box by clicking on OK instead of Submit, you will need to reopen it by selecting Graphics ▷ Twoway graph (scatter, line, etc.). We already have the information entered for *Plot 1*, the scattergram. Click on *Create....* Previously, we selected *Basic plots* and *Scatter*. This time we select *Fit plots* and make sure that *Linear prediction* is highlighted. We type `educ` and `paeduc` as the Y variable and X variable, respectively. Click on Submit, and the graph appears as shown in figure 8.4.

(*Continued on next page*)

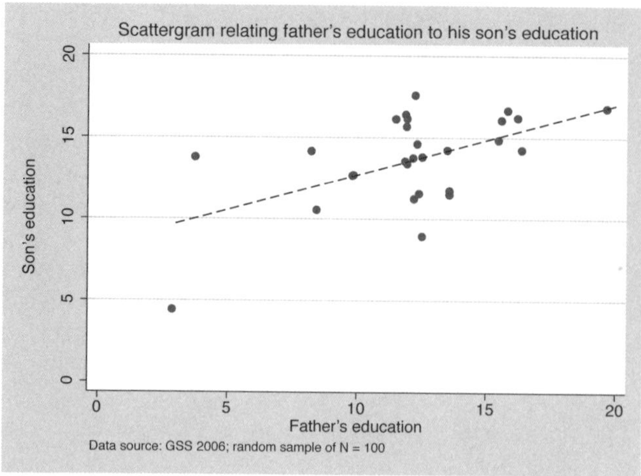

Figure 8.4. Scattergram of participant's education on father's education with a regression line

The linear regression line gives you a good sense of the relationship. It shows how the higher the father's education is, the higher his son's education is on average. The observations are not extremely close to the regression line, meaning that there is substantial intergenerational mobility. Sons whose fathers have little education are expected to have more education than their fathers, but sons whose fathers have a lot of education are expected to have a little less education than their fathers. Can you see this in the scattergram? Pick a father who has 5 years of education, $X = 5$. Project a line straight up to the regression line and then straight over to the Y axis, and you can see that we would expect his son to have about $Y = 10$ years of education. Pick a father who has 20 years of education, $X = 20$. Can you see that we would expect his son to have less education?

8.4 Correlation

Your statistics text gives you the formulas for computing correlation, and if you have done a few of these by hand, you will love using Stata. We will not worry about the formulas. Correlation measures how close the observations are to the regression line. We need to be cautious in interpreting a correlation coefficient. Correlation does not tell us how steep the relationship is; that comes from the regression. You may have a steep relationship or an almost flat relationship, and both relationships could have the same correlation. Suppose that the correlation between education and income is $r = 0.3$ for women and $r = 0.5$ for men. Does this mean that education has a bigger payoff for men than it does for women? We really cannot know the answer from the correlation. The correlation tells us almost nothing about the form of the relationship. The fact that the r is larger for men than for women is not evidence that men get a bigger payoff from an additional year of education. Only the regression line will tell us that. The $r = 0.5$ for men means that the observations are closer to the regression line for men than they are for women ($r = 0.3$), that the income of men is more predictable than that of women. Correlation also tells us whether the regression line goes up (r will be positive) or down (r will be negative). Strictly speaking, correlation measures the strength of the relationship only for how close the dots are to the regression line.

Bivariate correlation is used by social scientists in many ways. Sometimes we will be interested simply in the correlation between two variables. We might be interested in the relationship between calorie consumption per day and weight loss. If you discover that $r = -0.5$, this would indicate a fairly strong relationship. Generally, a correlation of $|r| = 0.1$ is a weak relationship, $|r| = 0.3$ is a moderate relationship, and $|r| = 0.5$ is a strong relationship. An r of -0.3 and an r of 0.3 are equally strong. The negative correlation means that as X goes up, Y goes down. The positive correlation means that as X goes up, Y goes up.

We might be interested in the relationship between several variables. We could compare three relationships between (a) weight loss and calorie consumption, (b) weight loss and the time spent in daily exercise, and (c) weight loss and the number of days per week a person exercises. We could use the three correlations to see which predictor is more correlated with weight loss.

Suppose that we wanted to create a scale to measure a variable, such as political conservatism. We would use several specific questions and combine them to get a score on the scale. We can compute a correlation matrix of the individual items. All of them should be at least moderately correlated with each other because they were selected to measure the same concept.

When we estimate a correlation, we also need to report its statistical significance level. The test of statistical significance of a correlation depends on the size or substantive significance of a correlation in the sample and depends on the size of the sample. An $r = 0.5$ might be observed in a very small sample, just by chance, even though there were no correlations in the population. On the other hand, an $r = 0.1$, although a weak substantive relationship, might be statistically significant if we had a huge sample.

(*Continued on next page*)

Statistical and substantive significance

It is easy to confuse statistical and substantive significance. Usually, we want to find a correlation that is substantively significant (r is moderate or strong in our sample) and statistically significant (the population correlation is almost certainly not zero). With a very large sample, we can find statistical significance even when $r = 0.1$ or less. What is important about this is that we are confident that the population correlation is not zero and that it is very small. Some researchers mistakenly assume that a statistically significant correlation automatically means that it is important when it may mean just the opposite—we are confident it is not very important.

With a very small sample, we can find a substantively significant $r = 0.5$ or more that is not statistically significant. Even though we observe a strong relationship in our small sample, we are not justified in generalizing this finding to the population. In fact, we must acknowledge that the correlation in the population might even be zero.

- Substantive significance is based on the size of the correlation.
- Statistical significance is based on the probability you could get the observed correlation by chance if the population correlation was zero.

Now let's look at an example that we downloaded from the UCLA Stata Portal. As mentioned at the beginning of the book, this is an exceptional source of tutorials, including movies on how to use Stata. They used data from a study called "High School and Beyond". Here you will download a part of this dataset used for illustrating how to use Stata to estimate correlations. Go to your command line, and enter the command

```
. use http://www.ats.ucla.edu/stat/stata/notes/hsb2, replace
```

You will get a message back that you have downloaded 200 cases, and your listing of variables will show the subset of the High School and Beyond dataset. If your computer is not connected to the Internet, you should use one that is connected, download this file, and save it to a flash disk. This dataset is also available from this book's web page.

Say that we are interested in the bivariate correlations between `reading`, `writing`, `math`, and `science` skills for these 200 students. We are also interested in the bivariate relationships between each of these skills and the students' socioeconomic status and between each of these skills and the students' gender. We believe that socioeconomic status is more related to these skills than gender is.

It is reasonable to treat the skills as continuous variables measured at close to the interval level, and some statistics books say that interval-level measurement is a critical assumption. It is problematic to treat socioeconomic status and gender as continuous variables. If we run a tabulation of socioeconomic status on gender, `tab1 female ses`, we will see the problem. Socioeconomic status has just three levels (low, middle, and high) and gender has just two levels (male and female). This dataset has all values labeled, so the default tabulation does not show the numbers assigned to these codes. We can run `codebook ses female` and see that `female` is coded `1` for girls and `0` for boys. Similarly, `ses` is coded `1` for low, `2` for middle, and `3` for high. If you have installed the `fre` command using `ssc install fre`, you will remember that the command `fre female ses` will show both the value labels and the codes. We will go ahead and compute the correlations anyway and see if they make sense.

Stata has two commands for doing a correlation: `correlate` and `pwcorr`. The `correlate` command runs the correlation using a casewise deletion (some books call this listwise deletion) option. Casewise deletion means that if any observation is missing for any of the variables, even just one variable, the observation will be dropped from the analysis. In some datasets that have problems with missing values, casewise deletion can introduce serious bias and greatly reduce the working sample size. Casewise deletion is a problem for external validity or the ability to generalize, when there is a lot of missing data. Many studies using casewise deletion will end up dropping 30% or more of the observations, and this makes generalizing a problem even though the total sample may have been representative.

The `pwcorr` command estimates each correlation based on all the people who answered each pair of items. For example, if Julia has a score on `write` and `read` but nothing else, she will be included in estimating the correlation between `write` and `read`. Pairwise deletion introduces its own problems. Each correlation may be based on a different subsample of observations, namely, those observations who answered both variables in the pair. We might have 500 people who answered both `var1` and `var2`, 400 people who answered both `var1` and `var3`, and 450 people who answered both `var2` and `var3`. Because each correlation is based on a different subsample, under extreme circumstances it is possible to get a set of correlations that would be impossible for a population.

To open the `correlate` dialog box, select Statistics ▷ Summaries, tables, and tests ▷ Summary and descriptive statistics ▷ Correlations and covariances. To open the `pwcorr` dialog box, select Statistics ▷ Summaries, tables, and tests ▷ Summary and descriptive statistics ▷ Pairwise correlations. Because the command is so simple, we can just enter the command directly.

(Continued on next page)

```
. correlate read write math science ses female
(obs=200)

             |     read    write     math  science      ses   female
-------------+------------------------------------------------------
        read |   1.0000
       write |   0.5968   1.0000
        math |   0.6623   0.6174   1.0000
     science |   0.6302   0.5704   0.6307   1.0000
         ses |   0.2933   0.2075   0.2725   0.2829   1.0000
      female |  -0.0531   0.2565  -0.0293  -0.1277  -0.1250   1.0000
```

We can read the correlation table going either across the rows or down the columns. The $r = 0.63$ between `science` and `read` indicates that these two skills are strongly related. Having good reading skills is probably helpful to having good science skills. All the skills are weakly to moderately related to socioeconomic status, `ses` ($r = 0.21$ to $r = 0.29$). Having a higher socioeconomic status does result in higher expected scores on all the skills for the 200 adolescents in the sample.

A dichotomous variable, such as gender, that is coded with a `0` for one category (man) and `1` for the other category (woman) is called a dummy variable or indicator variable. Thus `female` is a dummy variable (a useful standard is to name the variable to match the category coded as `1`). When you are using a dummy variable, the stronger the correlation is, the greater impact the dummy variable has on the outcome variable. The last row of the correlation matrix shows the correlation between `female` and each skill. The $r = 0.26$ between being a girl and writing skills means that girls (they were coded `1` on `female`) have higher writing skills than boys (they were coded `0` on `female`), and this is almost a moderate relationship. You have probably read that girls are not as skilled in math as boys. The $r = -0.03$ between `female` and `math` means that in this sample, the girls had just ever so slightly lower scores (remember an $|r| = 0.1$ is weak, so anything close to zero is very weak). If, instead of having 200 observations, we had 20,000, this small of a correlation would be statistically significant. Still, it is best described as very weak, whether it is statistically significant or not. The math advantage that is widely attributed to boys is very small compared with the writing advantage attributed to girls.

Stata's `correlate` command does not give us the significance of the correlations when using casewise deletion. To obtain these and to use all the available data, we need to use the `pwcorr` command. This command is a bit more complicated than `correlate` because it has more options. For this example, it includes the following options: `obs`, which gives the number of observations for that particular correlation; `sig`, which gives the significance level; and `star(5)`, which puts an asterisk by every correlation that is significant at the 5% level. In the following table, we add a new variable, `socst`.

```
. pwcorr read write math science socst ses female, obs sig star(5)
             |     read    write     math  science    socst      ses   female
-------------+---------------------------------------------------------------
        read |   1.0000
             |
             |      200
             |
       write |  0.5968*  1.0000
             |   0.0000
             |      200      200
             |
        math |  0.6623*  0.6174*  1.0000
             |   0.0000   0.0000
             |      200      200      200
             |
     science |  0.6302*  0.5704*  0.6307*  1.0000
             |   0.0000   0.0000   0.0000
             |      200      200      200      200
             |
       socst |  0.6215*  0.6048*  0.5445*  0.4651*  1.0000
             |   0.0000   0.0000   0.0000   0.0000
             |      200      200      200      200      200
             |
         ses |  0.2933*  0.2075*  0.2725*  0.2829*  0.3319*  1.0000
             |   0.0000   0.0032   0.0001   0.0000   0.0000
             |      200      200      200      200      200      200
             |
      female | -0.0531   0.2565* -0.0293  -0.1277   0.0524  -0.1250   1.0000
             |   0.4553   0.0002   0.6801   0.0714   0.4614   0.0778
             |      200      200      200      200      200      200      200
             |
```

Because this is a fairly large sample ($N = 200$), it is not surprising that all the moderate to large correlations are statistically significant. Beneath each correlation are two numbers. The first is the significance level; for example, the correlation between ses and write, $r = 0.21$, has a significance level of 0.0032, which we would write as $r = 0.21$, $p < 0.01$. The next number below the correlation is the number of observations that have a score on both variables; that is, they are not missing either item. There were no missing data in this sample—every correlation is based on $N = 200$ observations. In practice, there is often a wide variation in the number of observations. For example, income is often included as a variable, and about 30% of participants fail to report their income in national surveys, so any variable that is correlated with income will have many missing data and, hence, fewer observations.

(Continued on next page)

Multiple-comparison procedures with correlations

When you are estimating several correlations, the reported significance level given by the `sig` option can be misleading. If you made 100 independent estimates of a correlation that was zero in the population, you would expect to get five significant results by chance (using the 5% level). In this example, we had 21 correlations, and because we are considering all of them, we might want to adjust the probability estimate. One of the ways this is available in the `pwcorr` command is with the option `bon`; short for the Bonferroni multiple-comparison procedure. You can get this simply by adding the `bon` option at any point after the comma in the `pwcorr` command. The complete command would be

```
. pwcorr read write math science socst ses female, bon obs sig star(5)
```

Without this correction, the correlation between `write` and `ses`, $r = .21$, had a $p = 0.0032$ and was significant at the .01 level. With the Bonferroni adjustment, the $r = .21$ does not change, but the correlation now has a $p = 0.067$ and is no longer statistically significant. (An alternative multiple-comparison procedure uses the `sidak` option, which produces the Šidák-adjusted significance level.) It is difficult to give simple advice on when you should or should not use a multiple-comparison adjustment. If your hypothesis is that a certain pattern of correlations will be significant and this involves the set of all the correlations (here, 21), the adjustment is appropriate. If your focus is on individual correlations, as it probably is here, then this adjustment is not necessary.

I want the significance or multiple comparisons with listwise deletion

What happens if you want to use listwise/casewise deletion and want the significance level reported or the Bonferroni multiple-comparison adjustment? Because the significance level is available only with `pwcorr`, we need to tell Stata to do listwise deletion. The command to do this is

```
. pwcorr read write math science socst ses female, listwise
> bon obs sig star(5)
```

The `listwise` option means to keep only cases that are not missing any of the variables in the list.

8.5 Regression

Earlier you learned how to plot a regression line on a scattergram. Now we will focus on how to estimate the regression line itself. Suppose that you are interested in the relationship between how many hours per week a person works and how much occupational prestige they have. You expect that careers with high occupational prestige require more work rather than less. Therefore, you expect that the more hours respondents work, the more occupational prestige they will have. This is certainly not a perfect relationship, and we have all known people who work many hours, even doing two jobs, who do not have high occupational prestige. We will use the General Social Survey 2006 dataset (`gss2006_chapter8_selected.dta`) for this section. It has a variable called `prestg80`, which is a scale of occupational prestige, and `hrs1`, which is the number of hours respondents worked last week in their primary jobs.

Before doing the regression procedure, we should summarize the variables:

```
. summarize prestg80 hrs1
    Variable |       Obs        Mean    Std. Dev.       Min        Max
-------------+--------------------------------------------------------
    prestg80 |      4270    44.16745    13.99946         17         86
        hrs1 |      2739    42.07631    14.23166          1         89
```

This summary gives us a sense of the scale of the variables. The independent variable, `hrs1`, is measured in hours with a mean of $M = 42.08$ hours and a standard deviation of SD = 14.23 hours. The outcome variable, `prestg80`, has corresponding values of $M = 44.17$ and SD = 14.00. A scattergram does not help because there are so many cases that no pattern is clear. A correlation, `pwcorr prestg80 hrs1, obs sig`, tells us that $r = 0.16$, $p < 0.001$. We can interpret this as a fairly weak relationship that is statistically significant. To estimate the regression, open the `regress` dialog box by selecting Statistics ▷ Linear models and related ▷ Linear regression. This dialog box is shown in figure 8.5.

(*Continued on next page*)

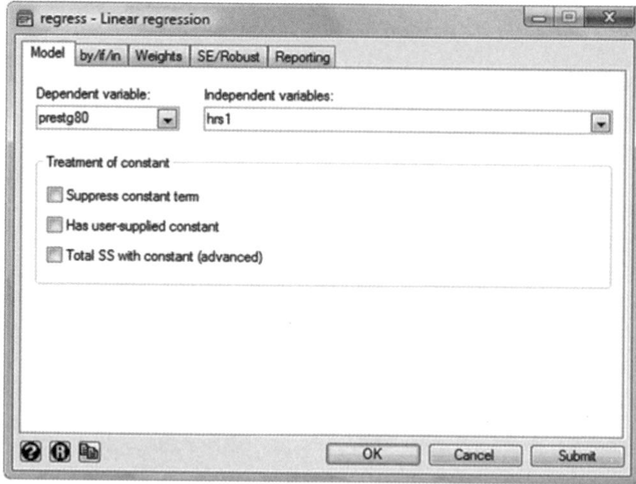

Figure 8.5. The Model tab of the regression analysis dialog box

Enter the *Dependent variable*, `prestg80`, and the *Independent variable*, `hrs1`. Click on the Reporting tab, and check *Standardized beta coefficients*. Click on Submit to obtain

```
. regress prestg80 hrs1, beta

      Source |       SS       df       MS              Number of obs =    2715
-------------+------------------------------           F(  1,  2713) =   67.32
       Model |  13125.035     1   13125.035            Prob > F      =  0.0000
    Residual | 528901.224  2713  194.950691            R-squared     =  0.0242
-------------+------------------------------           Adj R-squared =  0.0239
       Total | 542026.259  2714  199.714908            Root MSE      =  13.962

------------------------------------------------------------------------------
    prestg80 |      Coef.   Std. Err.      t    P>|t|                     Beta
-------------+----------------------------------------------------------------
        hrs1 |   .1546416   .0188468     8.21   0.000                 .1556109
       _cons |   38.14733   .8362129    45.62   0.000                        .
------------------------------------------------------------------------------
```

Understanding the format of this regression command is important because more-advanced procedures that generalize from this command follow the same command structure. The first variable after the name of the command, i.e., `prestg80`, is always the dependent variable. The second variable, `hrs1`, is the independent variable. When we do multiple regression in chapter 10, we will simply add more independent variables. When we do logistic regression in chapter 11, we will simply change the name of the command. After the comma, we have the option `beta`. This will give us beta weights, which are represented as β and will be interpreted below.

The table in the upper left of the output shows the `Source`, `SS`, `df`, and `MS`. This is an analysis of variance (ANOVA) table summarizing the results from an ANOVA perspective (ANOVA is covered in chapter 9), and we will ignore this table for now. In the upper right corner is the number of observations, $N = 2715$, representing the number of people who have been measured on both variables. We also have an F test, which we will cover more fully in the ANOVA chapter. The larger the F ratio is, the greater the significance. Like the t test and chi-squared, F also involves the idea of degrees of freedom. There are two values for the degrees of freedom associated with an F test: the number of predictors (here, 1) and $N - 2$ (here, 2,713). Just below the number of observations are $F(1, 2713) = 67.32$ and the probability level (Prob $> F = 0.0000$). Any probability less than 0.0001 is reported as 0.0000 by Stata. We could write this as $F(1, 2713) = 67.32$, $p < 0.001$. Thus there is a statistically significant relationship between hours worked and the prestige of the job.

Is this relationship strong? We have two values, namely, R^2 and the adjusted R^2, that serve to measure the strength of the relationship. When we are doing bivariate regression, R^2 is simply $r \times r$. Similarly, $r = \sqrt{R^2}$, but we need to decide on the sign of the r value—whether it is positive or negative. For our model, $R^2 = 0.02$, meaning that the hours worked explain 2% of the variation in the prestige rating. Because of the large sample, this R^2, although obviously weak because it does not explain 98% of the variation in prestige, is still statistically significant. When you have many predictors and a small sample, neither of which apply here, some report the adjusted R^2. This option removes the part of R^2 that would be expected just by chance. Whenever the adjusted R^2 is substantially smaller than the R^2 because there is a small sample relative to the number of predictors, it is good to report both values.

The root mean squared error, `Root MSE = 13.96`, has a strange name, but it is a useful piece of information. The `summarize` command showed SD = 14.00 for our dependent variable, `prestg80`. The root mean squared error is the standard deviation around the regression line. Recall when we did a plot of a regression line. If the observations were close to this line, the standard deviation around it should be much smaller than the standard deviation around the overall mean. It is not surprising that our R^2 is so small, given that the standard deviation around the regression line, 13.96, is nearly as big as the standard deviation around the mean, 14.00. In other words, the regression line does little to improve our prediction.

The bottom table gives us the regression results. The first column lists the outcome variable, `prestg80`, followed by the predictor, `hrs1`, and the constant called `_cons`. This last variable is the constant, or intercept. The equation would be written as

$$\text{Estimated prestige} = 38.15 + 0.15(\text{hours})$$

Remember that $M = 44.17$ and SD = 14.00 for prestige. If we did not know how many hours a person worked per week, we might guess that they had a prestige score of 44.17 given that 44.17 is the mean prestige score. The regression equation lets us estimate prestige differently depending on how many hours a person works per week. This equation tell us that a person who worked 0 hours would have a prestige score of 38.15 (the intercept or constant), and for each additional hour he or she worked, the prestige score would go up 0.15. For example, a person who worked 40 hours a week would have a predicted prestige score of (approximately, because we have rounded)

$38.15 + 0.15(40) = 44.15$, but a person working 60 hours a week would have a predicted prestige score of $38.15 + 0.15(60) = 47.15$. This shows a small payoff in prestige for working longer hours. The payoff is statistically significant but not very big.

The column labeled `Std. Err.` is the standard error. The `t` is computed by dividing the regression coefficient by its standard error; for example, $0.1546416/0.0188468 = 8.21$. The `t` is evaluated using $N - 2$ degrees of freedom. We do not need to look up the t value because Stata reports the probability as 0.000. We would report this as $t(2713) = 8.21$, $p < 0.001$.

The final column gives us a beta weight, $\beta = 0.16$. When we have just one predictor, the β weight is always the same as the correlation (this is not the case when there are multiple predictors). The β is a measure of the effect size and is interpreted much like a correlation with $\beta = 0.10$ being weak, 0.30 being moderate, and 0.50 being strong.

Your statistics book may also show you how to do confidence intervals. There are two types of these: a confidence interval around the regression coefficient and a confidence interval around the regression line. Stata gives you the former as an option for regression and the latter as an option for the scattergram. First, let's do the confidence interval around the regression coefficient. Reopen the `regress` dialog box, and click on the Reporting tab. This time, make sure that the *Standardized beta coefficients* box is not checked. By unchecking this option, you automatically get the confidence interval in place of β. The command is the same except there is no option, that is, we just have `regress prestg80 hrs1`.

```
. regress prestg80 hrs1
      Source |       SS       df       MS              Number of obs =    2715
-------------+------------------------------           F(  1,  2713) =   67.32
       Model |   13125.035     1   13125.035           Prob > F      =  0.0000
    Residual |  528901.224  2713  194.950691           R-squared     =  0.0242
-------------+------------------------------           Adj R-squared =  0.0239
       Total |  542026.259  2714  199.714908           Root MSE      =  13.962

------------------------------------------------------------------------------
    prestg80 |      Coef.   Std. Err.      t    P>|t|     [95% Conf. Interval]
-------------+----------------------------------------------------------------
        hrs1 |   .1546416   .0188468     8.21   0.000      .117686    .1915972
       _cons |   38.14733   .8362129    45.62   0.000     36.50765    39.78701
------------------------------------------------------------------------------
```

All of this is identical, except that where we had β, we now have a 95% confidence interval for each regression coefficient. The effect of an additional hour of work on prestige is $b = 0.15$. (Do not confuse $b = 0.15$ with $\beta = 0.16$ from the previous example.) The bs and βs are usually different values. The b has a 95% confidence as low as 0.09 and as high as 0.18. Because a value of zero is not included in the confidence interval (a zero value signifies no relationship), we know that the slope is statistically significant. It could have been as little as 0.12 or as much as 0.18. We would report this by stating we are 95% confident that the interval of 0.12 to 0.19 contains the true slope.

The second type of confidence interval is on the overall regression line. We will run into trouble using the regression line to predict cases that are very high or very low on the predictor. Usually there are just a few people at the tails of the distribution, so we do not have as much information there as we do in the middle. Also, we get to areas that make no sense or where there is likely a misunderstanding. For example, predicting the occupational prestige of a person who works 0 hours a week makes no sense, but it does not make much more sense to predict it for somebody who works just 1 or 2 hours a week. Similarly, if a person said they worked 140 hours a week, they probably misunderstood the question because hardly anybody actually works 140 hours a week on a regular basis, given that there are only 168 hours in a week. That would be 20 hours a day, 7 days a week! On the other hand, there are lots of people who work 30–50 hours a week, and here we have much more information for making a prediction. Thus, if we put a band around the regression line to represent our confidence, it would be narrowest near the middle of the distribution on the independent variable and widest at the ends.

This relationship is easy to represent with a scattergram. If we use the menu system, selecting Graphics ▷ Twoway graph (scatter, line, etc.) opens the dialog box we need. Next click on *Create....* Then click on *Fit plots* and highlight *Linear prediction w/CI.* Finally, we enter our two variables. The results are shown in figure 8.6.

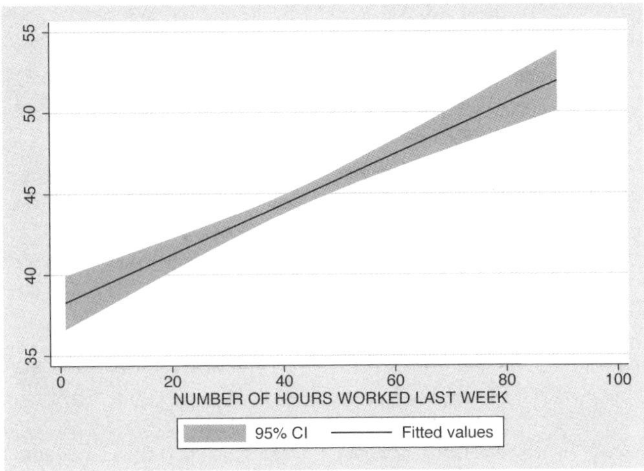

Figure 8.6. Confidence band around regression prediction

8.6 Spearman's rho: rank-order correlation for ordinal data

Spearman's rho, ρ_s, is the correlation between two variables that are treated as ranks. This procedure converts the variables to ranks (1st, 2nd, ..., nth) before estimating the correlation. For example, if the ages of five observations are 18, 29, 35, 61, and 20, these five participants would be assigned ranks of 1 for 18, 3 for 29, 4 for 35, 5 for 61, and 2 for 20. If we had a measure of liberalism for these five cases, we could convert that to a rank as well. Here are the data (`spearman.dta`) for this simple example:

```
. list
```

	age	liberal	rankage	ranklib
1.	18	90	1	4.5
2.	29	90	3	4.5
3.	35	80	4	2
4.	61	50	5	1
5.	20	89	2	3

Ideally, all the scores (and hence, the ranks) would be unique. Stata will assign an average value when there are ties. The two observations who have a score on `liberal` of 90 occupy the fourth and fifth (highest) rank. Because they are tied, they are both assigned a rank score of 4.5. Computing the correlation between `age` and `liberal` by using the command `corr age liberal`, we obtain an r of -0.97. Computing the correlation between the ranked data by using the command `corr rankage ranklib`, we obtain an r of -0.82. This is lower because the one extreme case (`age` = 61, `liberal` = 50) inflates the Pearson correlation, but this case is not extreme when ranked data are used.

Spearman's rho is a correlation of ranked data. To save the time of converting the variables to ranks and then doing a Pearson's correlation, Stata has a special command: `spearman age liberal`. Running this commands yields $\rho_s = -0.82$.

8.7 Summary

Scattergrams, correlations, and regressions are great ways to evaluate the relationship between two variables.

- The scattergram helps us visualize the relationship and is usually most helpful when there are relatively few observations.
- The correlation is a measure of the strength of the relationship between the two variables. Here it is important to recognize that it measures the strength of a particular form of the relationship. The other examples have used a linear regression line as the form of the relationship. Although it is not covered here, it is possible for regression to have other forms of the relationship.
- The regression analysis tells us the form of the relationship. Using this line, we can estimate how much the dependent variable changes for each unit change in the independent variable.
- The standardized regression coefficient, β, measures the strength of a relationship and is identical to the correlation for bivariate regression.

- You have learned how to compute Spearman's rho for rank-order data, and you now understand its relationship to Pearson's r.

8.8 Exercises

1. Use the gss2006_chapter8.dta dataset. Say you heard somebody say that there was no reason to provide more educational opportunities for women because so many of them just stay at home anyway. You have a variable measuring education, educ, and a variable measuring hours worked in the last week, hrs1. Do a correlation and regression of hours worked in the last week on years of education. Then do this separately for women and for men. Interpret the correlation and the slope for the overall sample and then for women and for men separately. Is there an element of truth to what you heard?

2. Use the gss2006_chapter8.dta dataset. What is the relationship between the hours a person works and the hours his or her spouse works? Do this for women and for men separately. Compute the correlation, the regression results, and the scattergrams. Interpret each of these. Next test if the correlation is statistically significant and interpret the results.

3. Use the gss2006_chapter8.dta dataset. Repeat figure 8.2 using your own subsample of 250 observations. Then repeat the figure using a jitter(3) option. Compare the two figures. Set your seed at 111.

4. Use the gss2006_chapter8.dta dataset. Compute the correlations between happy, hapmar, and health by using correlate and then again by using pwcorr. Why are the results slightly different? Then estimate the correlations by using pwcorr, and get the significance level and the number of observations for each case. Finally, repeat the pwcorr command so that all the Ns are the same (i.e., there is casewise or listwise deletion).

5. Use the gss2002_chapter8.dta dataset. There are two variables called happy7 and satfam7. Run the codebook command on these variables. Notice how the higher score goes with being unhappy or being dissatisfied. You always want the higher score to mean more of a variable, so generate new variables (happynew and satfamnew) that reverse these codes so that a score of 1 on happynew means very unhappy and a score of 7 means very happy. Similarly, a score of 1 on satfamnew means very dissatisfied and a score of 7 means very satisfied. Now do a regression of happiness on family satisfaction with the new variables. How correlated are these variables? Write the regression equation. Interpret the constant and the slope.

6. Use the spearman.dta dataset. Plot a scattergram, including the regression line for age and liberal, treating liberal as the dependent variable. Repeat this using the variables rankage and ranklib. Interpret this to explain why the Spearman's rho is smaller than the Pearson's correlation. Your explanation should involve the idea of one observation being an outlier.

9 Analysis of variance

9.1 The logic of one-way analysis of variance

In many research situations, a one-way analysis of variance (ANOVA) is appropriate, most commonly when you have three or more groups or conditions and want to see if they differ significantly on some outcome. This procedure is an extension of the two-sample t test.

You might have two new teaching methods you want to compare with a standard method. If you randomly assign students to three groups (standard method, first new method, and second new method), an ANOVA will show if at least one of these groups has a significantly different mean. ANOVA is usually a first step. Suppose you find that the three groups differ, but you do not know how they differ. The first new method may be best, the second new method may be second best, and the standard method may be worst. Alternatively, both new methods may be equal but worse than the standard method. When you do an ANOVA and find a statistically significant result, this begs a deeper look to describe exactly what the differences are. These follow-up tests often involve several specific tests (first new method versus standard method, second new method versus standard method, first new method versus second new method). When you do multiple tests like this, you need to make adjustments to the tests because they are not really independent tests.

ANOVA is normally used in experiments, but it can also be used with survey data. In a survey, you might want to compare Democrats, Republicans, independents, and noninvolved adults on their attitudes toward stem-cell research. There is no way you could do an experimental design because you could not randomly assign people to these different party identifications. However, in a national survey you could find many people belonging to each group. If your overall sample was random, then each of these subgroups would be a random sample as well. An ANOVA would let you compare the mean score on the value of stem-cell research (your outcome variable) across the four party identifications.

ANOVA makes a few assumptions:

- The outcome variable is quantitative (interval level).
- The outcome variable is normally distributed. This is problematic if we have a small sample.
- The observations represent a random sample of the population.
- The outcomes are independent. (We have repeated-measures ANOVA when this assumption is violated.)
- The variance of each group is equal. You can test this assumption.
- The number of observations in each group does not vary widely.

Violating combinations of these assumptions can be especially problematic. For example, unequal Ns for each group combined with unequal variances is far worse than unequal variances when the Ns are equal.

9.2 ANOVA example

People having different party identifications may vary in how much they support stem-cell research. You might expect Democrats to be more supportive than Republicans, on average. What about people who say they are independents? What about people who are not involved in politics? Stata allows you to do a one-way ANOVA and then do multiple-comparison tests of all pairs of means to answer two questions:

- The global question answered by ANOVA is whether the means are equal for all groups.
- Specific tests answer whether pairs of group means are different from one another.

There are two ways of presenting the data. Most statistics books use what Stata calls a wide format, in which the groups appear as columns and the scores on the dependent variables appear as rows under each column. An example appears in table 9.1.

Table 9.1. Hypothetical data—wide view

	`democrat`	`republican`	`independent`	`noninvolved`	overall
	9	5	7	5	
	7	9	8	4	
	9	4	6	6	
	6	6	6	4	
	9	3	7	5	
	8	1	7	4	
	9	5	8	6	
	7	9	7	4	
	9	4	8		
	6	6	6		
	9	3	6		
	8	1	7		
			7		
			8		
M	8.00	4.67	7.00	4.75	6.26
SD	1.21	2.61	.78	.88	2.09

Hypothetical data comparing views on stem-cell research for people with different party identifications.

Stata can work with data arranged like this, but it is usually easier to enter what Stata calls a long format. We could enter the data as shown in table 9.2.

(*Continued on next page*)

Table 9.2. Hypothetical data—long view

`stemcell`	`partyid`
9	1
6	1
6	1
9	1
9	1
7	1
8	1
7	1
9	1
8	1
9	1
9	1
6	2
5	2
4	2
1	2
6	2
4	2
1	2
3	2
5	2
9	2
3	2
9	2
...	...
5	4

There are 46 observations altogether, and table 9.2 shows 25 of them. We know the party membership of each person by looking at the `partyid` column. We have coded Democrats with a `1`, Republicans with a `2`, independents with a `3`, and those not involved with a `4`. If you want to see the entire dataset, you can open up `partyid.dta` and enter the `list, nolabel` command. This is similar to entering data for a two-sample t test, with one important exception. This time, we have four groups rather than two groups, and the grouping variable is coded from `1` to `4` instead of from `1` to `2`.

To perform a one-way ANOVA, select Statistics ▷ Linear models and related ▷ ANOVA/MANOVA ▷ One-way ANOVA, which opens the dialog box shown in figure 9.1.

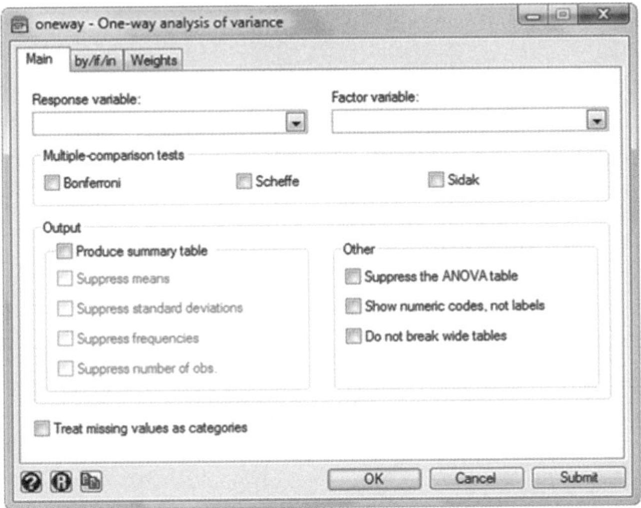

Figure 9.1. One-way analysis of variance dialog box

Under the Main tab, indicate that the *Response variable* is `stemcell` and the *Factor variable* is `partyid`. Many analysis-of-variance specialists call the response variable the dependent or outcome variable. They call the categorical factor variable the independent, predictor, or grouping variable. In this example, `partyid` is the factor that explains the response, `stemcell`. Stata uses the names *response variable* and *factor variable* to be consistent with the historical traditions of analysis of variance. Many sets of statistical procedures developed historically, in relative isolation, and produced their own names for the same concepts. Remember, the response or dependent variable is quantitative, and the independent or factor variable is categorical.

Stata asks if we want multiple-comparison tests, and we can choose from three: *Bonferroni*, *Scheffe*, and *Sidak*. These are three multiple-comparison procedures for doing the follow-up tests that compare each pair of means. A comparison of these three approaches is beyond the purpose of this book. We will focus on the Bonferroni test, so click on *Bonferroni*. Finally, check the box to *Produce summary table* to get the mean and standard deviation on support for stem-cell research for members grouped by `partyid`. Here are the results:

(*Continued on next page*)

```
. oneway stemcell partyid, bonferroni tabulate
      party |  Summary of support for stem cell
 identificat |            research
         ion |        Mean   Std. Dev.       Freq.
 ------------+------------------------------------
    democrat |           8   1.2060454          12
   republica |   4.6666667   2.6053558          12
   independe |           7   .78446454          14
   noninvolv |        4.75   .88640526           8
 ------------+------------------------------------
       Total |   6.2608696   2.0916212          46
                        Analysis of Variance
    Source              SS         df      MS            F     Prob > F
------------------------------------------------------------------------
Between groups      92.7028986      3   30.9009662     12.46     0.0000
 Within groups      104.166667     42   2.48015873
------------------------------------------------------------------------
    Total           196.869565     45   4.37487923
Bartlett's test for equal variances:  chi2(3) =  20.1167  Prob>chi2 = 0.000
     Comparison of support for stem cell research by party identification
                                (Bonferroni)
Row Mean-|
Col Mean |   democrat   republic   independ
---------+---------------------------------
republic |   -3.33333
         |      0.000
         |
independ |         -1    2.33333
         |      0.684      0.003
         |
noninvol |      -3.25    .083333      -2.25
         |      0.000      1.000      0.015
```

This is a lot of output, so we need to go over it carefully. Just below the command is a tabulation showing the mean, standard deviation, and frequency on support for stem-cell research by people in each group. Notice that Democrats and independents both have relatively high means, $M_{\text{democrat}} = 8.00$ and $M_{\text{independent}} = 7.00$, although Republicans and those who are not involved have relatively low means, $M_{\text{republican}} = 4.67$ and $M_{\text{noninvolved}} = 4.75$. (These are hypothetical data.)

Looking at the standard deviations, we can see a potential problem. In particular, the Republicans have a much larger standard deviation than that of any of the other groups. This could be a problem because we assume that the standard deviations are equal for all groups. Finally, the table gives the frequency of each group. There are relatively fewer noninvolved people than people in the other groups, but the differences are not dramatic.

Next is the ANOVA table. This is undoubtedly discussed in your statistics textbook, so we will go over it only briefly. The first column, `Source`, has two sources. There is variance between the group means, which should be substantial if the groups are really different. There is also variance within groups, which should be relatively small if the groups are different from each other, but should be homogeneous within each group. Think about this a moment. If the groups are really different, their means will be spread out, but within each group the observations will be homogeneous. The column labeled `SS` is the sum of squared deviations for each source, and when we divide this by the degrees of freedom (labeled `df`), we get the values in the column labeled `MS` (mean squares). This label sounds strange if you are not familiar with ANOVA, but it has a simple meaning. The between-group mean square is the estimated population variance based on differences between groups—this should be large when there are significant differences between the groups. The within-group mean square is the estimated population variance based on the distribution within each group—this should be small when most of the differences are between the groups. The test statistic, F, is computed by dividing the MS(between) by the MS(within). F is the ratio of two variance estimates. The $F = 30.90/2.48 = 12.46$. This ratio is evaluated using the degrees of freedom. We have 3 degrees of freedom for the numerator (30.90) and 42 degrees of freedom for the denominator (2.48). You can look this up in a table of the F distribution. However, Stata gives you the probability as 0.0000. We would never report a probability as 0.0000 but would say $p < 0.001$.

Can Stata give me an F table instead of looking it up in a book?

We get the probability directly from the Stata output, so we do not need to look it up in an F table. However, if you ever need to use an F table, you can get one from Stata without having to find the table in a statistics textbook. Type the command `findit ftable`, which opens a window with a link to a web page. Go to that web page, and click on `install` to install several tables. You may have already downloaded these tables when you were obtaining z tests or t tests. From now on, whenever you want an F table, you need only enter the command `ftable`.

When you do this, you also have `ttable`, `ztable`, and `chitable`.

Years ago, these ANOVA tables appeared in many articles. Today, we simplify the presentation. We could summarize the information in the ANOVA table as $F(3, 42) = 12.46$, $p < 0.001$, meaning that there is a statistically significant difference between the means.

Stata computes Bartlett's test for equal variances and tells us that the χ^2 with 3 degrees of freedom is 20.12, $p < 0.001$. (Do not confuse this χ^2 test with the χ^2 test for a frequency table. The χ^2 distribution is used in many tests of significance.) One of the assumptions of ANOVA is that the variances of the outcome variable, `stemcell`, are equal in all four groups. The data do not meet this assumption. Some researchers discount this test because it will usually not be significant, even when there are substantial differences in variances, if the sample size is small. By contrast, it will usually be significant when there are small differences, if the sample size is large. Because unequal variances are more problematic with small samples, where the test lacks power, and less important with large samples, where the test may have too much power, the Bartlett test is often ignored. Be careful when the variances are substantially different and the Ns are also substantially different: this is a serious problem. One thing we might do is go ahead with our ANOVA but caution our readers in our reports that the Bartlett test of equal variances was statistically significant.

Given that the overall F test is statistically significant, we can proceed to compare the means of the groups. This is a multiple comparison involving six tests of significance: Democrat versus Republican, Democrat versus independent, Democrat versus noninvolved, Republican versus independent, Republican versus noninvolved, and independent versus noninvolved. Stata provides three different options for this, but here we consider only the Bonferroni. Multiple comparisons involve complex ideas, and we will mention only the principal issues here. The traditional t test worked fine for comparing two means, but what happens when we need to do six of these tests? Stata does all the adjustments for us, and we can interpret the probabilities Stata reports in the way we always have. Thus, if the reported p is less than 0.05, we can report $p < 0.05$ using the Bonferroni multiple-comparison correction.

Below the ANOVA table is the comparison table for all pairs of means. The first number in each cell is the difference in means (`Row Mean - Col Mean`). Because political independents had a mean of 7 and Democrats had a mean of 8, the table reports the difference, $7 - 8 = -1$. Independents, on average, have a lower mean score on support for stem-cell research. Is this statistically significant? No; the $p = 0.684$ far exceeds a critical level of $p < 0.05$. Notice that the noninvolved people and the Republicans have similar means and the small difference, 0.08, is not statistically significant. However, all the other comparisons are statistically significant. For example, Republican support for stem-cell research has a mean that is 3.33 points lower than that of Democrats, $p < 0.001$. One way to read the table is to remember that a negative sign means that the row group mean is lower than the column group mean and that a positive sign means that the row group mean is higher than the column group mean.

How strong is the difference between means? How much of the variance in `stemcell` is explained by different party identification? We can compute a measure of association to represent the effect size. Eta-square (η^2) is a measure of explained variation. Some refer to this as r^2 because it is the ANOVA equivalent of r^2 for correlation and regression. Knowing the party identification improves our ability to predict the respondent's attitude toward stem-cell research. The η^2 or r^2 in the context of ANOVA is the ratio of the between-groups sum of squares to the total sum of squares. Stata does not compute this for you, but you can compute it by using a simple division:

$$\eta^2 = r^2 = \frac{\text{Between group SS}}{\text{Total SS}} = \frac{92.703}{196.870} = 0.471$$

Do not forget about Stata's calculator. You can enter `display 92.703/196.870`, and Stata will report back the answer of 0.471. This result means that 47.1% of the variance in attitude toward stem-cell research is explained by differences in party identification. Some researchers prefer a different measure that is called ω^2 (pronounced omega-squared), and you will learn how to obtain this later in the chapter.

9.3 ANOVA example using survey data

You will find several examples of studies like the one we just did in standard statistics textbooks. Our second example uses data from a large survey, the 2006 General Social Survey (`gss2006_chapter9.dta`) and examines occupational prestige. For our one-way ANOVA, we will see if people who are more mobile have the benefit of higher occupational prestige. We will compare three groups (the factor or independent variable must be a grouping variable). One group includes people who are living in the same town or city they lived in at the age of 16. We might think that these people have lower occupational prestige, on average, because they did not or could not take advantage of a broader labor market that extended beyond their immediate home city. The second group comprises those who live in a different city but still the same state. Because they have a larger labor market, one that includes areas outside of their city of origin, they may have achieved higher prestige. The third group contains those who live in a different state. By being able and willing to move this far, they have the largest labor market available to them. We will restrict the age range so that age is not a second factor; the sample is restricted to adults who are between 30 and 59 years of age.

Make sure that you open the *One-way analysis of variance* dialog box shown in figure 9.1. There is another option for analysis of variance and covariance that we will cover later in this chapter. In the Main tab, we are asked for a *Response variable*. The dependent variable is `prestg80`, so enter this as the response variable. This is a measure of occupational prestige that was developed in 1980 and applied to the 2006 sample. The dialog box asks for the *Factor variable*, so type `mobile16` (the independent variable). This variable is coded `1` for respondents who still live in the same city they lived in when they were 16; `2` if they live in a different city, but still in the same state; and `3` if they live in a different state. Also click on the box by *Produce summary table* to get the means for each group on the factor variable. Finally, click on the box by *Bonferroni* in the section labeled *Multiple-comparison tests*. Because we have three groups, we can use this test to compare all possible pairs of groups, i.e., same city to same state, same city to different state, and same state to different state. On the by/if/in tab, we enter the following restriction under *If: (expression)*: `age > 29 & age < 60 & wrkstat==1`. This restriction limits the sample to those who are 30–59 years old and have a `wrkstat` of `1`, signifying that they work full time. Remember to use the symbol `&` rather than "and".

In our results, we first get a table of means and standard deviations. Notice that the first two means are in the direction predicted, but the third mean is not. Those who live in the same city have a mean of 44.12, those who have moved from the city but are still in the same state have a mean of 48.67, and those who have moved out of the state have a mean of 47.19. The standard deviations are similar (we can use the Bartlett test of equal variances to test this assumption). Also the N for each group varies from 269 to 425. Usually, the unequal Ns are not considered a serious problem unless they are extremely unequal and one or more groups have few observations.

```
. oneway prestg80 mobile16 if age > 29 & age < 60 & wrkstat==1, bonferroni
> tabulate
 GEOGRAPHIC
   MOBILITY  |  Summary of RS OCCUPATIONAL PRESTIGE
  SINCE AGE  |            SCORE  (1980)
         16  |        Mean   Std. Dev.       Freq.
-------------+------------------------------------
  SAME CITY  |   44.116162    13.73443         396
  SAME ST,D  |   48.672862   13.002183         269
  DIFFERENT  |   47.185882   13.959368         425
-------------+------------------------------------
      Total  |   46.437615   13.758919        1090
                        Analysis of Variance
    Source              SS         df      MS            F     Prob > F
------------------------------------------------------------------------
Between groups      3716.07404      2   1858.03702      9.98     0.0001
 Within groups      202440.184   1087    186.23752
------------------------------------------------------------------------
    Total           206156.258   1089   189.307858
Bartlett's test for equal variances:  chi2(2) =   1.6873  Prob>chi2 = 0.430
             Comparison of RS OCCUPATIONAL PRESTIGE SCORE  (1980)
                    by GEOGRAPHIC MOBILITY SINCE AGE 16
                               (Bonferroni)
Row Mean-|
Col Mean |   SAME CIT   SAME ST,
---------+----------------------
SAME ST, |     4.5567
         |      0.000
         |
DIFFEREN |    3.06972   -1.48698
         |      0.004      0.487
```

We can summarize the second table (the ANOVA table) by writing $F(2, 1087) = 9.98$, $p < 0.001$. Just beneath the ANOVA table is Bartlett's test for equal variances. Because $\chi^2(2) = 1.69$, this finding is not statistically significant ($p = 0.43$). This is good because equal variances is an assumption of ANOVA. The F test is an overall test of significance for any differences between the group means. You can have a significant F test when the means are different but in the opposite direction of what you expected.

Now, let's perform multiple comparisons using the Bonferroni adjustment. Those adults who moved to a different state scored 3.07 points higher on prestige, on average, than those who stayed in their home town ($47.19 - 44.12 = 3.07$), $p < 0.01$. Those who moved but stayed in the same state scored 4.56 points higher on the prestige scale ($48.67 - 44.12 = 4.55$), $p < 0.001$. Although these two comparisons are significant, those who moved but stayed in the same state are not significantly different from those who moved to a new state.

We can do a simple one-way bar chart of the mean. The command is

```
. graph bar (mean) prestg80 if age > 29 & age < 60 & wrkstat==1, over(mobile16)
```

You can enter this command as one line in the Command window or in a do-file using /// as needed for a line break. Because of the length of graph commands, it is probably a good idea to use the dialog system as discussed in chapter 5. Select Graphics ▷ Bar chart to open the appropriate dialog box. On the Main tab, select *Graph by calculating summary statistics* and *Vertical* in the *Orientation* section. Under *Statistics to plot*, check the first box, select *Mean* from the drop-down menu and, to the right of this, type `prestg80` as the variable. Next go the Categories tab and check *Group 1* and select `mobile16` as the *Grouping variable*. Finally, open the if/in tab and, under the *If: (expression)*, type `age > 29 & age < 60 & wrkstat==1`.

The resulting bar chart provides a visual aid for showing that the mean is highest for those with some geographical mobility, i.e., same state but different city. Although all three means are different in our sample, only two comparisons (staying in your home town versus moving to a different city within your home state and staying in your home town versus moving to a different state) reach statistical significance.

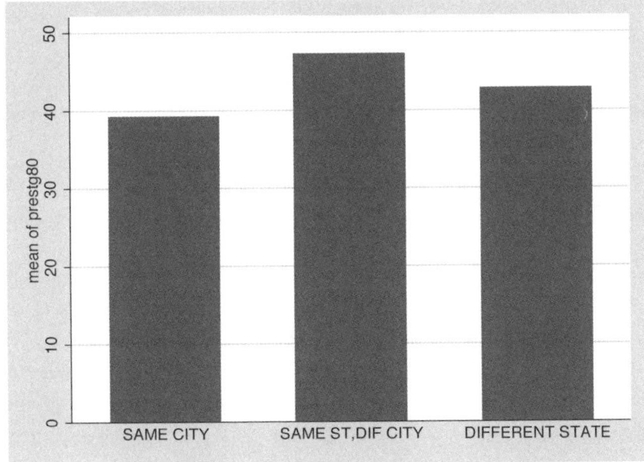

Figure 9.2. Bar graph of relationship between prestige and mobility

We can compute $\eta^2 = r^2 = 3716.074/206156.258 = 0.02$ as described in the previous section. Thus, even though the overall result is significant and two of the three comparisons are significant in the direction we predicted, the mobility variable does not explain much of the variance in prestige. Why is this? To get an answer, look at the means and standard deviations. In the previous example, the standard deviations for groups tended to be much smaller than the overall standard deviation. This is not the case in this example. The standard deviations for each group are between 13.00 and 13.96, whereas the overall standard deviation is 13.76. Although the means do differ, most of the variance still remains within each group. When the variance within the groups is not substantially smaller than the overall variance, the group differences are not explaining much variance.

Using Stata's calculator to compute η^2

> Stata saves many of the statistics that are computed so that you can use them. Regular statistics produced by a command are listed by the command `return list`, and statistics that are estimated by a command are listed by the command `ereturn list`. Running the command `return list` displays the names of the statistics the one-way ANOVA computed. Two of these are the sum of squares between groups, `r(mss)`, and the sum of squares residuals, `r(rss)`. Stata does not save the total sum of squares—that is, the sum of `r(mss)` + `r(rss)`. Looking at the formula for the correlation ratio, we can calculate it using the display calculator, `display "eta-squared = " r(mss) / (r(mss) + r(rss))`.
>
> It would be easier here to enter the values by hand. The advantage of using the returned values is that they keep all the decimal places. Also, these values are useful when you develop your own programs to do specialized analyses that are not available in Stata.

9.4 A nonparametric alternative to ANOVA

Sometimes treating the outcome variable as a quantitative, interval-level measure is problematic. Many surveys have response options, such as agree, don't know, and disagree. In such cases, we can score these so that `1` is "agree", `2` is "don't know", and `3` is "disagree", and then we can do an ANOVA. However, some researchers might say that the score was only an ordinal-level measure, so we should not use ANOVA. The Kruskal–Wallis rank test lets us compare the median score across the groups. If we are interested in party identification and differences in support for stem-cell research, we might use Kruskal–Wallis instead of the one-way ANOVA. If we want to use only ordinal information, it may be more appropriate to compare the median rather than the mean.

For this example, we will use the `partyid.dta` dataset. To open the dialog box for the Kruskal–Wallis test, select Statistics ▷ Summaries, tables, and tests ▷ Nonparametric tests of hypotheses ▷ Kruskal-Wallis rank test. In the resulting screen, there are only two options. Type `stemcell` under *Outcome variable* and `partyid` under *Variable defining groups*. As with ANOVA, the variable defining the groups is the independent variable, and the outcome variable is the dependent variable.

```
. kwallis stemcell, by(partyid)
Kruskal-Wallis equality-of-populations rank test

  +-----------------------------+
  |     partyid | Obs | Rank Sum |
  |-------------+-----+----------|
  |    democrat |  12 |   422.00 |
  |  republican |  12 |   174.00 |
  | independent |  14 |   391.00 |
  | noninvolved |   8 |    94.00 |
  +-----------------------------+

chi-squared =    22.115 with 3 d.f.
probability =     0.0001

chi-squared with ties =    22.696 with 3 d.f.
probability =     0.0001
```

This test ranks all the observations from the lowest to the highest score. With a scale that ranges from 1 to 9, there are many ties, and the program adjusts for that. If the groups were not different, the rank sum for each group would be the same, assuming an equal number of observations. Notice from the output that the `Rank Sum` for Democrats is 422 and for Republicans it is just 174, even though there are 12 observations in each group. This means that Democrats must have higher scores than Republicans. The output gives us two chi-squared tests. Because we have people who are tied on the outcome variable (have the same score on `stemcell`), use the chi-squared with ties: chi-squared(3) = 22.696, $p < 0.001$. Thus there is a highly significant difference between the groups in support for stem-cell research.

Because we are treating the data as ordinal, it makes sense to use the median rather than the mean. Now run a `tabstat` to get the median. It is easiest to type the command in the Command window, but if you prefer to use the dialog box, select Statistics ▷ Summaries, tables, and tests ▷ Tables ▷ Table of summary statistics (tabstat).

```
. tabstat stemcell, statistics(mean median sd) by(partyid)

Summary for variables: stemcell
     by categories of: partyid (party identification)

    partyid |      mean       p50        sd
------------+------------------------------
   democrat |         8       8.5  1.206045
 republican |  4.666667       4.5  2.605356
independent |         7         7  .7844645
noninvolved |      4.75       4.5  .8864053
------------+------------------------------
      Total |   6.26087         6  2.091621
-------------------------------------------
```

We have the same pattern of results we had when we ran an ANOVA on these data. The `p50` is the median. Notice that the medians are in the same relationship as the means with Democrats having the highest median, Mdn = 8.5; followed by the independents, Mdn = 7.0; and the Republicans and noninvolved having the lowest, Mdn = 4.5.

There are two graphs we can present. We can do a bar chart like the one we did for the ANOVA, only this time we will do it for the median rather than for the mean. Select Graphics ▷ Bar chart to open the appropriate dialog box, or type the command

```
. graph hbar (median) stemcell, over(partyid)
> title(Median stem cell attitude score by party identification)
> ytitle(Median score on stem cell attitude)
```

which produces the following graph (figure 9.3).

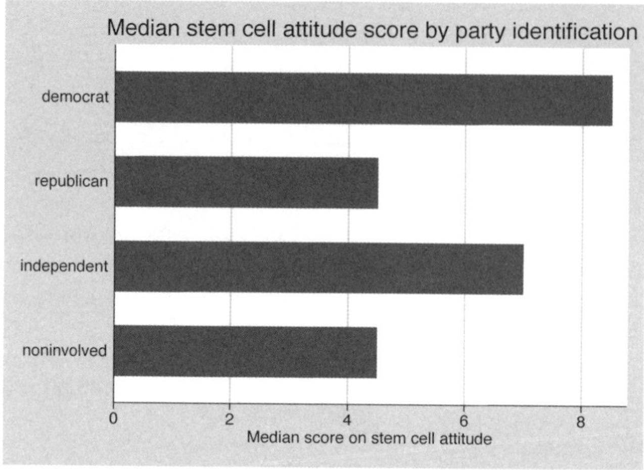

Figure 9.3. Bar graph of support for stem-cell research by party identification

Because we are working with the median, we can use a box plot. Open the dialog box by selecting Graphics ▷ Box plot. Under the Main tab, we enter `stemcell` under *Variables*. Switching to the Categories tab, we check the *Group 1* box and enter `partyid` as the variable under *Grouping variable*. Clicking on OK produces the following command and figure 9.4.

```
. graph box stemcell, over(partyid)
```

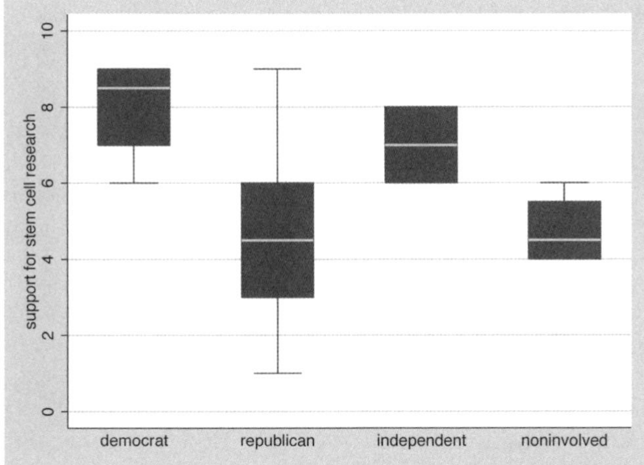

Figure 9.4. Box plot of support for stem-cell research by party identification

This is a much nicer graph than the bar chart because it not only shows the median of each group but also gives us a sense of the within-group variability. With the ANOVA, party identification explained nearly half (0.471) of the variance in support for stem-cell research. This box plot shows that the medians are not only different but that there is a lot of homogeneity within each group, except for the Republicans. (Again, these are hypothetical data.)

9.5 Analysis of covariance

We will discuss analysis of covariance (ANCOVA) as an extension of multiple regression in chapter 10. However, we will present an example of it here because it has a long history as a type of analysis of variance. ANCOVA has a categorical predictor (factor) and a quantitative dependent variable (response) like one-way ANOVA. ANCOVA adds more covariates that are quantitative variables that need to be controlled.

One of the areas in which ANCOVA developed was education, where the researchers could not assign children to classrooms randomly. Researchers needed to adjust for the fact that without random assignment, there might be systematic differences between the classrooms before the experiment started. For instance, one classroom might have students who have better math backgrounds than those of students in another classroom, so they might seem to do better because they were ahead of the others before the experiment started. We would need to control for initial math skills of students. People who volunteer to participate in a nutrition education course might do better than those in a control group, but this might happen because volunteers tend to be more motivated. We would need to control for the level of motivation. The idea of ANCOVA is to statistically control for group differences that might influence the result when you cannot rule out these possible differences through randomization. We might think of

ANCOVA as a design substitute for randomization. Randomization is ideal, but when it is impossible to randomly assign participants to conditions, we can use ANCOVA as a fallback.

I do not mean to sound negative. ANCOVA allows us to make comparisons when we do not have randomization. Researchers often need to study topics for which randomization is out of the question. If we can include the appropriate control variables, ANCOVA does a good job of mitigating the limitations caused by the inability to use randomization.

The one-way ANOVA showed that people who moved out of their hometown had higher prestige than people who did not. Perhaps there is a benefit to opening up to a broader labor market among people who move away from their hometown. They advance more because there are more opportunities when they cast their job search more broadly than if they never moved out of their home city. On the other hand, there might be a self-selection bias going on. People who decide to move may be people who have more experience in the first place. A person who is 30 years old may be less likely to have moved out of state than a person who is 40 years old simply because each year adds to the time in which such a move could occur. Also, people who are 40 or 50 years old may have higher occupational prestige simply because they have had more time to gain experience. Because age could be related to both the chances of living in the same place that respondents did at age 16 and to their prestige, we should control for age. Ideally, we would identify several other variables that need to be controlled, but for now we will just use age.

Instead of using the one-way ANOVA, use the full ANOVA procedure with the dataset `gss2006_chapter9.dta`. Open the dialog box by selecting Statistics ▷ Linear models and related ▷ ANOVA/MANOVA ▷ Analysis of variance and covariance. This dialog box is more complicated than the one-way ANOVA dialog box. First, type `prestg80` as the *Dependent variable*. In the *Model* box, type `mobile16 age` because these are the predictor and the covariate. If we had more covariates, we would add them here. If you click on the *Examples...* button, Stata will give you several examples of things you can enter in the *Model* box. Under *Model variables*, click on *Categorical except the following continuous variables*, and type `age`, which is the quantitative covariate. Also make sure that *Partial* is checked under *Sums of squares*. The partial sum of squares automatically allocates to each predictor and covariate the portion of the variance that it uniquely explains. We will discuss the sequential approach (often called hierarchical or nested) when we cover multiple regression. The filled-in dialog box is shown in figure 9.5.

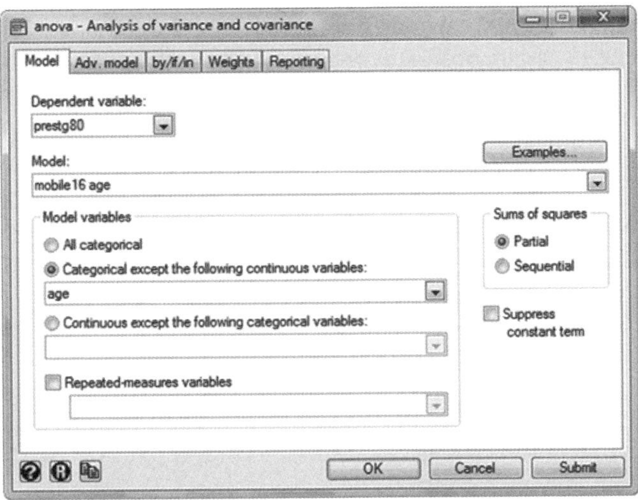

Figure 9.5. Analysis of variance and covariance dialog box

Under the by/if/in tab, enter the restriction `age > 29 & age < 60 & wrkstat==1`. This restricts the analysis to adults who are between 30 and 60 years old and who are working full time. Here is what we get:

```
. anova prestg80 mobile16 age if age > 29 & age < 60 & wrkst==1,
> continuous(age) partial
                         Number of obs =     1090     R-squared     =  0.0183
                         Root MSE      =  13.6511     Adj R-squared =  0.0156

                Source   Partial SS    df       MS            F     Prob > F

                 Model    3777.3325     3   1259.11083       6.76     0.0002

              mobile16   3750.73122     2   1875.36561      10.06     0.0000
                   age   61.2584614     1   61.2584614       0.33     0.5665

              Residual   202378.925  1086   186.352602

                 Total   206156.258  1089   189.307858
```

Notice what happens when we add `age` as a covariate. The table has different names for the `Sources` from those it did for the one-way ANOVA. What was called the `Between groups` is now given the name of the categorical variable, `mobile16`. In the ANCOVA table, we also have the covariate, `age`, listed as a source. Finally, we have what was called `Within Group`, now called `Residual`. A lot of different terminology has been applied to ANOVA. Even within a package like Stata, the same thing can be given different names.

The overall model is statistically significant: $F(3, 1086) = 6.76$, $p < 0.001$. The model includes both the predictor (mobility) and the covariate (age). The predictor—mobility—is still significant when we control for age. In the one-way ANOVA, $F(2, 1087)$

$= 9.98$, $p < 0.001$. In our ANCOVA, $F(2, 1086) = 10.06$, $p < 0.001$. The covariate, `age`, is not significant: $F(1, 1086) = 0.33$, p not significant. With hindsight, we can see that we did not need to control for age. The important thing is that the effect of mobility is robust in that we can control for age and the mobility effect is still significant.

When we discussed one-way ANOVA, we said that you can compute a measure of association called η^2 or R^2 by dividing the between-group sum of squares by the total sum of squares. Notice that whenever there is more than one predictor, we use a capital R^2 rather than a lowercase r^2. This is just a convention, but it is important to follow, or readers will be confused about whether you have one predictor or a more complicated model. We can compute this measure of association here. Stata reports $R^2 = 0.02$, but this is *not* what we usually want. The R^2 that Stata reports is for the entire model and includes both the effect of the predictor and the effect of the covariate. Because we really want to know if the predictor (mobility) has an effect after we control for the covariate, the ANCOVA table has the column labeled `Partial SS`, which is the sum of squares contributed uniquely by each source. To get the η^2 or R^2 for mobility, we divide the partial SS for mobility by the total SS, $3750.73/206156.26 = 0.02$. The results are the same as the overall R^2 after rounding because the covariate has little impact. If the covariate were significant, there might be a big difference between the R^2 Stata reports for the overall model and the R^2 we get for the unique (partial) effect of mobility, controlling for age.

Estimating the effect size and omega-squared, ω^2

Some fields use a measure of effect size based on ω^2 (omega-squared) as a measure of association when doing analysis of variance. They also compute

$$\text{Effect size} = \frac{\omega^2}{1-\omega^2}$$

Stata does not estimate either ω^2 or effect size for us. You can obtain them by using the user-written command `omega2`. Type the command `findit omega2`, and install the command `omega2`. After doing an `anova` command, you run the command `omega2`. This does not work after running the `oneway` command. Omega-squared is often small: A value of 0.01 is a small effect, 0.06 is a moderate effect, and 0.14 is a strong effect. Corresponding values for the effect size are 0.10, 0.25, and 0.40.

`omega2` is fairly simple to use. When you install this command, Stata saves it to a specific directory, `c:\ado\plus\o\omega2.ado`. If you open this file in a text editor such as Notepad, you will realize that we have not given you enough information yet to write your own programs and that the command is using the saved statistics we discussed before in a box about computing the correlation ratio. The authors were able to add a useful command to Stata with what is actually a simple program. If you become a regular user of Stata, you will soon be writing your own commands!

Immediately after running the ANCOVA on page 199, enter the command `omega2 mobile16`, which produces

```
. omega2 mobile16
omega squared for mobile16 = 0.0164
fhat effect size = 0.1290
```

This gives you ω^2 for mobility and the effect size for mobility.

Although `omega2` does not work for the one-way ANOVA command, you can always do a one-way ANOVA by using the full `anova` command rather than the `oneway` command. Then, after doing the one-way ANOVA using the `anova` command, you can run the `omega2` command.

A final thing we can do with ANCOVA is to estimate adjusted means. This is an extremely important thing to do when the covariate or covariates are statistically significant. Adjusted means for the dependent variable are computed for each level of the predictor. These adjusted means adjust for the covariate or covariates. They tell us what the means on the dependent variable would be for each category of our independent variable if the participants were equal on the covariates. Saying "if the participants were equal" begs the question of equal at what level. By default, Stata makes them equal at the mean of the covariate. The overall mean age is $M = 47.1$, so Stata makes a linear estimation of what the mean would be if people in all categories of the independent variable were 47.1 years old.

Immediately after running the command to do the ANCOVA, open the dialog box for postestimation of adjusted means.We open the dialog box by selecting Statistics ▷ Postestimation ▷ Adjusted means and proportions. Under *Compute and display predictions for each level of variables*, enter the categorical independent variable, `mobile16`. Under *Variables to be set to their overall mean value*, enter the quantitative covariate, age. The resulting dialog box is shown in figure 9.6.

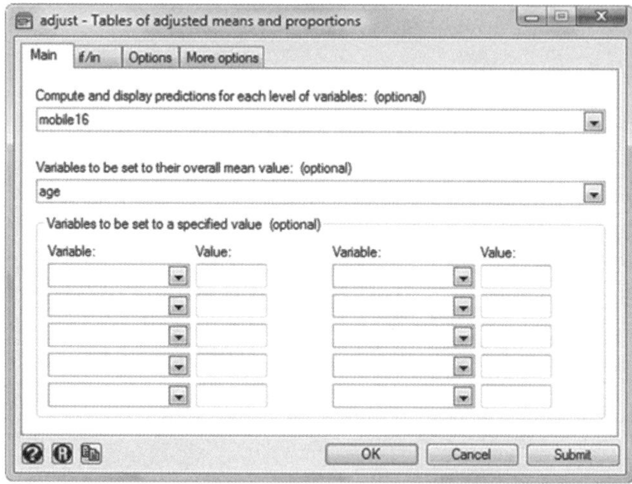

Figure 9.6. The Main tab for estimating adjusted means

Switching to the Options tab, check *Linear prediction* and *Confidence or prediction intervals*. Under *Generate new variables*, click on *Prediction variable* and type `adj_mean` in the box next to it; this is a new variable we will use later. Next, switching to the More options tab, enter a *Prediction label*, `Adjusted Means`; enter a *Confidence-interval label*, `Confidence Interval`; and check *Left-align column labels*.

The resulting command creates a new variable, `adj_mean`, that is the estimated mean on the dependent variable for each category of the predictor variable, after you adjust for the effect of the covariate. This is summarized in the following output:

```
. adjust age, by(mobile16) xb ci generate(adj_mean) label(Adjusted Means)
> cilabel(Confidence Interval) left
```

```
     Dependent variable: prestg80     Command: anova
       Created variable: adj_mean
  Covariate set to mean: age = 47.134835
```

GEOGRAPHIC MOBILITY SINCE AGE 16	xb	lb	ub
SAME CITY	44.1993	[42.8235	45.5751]
SAME ST,DIF CITY	48.7834	[47.107	50.4597]
DIFFERENT STATE	47.2828	[45.9419	48.6238]

```
     Key:  xb          =  Adjusted Means
           [lb , ub]   =  [95% Confidence Interval]
```

The first column of means, `xb`, is the estimated adjusted mean, as defined in the `Key` at the bottom of the table. The columns labeled `lb` and `ub` are the lower and upper bounds of the confidence intervals for the adjusted means. The group differences in prestige are consistent with what we had before when we made the adjustment for age. This table may be difficult to read, and you may want to compare the adjusted means with the unadjusted means. We might want to create a better table using Stata's `table` command. Open the dialog box by selecting Statistics ▷ Summaries, tables, and tests ▷ Tables ▷ Table of summary statistics (table). On the Main tab, enter the categorical independent variable, `mobile16`, under *Row variable*. Under *Statistics*, select `Mean` in the first row, and enter `adj_mean` under *Variable* (the postestimation command created `adj_mean` as a new variable). In the second row, select `Mean`, and under *Variable*, enter the unadjusted dependent variable, `prestg80`. Because we restricted our analysis to people based on age and work status, we need to include those restrictions in this dialog box. Click on the by/if/in tab and enter the following restriction: `age > 29 & age < 60 & wrkstat == 1`. If any of our variables had a name with more than eight characters, we would need to use the options on the Options tab to make the columns wider. The resulting table is

```
. table mobile16 if age > 29 & age < 60 & wrkstat==1,
> contents(mean adj_mean mean prestg80)
```

GEOGRAPHIC MOBILITY SINCE AGE 16	mean(adj_mean)	mean(prestg80)
SAME CITY	44.19931	44.11616
SAME ST,DIF CITY	48.78338	48.67286
DIFFERENT STATE	47.28284	47.18588

Although the adjusted means (`adj_mean`) are different from the unadjusted means, `prestg80`, their ordering is preserved. The difference between the adjusted means and the unadjusted means is trivial in this example. Can you guess why? The age of a person is not an important covariate. Indeed, correlating age with prestige results in an $r = -0.03$. However, if we had controlled for an important covariate, namely, a covariate that was highly correlated with prestige and mobility classification, then there would be substantial differences between the adjusted and the unadjusted means.

How do I control for a binary covariate?

In this example, we controlled for `age`, which is a quantitative, interval-level covariate. Sometimes we may want to control for a binary variable such as gender. Stata assumes that all covariates are quantitative variables, so we would code gender using 0s and 1s for men and women, respectively, and then we would enter it in the dialog box as a continuous variable.

There is one problem with this approach. The adjusted means are adjusting for the covariates by fixing them at their means. It makes sense to fix a variable, such as age, at its mean to adjust for differences in the average age of the groups being compared. It makes less sense to fix a dichotomous variable at its mean. Still, many researchers will use this strategy, and when you do not have the ability to randomly assign people to your groups, your options are limited.

Another way of computing adjusted means is available. In figure 9.6, there is the option to fix some of the variables at specific values. If we were controlling for gender as well as age, we would change the information in the Main tab by first fixing gender at `1` and doing a table and then by rerunning the command fixing gender at `0`. We would then have two tables, one for women and one for men.

9.6 Two-way ANOVA

It is possible to extend one-way ANOVA to two-way ANOVA, in which you have a pair of categorical predictors. Using our example of the relationship between prestige and mobility, we might think this relationship varies by gender. With two categorical predictors, we have three hypotheses. First, we want to continue testing whether prestige is related to mobility. Second, we want to test if prestige is related to gender. That is, do men or do women have higher occupational prestige? Third, and sometimes most importantly, we might want to see if there is an interaction effect, which occurs when the effect of one variable, say, mobility, is contingent on another, say, gender. For example, interaction would be evident if mobility had a stronger or weaker effect for men than it had for women.

First, let's look at the mean score on prestige by mobility and by gender. Open the dialog box by selecting Statistics ▷ Summaries, tables, and tests ▷ Tables ▷ One/two-way table of summary statistics. Type `mobile16` as *Variable 1* (this will be the row variable) and `sex` as *Variable 2* (this will be the column variable). Then type `prestg80` as the *Summarize variable*. The resulting dialog box is shown in figure 9.7.

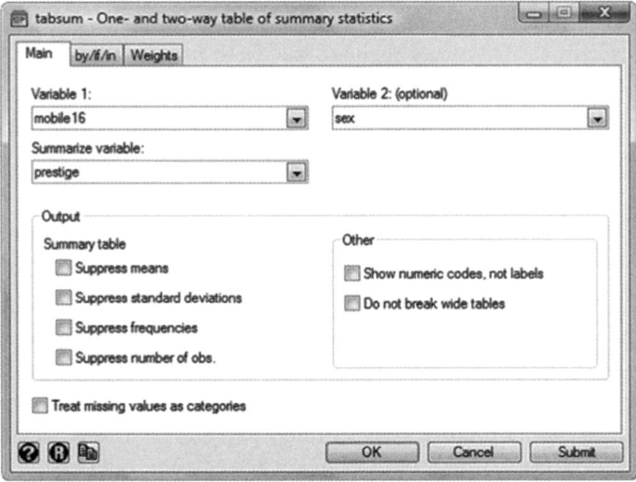

Figure 9.7. The Main tab for one- and two-way tables of means

Click on the by/if/in tab, and make sure to restrict the sample to people who are age 30–59 and who work full time (`age > 29 & age < 60 & wrkstat==1`). Clicking on OK results in the following command and output:

```
. tabulate mobile16 sex if age > 29 & age < 60 & wrkstat==1, summarize(prestg80)
                    Means, Standard Deviations and Frequencies
                     of RS OCCUPATIONAL PRESTIGE SCORE  (1980)

GEOGRAPHIC |
  MOBILITY |
 SINCE AGE |   RESPONDENTS SEX
        16 |      MALE     FEMALE |     Total
-----------+----------------------+----------
 SAME CITY | 43.521505  44.642857 | 44.116162
           | 13.179512  14.218428 |  13.73443
           |       186        210 |       396
-----------+----------------------+----------
 SAME ST,D |  47.70922  49.734375 | 48.672862
           | 13.843275   11.97207 | 13.002183
           |       141        128 |       269
-----------+----------------------+----------
 DIFFERENT | 46.697581  47.870056 | 47.185882
           | 14.124134  13.735739 | 13.959368
           |       248        177 |       425
-----------+----------------------+----------
     Total | 45.918261  47.017476 | 46.437615
           | 13.837607   13.66062 | 13.758919
           |       575        515 |      1090
```

This table provides descriptive answers to the three research hypotheses. First, mobility does seem to be related to prestige. The `Total` column on the right shows that the mean prestige grows from 44.12 to 48.67 and finally to 47.19, depending on the person's level of mobility. There also appears to be a gender effect, with women having higher prestige, on average, than men. The mean prestige for women is 47.02 and for men the mean is 45.92. Is the difference sufficient to be statistically significant?

Finally, there may be some interaction. For both women and men, the greatest prestige comes with moderate mobility. However, the effect of mobility appears to be slightly greater for women than it is for men. For example, women who live in the same city score just over 5 points below women who moved to a different city in the same state. For men this difference is just over 4 points. Is this enough of a difference to say there is interaction? This seems like little interaction, but we need to test to see if this assessment is correct.

To run the two-way ANOVA, open the same *Analysis of Variance and Covariance* dialog box used previously for ANCOVA (figure 9.5). Type `prestg80` for the *Dependent variable* and `mobile16 sex mobile16*sex` under *Model*. These correspond to the three hypotheses that (a) mobility has an effect, (b) gender has an effect, and (c) the interaction between mobility and gender has an effect. `mobile16*sex` is the interaction term. We do not need to generate a new variable for `mobility*sex`; we just need to list it this way in the dialog box. Also click on the by/if/in tab, and enter the restriction that `age > 29 & age < 60 & wrkstat==1`. The results are

```
. anova prestg80 mobile16 sex mobile16*sex if age > 29 & age < 60 & wrkstat==1,
> partial

                           Number of obs =     1090     R-squared     =  0.0207
                           Root MSE      = 13.6475     Adj R-squared =  0.0161

                  Source |  Partial SS    df       MS           F     Prob > F
              -----------+----------------------------------------------------
                   Model |  4257.25292     5  851.450584       4.57     0.0004
                         |
                mobile16 |  3883.99466     2  1941.99733      10.43     0.0000
                     sex |  537.182245     1  537.182245       2.88     0.0897
            mobile16*sex |  38.9227025     2  19.4613512       0.10     0.9008
                         |
                Residual |  201899.005  1084  186.253695
              -----------+----------------------------------------------------
                   Total |  206156.258  1089  189.307858
```

We can interpret this table in the same way we interpreted the ANCOVA table. Mobility is statistically significant: $F(2,1084) = 10.43$, $p < 0.001$. Gender is not significant: $F(1,1084) = 2.88$. Thus the differences in means by level of mobility are significant, but the gender differences are not. The interaction between mobility and gender is also not significant: $F(2,1084) = 0.10$. Even though inspecting the table of means made us think that both gender and the interaction might be statistically significant, Stata is telling us that they are not significant. Remember, an F statistic has two separate degrees of freedom, one for the numerator and one for the denominator. For mobility, the numerator is the MS for `mobile16` (1,941.997), and this has 2 degrees of freedom. The denominator is the MS `Residual` of 186.254, and this has 1,084 degrees of freedom. The F is simply the ratio of these two MSs, that is $F = 1941.997/186.254 = 10.43$ for `mobile16`.

I will now show you how to create a graph to represent the results of our two-way ANOVA. Normally, this would be done if the interaction were significant to show the nature of the interaction. We do it here solely to demonstrate how to do it. To represent the relationships in a graph, you first have to compute the predicted prestige score. Open the postestimation dialog box by selecting Statistics ▷ Postestimation ▷ Adjusted means and proportions. Leave the Main tab blank. In the Options tab, check *Linear prediction*. Near the bottom, check that there is a *Prediction variable*, and enter the name of the variable we are predicting, `prestige`. Click on Submit, and Stata will compute the appropriate values for the graph. This shows up as a new variable, `prestige`, in your Variables window. The resulting command is

```
. adjust, xb generate(prestige)
```

The graph can be a powerful tool, but this graph involves several steps. First, open the appropriate dialog box by selecting Graphics ▷ Twoway graph (scatter, line, etc.). This dialog allows us to plot the relationship between mobility and prestige for men and then to create a second plot of the same relationship for women, which we will overlay on the first graph. Click on *Create....* From the *Basic plots* options, select *Connected*. Under the *Y variable*, enter the variable we just generated, `prestige`. Under the *X variable*, enter the independent variable, `mobile16`. Make sure to check the box for *Sort on x variable*. Under the if/in tab, for the *If: (expression)*, enter the restriction we had for our analysis plus `sex==1`: `age > 29 & age < 60 & wrkstat==1 & sex==1`. Click on Accept to return to the Plots tab. Click on *Create...* to create *Plot 2*, doing everything the same as you did for *Plot 1*, except this time use `sex==2` in the restrictions.

At this point you can either continue to work with the dialog box options for titles, the legend, overall, markers, etc., or you can Submit what you have done and use the Graph Editor. The advantage of continuing to work with the dialog box is that it will produce the Stata graph command for this graph that you can use later, if needed. The advantage of working with the Graph Editor is that it is probably easier to use and has more flexibility.

The resulting graph after using the Graph Editor to make it easier to understand is shown in figure 9.8. This graph may be easier to understand than the table of means. You should present the table of means, this graph, or both based on your target audience. The graph shows that women have higher prestige than men, regardless of how mobile they are, but be cautious because this difference is not statistically significant. The gender gap is greatest for people who have moved to a different city, but still live in their state of origin. With extreme interaction, the lines might even cross. This would happen if, for example, men who were more mobile had greater prestige than women who were more mobile, but men who were not mobile had less prestige than women who were not mobile. It would be helpful to do a hand drawing of such an interaction to help clarify the meaning of the interaction.

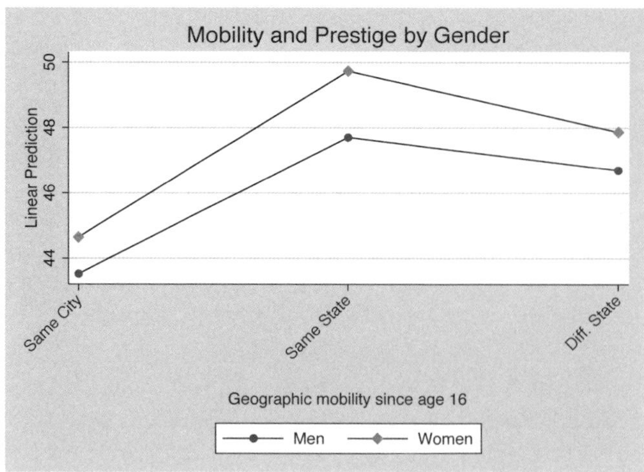

Figure 9.8. Two-way overlay graph of the relationship between prestige and mobility for women and men

9.7 Repeated-measures design

Chapter 7 included a section on repeated-measures t tests. These were useful when you had a before–after design in which participants were measured before an intervention and then the same participants were measured again after the intervention. We use repeated-measures ANOVA when there are more than two measurements. For example, we might want to know if the effect of the intervention was enduring. With only a before and after measurement, we do not know what happens in the weeks or months following the experiment. We might have three time points, namely, pretest (before the intervention), posttest (shortly after the intervention is completed), and follow-up (another measurement a couple of months later). Students can think of many subjects where the follow-up measure would be revealing. They would have scored poorly on the final examination if they took it before taking the class. They would score much better when they took the final examination after taking the course. However, they might score poorly again if they took the same test again a few months after completing the course. This is a special problem with courses that emphasize memorization. A repeated-measures ANOVA allows us to test what happens over time.

Here are some hypothetical data for students in a small class as they might be presented in a textbook (`wide9.dta`).

```
. list

     +-------------------------------+
     | id   test1   test2   test3 |
     |-------------------------------|
  1. |  1      55      85      80 |
  2. |  2      65      90      85 |
  3. |  3      34      70      71 |
  4. |  4      55      75      65 |
  5. |  5      61      59      65 |
     |-------------------------------|
  6. |  6      79      94      85 |
  7. |  7      63      59      59 |
  8. |  8      45      65      50 |
  9. |  9      54      70      60 |
 10. | 10      69      90      82 |
     +-------------------------------+
```

Stata refers to this arrangement as a "wide" layout of the data, but Stata prefers a "long" arrangement. You could make this conversion by hand with a small dataset like this, but it's easier to use the `reshape` command. Many find this command confusing, and I will illustrate only this one application of it. A detailed explanation appears in http://www.ats.ucla.edu/stat/stata/modules/reshapel.htm. For our purposes, the command to reshape the data into a long format is `reshape long test, i(id) j(time)`.

This looks strange until we analyze what is being done. The `reshape long` lets Stata know that we want to change from the wide layout to a long layout. This is followed by a variable name, `test`, that is not a variable name in the wide dataset. However, `test` is the prefix for each of the three measures: `test1`, `test2`, and `test3`. If we had used names like `pretest`, `posttest`, and `followup`, we would have needed to rename the variables so that they would have had the same stubname and then a number. We can do this by opening the Data Editor, double-clicking on the variable name we want to change, and then generating the commands `rename pretest test1`, `rename posttest test2`, and `rename followup test3`. We could also right-click on the variable name in the Variables window, and this would give us an option to rename the variable.

After the comma in the `reshape` command are two required options. (Stata calls everything after the comma an option, even if it is required for a particular purpose.) The `i(id)` will always be the ID number for each observation. The parentheses would contain whatever variable name was used to identify the individual observations. Here we use `id`, but another dataset might use a different name, such as `ident` or `case` (if you do not have an identification variable, run the command `gen id` = _n before using the `reshape` command). The `j(time)` option refers to the time of the measurement. This option creates a new variable, `time`, that is scored a `1` if the score is for `test1`, a `2` for `test2`, or a `3` for `test3`. The choice of the name `time` is arbitrary. Here are the command and its results:

(Continued on next page)

```
. reshape long test, i(id) j(time)
(note: j = 1 2 3)

Data                                    wide   ->   long
-----------------------------------------------------------------------------
Number of obs.                            10   ->      30
Number of variables                        4   ->       3
j variable (3 values)                          ->   time
xij variables:
                             test1 test2 test3  ->   test
-----------------------------------------------------------------------------

. list

     +-----------------+
     | id   time   test |
     |-----------------|
  1. |  1      1     55 |
  2. |  1      2     85 |
  3. |  1      3     80 |
  4. |  2      1     65 |
  5. |  2      2     90 |
     |-----------------|
  6. |  2      3     85 |
  7. |  3      1     34 |
  8. |  3      2     70 |
  9. |  3      3     71 |
 10. |  4      1     55 |
     |-----------------|
 11. |  4      2     75 |
 12. |  4      3     65 |
 13. |  5      1     61 |
 14. |  5      2     59 |
 15. |  5      3     65 |
     |-----------------|
 16. |  6      1     79 |
 17. |  6      2     94 |
 18. |  6      3     85 |
 19. |  7      1     63 |
 20. |  7      2     59 |
      (output omitted)
 26. |  9      2     70 |
 27. |  9      3     60 |
 28. | 10      1     69 |
 29. | 10      2     90 |
 30. | 10      3     82 |
     +-----------------+
```

If you are uncomfortable with the `reshape` command, you can always enter the data yourself using the long format. `id` is repeated three times for each observation, corresponding to the times `1`, `2`, and `3` that appear under `time`. The last column is the variable stub `test` with `55` being how person `1`, `id` = 1, did on `test1` at `time` = 1; `85` being how person `1` did at time `2`; and `80` being how person `1` did at time `3`. You can see how the long format appears as a wide format to check that you have either entered the data or done the `reshape` command correctly. In this example, the command is `tabdisp id time, cellvar(test)` (the results are not shown here).

Once the data are in this long arrangement, doing the repeated-measures ANOVA is fairly simple. Open the dialog box by selecting Statistics ▷ Linear models and related ▷ ANOVA/MANOVA ▷ Analysis of variance and covariance. Under *Dependent variable*, type `test`. Under *Model*, type `id time` because we want to estimate individual differences using the variable `id` and differences between the three measures using the variable `time`. Individual differences are important because some people will do better than others regardless of whether it is a pretest, posttest, or follow-up test. However, say that we are most interested in the differences between the three measurements represented by time. We think that people will do poorly the first time (pretest), much better the second time (posttest), and then drop back down somewhat the third time (follow-up).

Check the *All categorical* option because the independent variables, `id` and `time`, should be treated as categorical. For example, the second person is not higher than the first person, which would be true if `id` were continuous. At the bottom of the dialog, check the box by *Repeated-measures variables*, and enter the variable `time` because the measures are repeated over the three times. The completed dialog appears in figure 9.9.

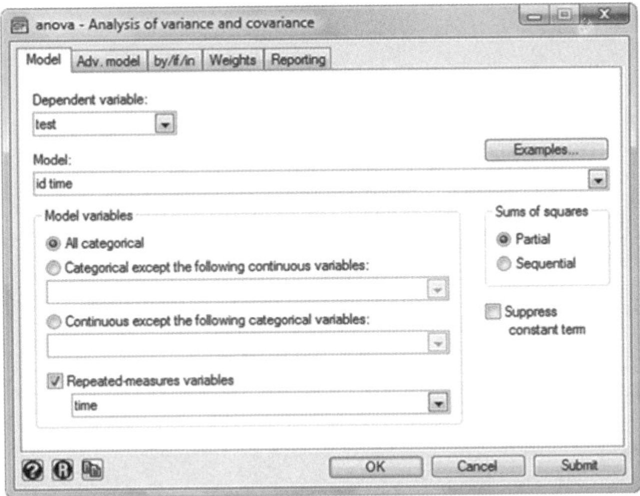

Figure 9.9. The Model tab for repeated-measures ANOVA

Submit this dialog to receive the following results:

(*Continued on next page*)

```
. anova test id time, repeated(time) partial
                          Number of obs =      30     R-squared     =  0.8284
                          Root MSE      = 7.56233     Adj R-squared =  0.7235

                  Source |  Partial SS    df       MS           F     Prob > F
              -----------+----------------------------------------------------
                   Model |  4969.56667    11  451.778788       7.90     0.0001
                         |
                      id |      3328.3     9  369.811111       6.47     0.0004
                    time |  1641.26667     2  820.633333      14.35     0.0002
                         |
                Residual |      1029.4    18  57.1888889
              -----------+----------------------------------------------------
                   Total |  5998.96667    29   206.86092

Between-subjects error term:  id
                     Levels:  10        (9 df)
     Lowest b.s.e. variable:  id

Repeated variable: time
                                          Huynh-Feldt epsilon        =  0.7848
                                          Greenhouse-Geisser epsilon =  0.6969
                                          Box's conservative epsilon =  0.5000

                                            ------------ Prob > F ------------
                  Source |     df      F    Regular    H-F      G-G      Box
              -----------+----------------------------------------------------
                    time |      2   14.35   0.0002   0.0007   0.0012   0.0043
                Residual |     18
              -----------+----------------------------------------------------
```

You should ignore the results for `id` because we are interested in the `time` variable instead. The test for the `time` variable provides the test of whether the three scores for each person differ significantly. We can see that $F(2,18) = 14.35$, $p < 0.001$.

We can obtain a listing of the means for each time by using the following command:

```
. table time, contents(mean test)

----------------------
     time | mean(test)
----------+-----------
        1 |         58
        2 |       75.7
        3 |       70.2
----------------------
```

There are serious limitations of the repeated-measures ANOVA. This results from the scores on the test for each time not being independent. A person who has a high score on `test1` is likely to also have relatively high scores on `test2` and `test3`. Similarly, a person who has a low score on one of these is likely to have a low score on all three. Stata provides three adjustments for this lack of independence: the Huynh–Feldt ϵ (pronounced epsilon), the Greenhouse–Geisser ϵ, and the conservative Box ϵ. See the *Stata Base Reference Manual* under the `anova` command for more information. We see four probability values for the F test. The `Regular` $p = 0.0002$, the Huynh–Feldt ϵ has a $p = 0.0007$, the Greenhouse–Geisser ϵ has a $p = 0.0012$, and the Box conservative ϵ has a $p = 0.0043$. Each of these is statistically significant, at least at the $p < 0.01$ level.

There are several assumptions that can be made about the covariances of the measures across time, and these may be used explicitly in a more complex analysis of repeated measures. One approach is using multivariate analysis of variance, `manova`, in which the three tests are treated as three dependent variables. The best solution is to use the `xtmixed` command. Unfortunately, a discussion of MANOVA and mixed models is beyond the scope of this book. It will be necessary to use the `manova` or `xtmixed` command if you have a large sample.

9.8 Intraclass correlation—measuring agreement

Intraclass correlation, ρ_I (pronounced rho-sub i), is a measure of agreement within a group of people. The group might be just two people, such as a wife and her husband, and you want to know how much spouses agree. A correlation coefficient, r, does not measure agreement in an absolute sense. If each husband wanted to spend exactly twice as much time watching televised football games on TV as his wife, the correlation would be $r = 1$ because each husband's score is perfectly predictable from his wife's score. However, they do not agree with each other. Stata has a special ANOVA command that produces the intraclass agreement for however many members there are in a group.

Suppose that we are doing a study of group dynamics. We randomly assign three people to each of 10 groups and have them discuss reforming Medicare. After 30 minutes of discussing the issue, each person completes a questionnaire that includes a scale measuring attitude toward welfare reform. Assume that one member of group 7 got sick, so we would have only two scores for group 7 on their attitudes toward reforming Medicare. You enter your data using the "long" format, and the data look like this (`intraclass.dta`):

(Continued on next page)

```
. list

     +------------------+
     | medicare   group |
     |------------------|
  1. |       21       1 |
  2. |       22       1 |
  3. |       22       1 |
  4. |       17       2 |
  5. |       16       2 |
     |------------------|
  6. |       15       2 |
  7. |       15       3 |
  8. |       16       3 |
  9. |       18       4 |
 10. |       19       4 |
     |------------------|
 11. |       20       4 |
 12. |       12       5 |
 13. |       12       5 |
 14. |       14       5 |
 15. |       14       6 |
     |------------------|
 16. |       21       6 |
 17. |       24       6 |
 18. |       23       7 |
 19. |       22       7 |
 20. |       26       8 |
  (output omitted)
 26. |       35      10 |
 27. |       33      10 |
 28. |       45      10 |
     +------------------+
```

The groups, numbered 1 to 10, each appear three times, with the exception of group 7 because one of its members got sick and did not complete the questionnaire. The three people in the first group have scores of 21, 22, and 22. The order of members is arbitrary in this example. If the group process leads to agreement among group members, we would expect the variance within each group to be small—all of them would have similar scores. By contrast, we would expect there to be considerable variance across groups. Compare the three people in group 1 with the three people in group 2, and you can see that most of the variance is between groups. If everybody within each group agrees with his or her other group members and all the differences are between groups, then the intraclass correlation will be 1.0.

The command for estimating the intraclass correlation is `loneway medicare group`. The lowercase letter "l" (not the number one) begins the command's name. The quantitative outcome variable, attitude toward Medicare, appears next. Finally, we have the grouping variable, `group`. Remember, the grouping variable may be a family ID, a group ID, a cluster of cases, or anything else that divides the sample into groups. We simply have the data entered so everybody in the first group appears at the top, everybody in the second group appears just below them, and so on. Submitting this command, we obtain these results:

```
. loneway medicare group
  One-way Analysis of Variance for medicare: Attitude toward Medicare Reform
                                              Number of obs =        28
                                                  R-squared =    0.8887

    Source                    SS          df       MS          F     Prob > F
-------------------------------------------------------------------------------
Between group              1285.9643       9    142.88492     15.97     0.0000
Within group                     161      18    8.9444444
-------------------------------------------------------------------------------
Total                      1446.9643      27     53.59127

         Intraclass        Asy.
         correlation       S.E.       [95% Conf. Interval]
         ------------------------------------------------
            0.84277      0.08207       0.68192     1.00363

         Estimated SD of group effect               6.924204
         Estimated SD within group                  2.990726
         Est. reliability of a group mean            0.93740
              (evaluated at n=2.79)
```

The only interesting part of this output is the intraclass correlation. Stata reports $\rho_I = 0.843$ and reports a confidence interval. Because the value of zero is not included in the confidence interval, we know that ρ_I is significantly greater than zero.

9.9 Summary

ANOVA is an extremely complex and powerful approach to analyzing data. We have touched only the surface here. Stata offers a powerful collection of advanced procedures that extend traditional ANOVA. In this chapter, you have learned how to

- Conduct a one-way ANOVA as an extension of the independent t test to three or more groups
- Use nonparametric alternatives to ANOVA for comparing distributions or comparing medians when you have three or more groups
- Control for a covariate through ANCOVA when you do not have randomization but know how the groups differ on background variables
- Use two-way ANOVA for situations in which you have two categorical predictors that might or might not interact with one another
- Graph interaction effects for two-way ANOVA
- Perform repeated-measures ANOVA as an extension of the dependent/paired t test for when you have three or more measurements on each observation
- Calculate the intraclass correlation coefficient to measure agreement or homogeneity within groups

The next chapter presents multiple regression. Many of the models we have fitted within an ANOVA framework can be done equally well or better within a multiple-regression format. Multiple regression is an extremely general approach to data analysis.

9.10 Exercises

1. Suppose that you have five people in condition A, five in condition B, and five in condition C. On your outcome variable, the five people in condition A have scores of 9, 8, 6, 9, and 5, respectively. The five in condition B have scores of 5, 9, 6, 4, and 8, respectively. The five in condition C have scores of 2, 5, 3, 4, and 6, respectively. Show how you would enter these data using a wide format. Show how you would enter them using a long format.

2. Using the data in the long format from exercise 1, do an ANOVA. Compare the means and the standard deviations. What does the F test tell us? What do the multiple-comparison tests tell us?

3. Use the `gss2002_chapter9.dta` dataset. Does the time an adult woman watches TV (`tvhours`) vary depending on her marital status (`marital`)? Do a one-way ANOVA, including a tabulation of means; do a Bonferroni multiple-comparison test; and then present a bar chart showing the means for hours spent watching TV by marital status. Carefully interpret the results. Can you explain the pattern of means and the differences of means? Do a tabulation of your dependent variable, and see if you find a problem with the distribution.

4. Use the `gss2002_chapter9.dta` dataset. Does the time an adult watches TV (`tvhours`) depend on his or her political party identification (`partyid`)? Do a tabulation of `partyid`. Drop the 48 people who are Other party. Combine the strong Democrats, not strong Democrats, and independent near Democrats into one group and label it Democrats. Combine the strong Republicans, not strong Republicans, and independent near Republicans into one group and label it Republicans. Keep the independents as a separate group. Now do a one-way ANOVA with a tabulation and Bonferroni test to answer the question of whether how much time adults spend watching TV depends on their party identification. Carefully interpret the results.

5. Use the `gss2002_chapter9.dta` dataset. Repeat exercise 3 using the Kruskal–Wallis rank test and a box plot. Interpret the results, and compare them with the results in exercise 3.

6. Use the `nlsy97_selected_variables.dta` dataset. A person is interested in the relationship between having fun with your family (`fun97`) and associating with peers who volunteer (`pvol97`). She thinks adolescents who frequently have fun with their families will associate with peers who volunteer. The fun-with-family variable is a count of the number of days a week an adolescent has fun with his or her family. Although this is a quantitative variable, treat this as eight categories,

0–7 days per week. Because of possible differences between younger and older adolescents, control for age (`age97`). Do an analysis of covariance, and compute adjusted means. Present a table showing both adjusted and unadjusted means. Interpret the results.

7. In figure 9.8, there was no statistically significant interaction. Do a freehand drawing for each of the following hypothetical results: (a) women who are not mobile have higher prestige than men who are not mobile, but men have higher prestige for both of the other categories on mobility; (b) women have less prestige for each level of mobility, and there is absolutely no interaction; and (c) women and men have the same level of prestige, except for men who move to a different state having much higher prestige than women who move to a different state.

8. Use the `partyid.dta` dataset. Do a two-way ANOVA to see if there is (a) a significant difference in support for stem-cell research (`stemcell`) by party identification (`partyid`), (b) a significant difference in support by gender, and (c) a significant interaction between gender and party identification. Construct a two-way overlay graph to show the interaction, and interpret it.

9. Use the data you entered in the wide format in the first exercise. You will need to add an identification variable to the wide format, label the three tests appropriately, and then use `reshape` to transform the data to the long format. Do a repeated-measures ANOVA, and interpret the results.

10. You have five two-parent families, each of which has two adolescent children. You want to know if the parents agree more with each other about the risks of premarital sex or if the siblings agree more. The scores for the five sets of parents are (9, 3) (5, 5) (4, 2) (6, 7) (8, 10). The scores for the five sets of siblings are (9, 8) (7, 4) (8, 10) (3, 2) (8, 8). Enter these data in a long format, and compute the intraclass correlation for parents and the intraclass correlation for siblings. Which set of pairs has greater agreement regarding the risks of premarital sex?

10 Multiple regression

10.1 Introduction to multiple regression

Multiple regression is an extension of bivariate correlation and regression that opens a huge variety of new applications. Multiple regression and its extensions provide the core statistical technique for most publications in social science research journals. Stata is an exceptional tool for doing multiple regression because regression applications are at the heart of the original conceptualization and subsequent development of Stata. In this chapter, I introduce multiple regression and regression diagnostics. In the following chapter, I will introduce logistic regression.

10.2 What is multiple regression?

In bivariate regression, you had one outcome variable and one predictor. Multiple regression expands this by letting you have any number of predictors. This makes sense because few outcomes have just one cause. Why do some people have more income than others? Clearly, education is an important predictor, but education is only part of the story. Some people inherit great wealth and would have substantial income whether they had any education. The income people have tends to increase as they get older, at least up to some age where their income may begin to decline. The career people select will influence their income. A person's work ethic may also be important because those who work harder may earn more income. There are many more variables that influence a person's income, e.g., race, gender, and marital status. You need more than one variable to adequately predict and understand the variation in income.

Predicting and explaining income can be even more complicated than this. Some of the predictors may interact such that combinations of independent variables have unique effects. For example, a physician with a strong work ethic may make much more than a physician who is lazy. By contrast, an assembly-line worker who has a strong work ethic has little advantage over an assembly-line worker who is working only at the minimum required work rate. Hence, the effect of work ethic on income can vary depending on your occupation. Similarly, the effect of education on income may be different for women and men. Clearly, social science requires a statistical strategy that allows you to study many variables working simultaneously to produce an outcome; multiple regression provides that strategy.

This chapter will use data from a 2004 survey of Oregon residents, `ops2004.dta`. There is a series of 11 items concerning views on how serious a concern the person has about different environmental issues, such as pesticides, noise pollution, food contamination, air quality, and water quality. Using the alpha reliability procedures and factor analysis that is covered in chapter 12, I constructed a scale called `env_con` that has a strong alpha of 0.89. This will serve as our dependent variable; it is the outcome we want to predict.

Say that you are interested in several independent variables (predictors). These might include variables related to the person's health, education, income, and community identification. For example, we might expect people who have higher education and community identification to score higher on concern for environmental issues. Perhaps the expectation is less clear for how a person's health or income will be related to concerns for the environment, but we think they both might be important. Clearly, we have left out some important predictors, but these predictors will be sufficient to show how multiple regression analysis is done using Stata.

10.3 The basic multiple regression command

Select Statistics ▷ Linear models and related, and you will see a long list of regression-related commands. Stata gives us more options than we could possibly cover in this book. In this chapter, we focus on the ▷ Linear regression menu item. The basic dialog box for linear regression asks for the dependent variable in one dialog box and a list of our independent variables in another dialog box. Type `env_con` as the dependent variable, and then type the following independent variables: `educat` (years of education), `inc` (annual income in dollars), `com3` (identification with the community), `hlthprob` (health problems), and `epht3` (impact of the environment on the person's own health). The resulting dialog box appears in figure 10.1.

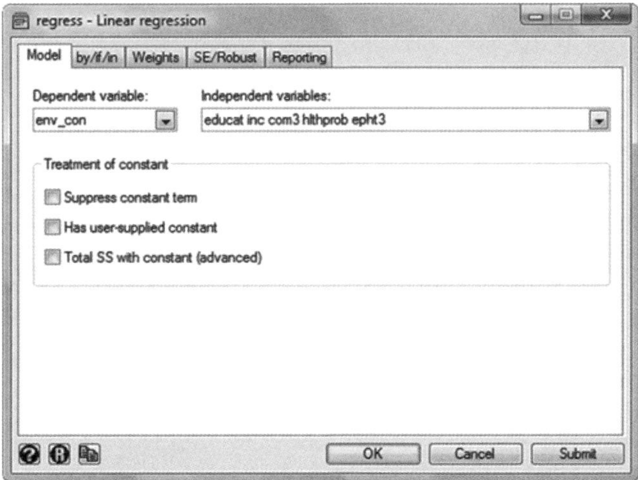

Figure 10.1. The Model tab for multiple regression

The by/if/in tab lets you fit the regression model by some grouping variable (you might estimate it separately for women and men), restrict your analysis to cases if a condition is met (you might limit it to people 18–40 years old), or apply your regression model to a subset of observations. We will go over the Weights tab and the SE/Robust tab later. The Reporting tab has several options for what is reported. Click on that tab, and check the option to show *Standardized beta coefficients*. This produces the simple command `regress env_con educat inc com3 hlthprob epht3, beta`. The syntax of this command is simple, and unless you want special options, you can just enter the command directly in the Command window. The command name, `regress`, is followed by the dependent variable, `env_con`, and then a list of the independent variables. After the comma comes the only option you are using, namely, `beta`. Here are the results:

```
. regress env_con educat inc com3 hlthprob epht3, beta

      Source |       SS       df       MS              Number of obs =    3769
-------------+------------------------------           F(  5,  3763) =  320.15
       Model |  647.67794     5  129.535588           Prob > F      =  0.0000
    Residual | 1522.55872  3763  .404613001           R-squared     =  0.2984
-------------+------------------------------           Adj R-squared =  0.2975
       Total | 2170.23666  3768  .575965144           Root MSE      =  .63609

------------------------------------------------------------------------------
     env_con |      Coef.   Std. Err.      t    P>|t|                     Beta
-------------+----------------------------------------------------------------
      educat |  -.0011841    .004077    -0.29   0.772                -.0044584
         inc |  -5.51e-08   3.62e-07    -0.15   0.879                -.0023317
        com3 |   .0503162   .0092717     5.43   0.000                  .074352
    hlthprob |  -.2974035   .0248129   -11.99   0.000                 -.172927
       epht3 |  -.4020741    .012687   -31.69   0.000                -.4575999
       _cons |   3.726345   .0651735    57.18   0.000                        .
------------------------------------------------------------------------------
```

The results have three sections. The upper left block of results is similar to what you saw in chapter 9 on analysis of variance. This time, the source called `Model` refers to the regression model rather than the between-group source. The regression model consists of the set of all 5 predictors. The source called `Residual` corresponds to the error component in analysis of variance. You have 5 degrees of freedom for the model. The degrees of freedom will always be k, the number of predictors (here `educat`, `inc`, `com3`, `hlthprob`, and `epht3`) in the model.

The analysis of variance does not show the F ratio like the tables in chapter 9. The F appears in the block of results in the upper right section. The output says $F(5, 3763) = 320.15$. The F of 320.15 is the ratio of the mean square for the model to the mean square for the residual. It is $129.535588/0.404613001 = 320.15$ in this example. The degrees of freedom for the numerator (`Model`) are 5, and for the denominator (`Residual`) they are 3,763. The probability of this F ratio appears just below the F value, Prob $> F = 0.0000$. When this probability is less than 0.05, you can say $p < .05$; when it is less than 0.01, you can say $p < 0.01$; and when it is less than 0.001, you can say $p < 0.001$. In a report, you would write this as $F(5, 3763) = 320.15$, $p < 0.001$. There is a highly significant relationship between environmental concerns and the set of 5 predictors.

How well does the model fit the data? This question is usually answered by reporting R^2. The regression model explains 29.8% of the variance in environmental concerns. Stata reports this as `R-squared = 0.2984`. In bivariate regression, the r^2 measured how close the observations were to the straight line used to make the prediction. With multiple regression, you use the capital R^2, which has the same meaning, except that this time it measures how close the observations are to the predicted value, based on the set of predictors. What is a weak or a strong value for R^2 varies by the topic being explained. If you are in a fairly exploratory area, an R^2 near 0.3 is considered reasonably good. A rule of thumb some researchers use is that an R^2 less than 0.1 is weak, between 0.1 to 0.2 is moderate, and greater than 0.3 is strong. Be careful in applying this rule of thumb because some areas of research require higher values and others require lower values. You might report our results as showing that we can explain

29.8% of the variance in environmental concern using our set of predictors, and this is a moderate-to-strong relationship.

On a small sample with several predictors, the value of R^2 can exaggerate the strength of the relationship. Each time you add a variable, you expect to increase R^2 just by chance. R^2 cannot get smaller as you add variables. When you have many predictors and a small sample, you may get a big R^2 just by chance. To offset this bias, some researchers report the adjusted R^2. This will be smaller than the R^2 because it attempts to remove the chance effects. When you have a large sample and relatively few predictors, R^2 and the adjusted R^2 will be similar, and you might report just the R^2. However, when there is a substantial difference between the two, you should report both values.

Across the bottom of the results is a block that has six columns containing the key regression results. A formal multiple regression equation is written as

$$\widehat{Y} = b_0 + b_1 X_1 + b_2 X_2 + \cdots + b_k X_k$$

$\widehat{Y}$ is the predicted value of the dependent variable. b_0 is the intercept or constant. Stata calls b_0 the `_cons`, which is an abbreviation for the constant. You can think of this as the base prediction of Y when all the X variables are fixed at zero. b_1 is the regression coefficient for the effect of X_1, b_2 is the regression coefficient for the effect of X_2, and b_k is the regression coefficient for the last X variable. You may see these coefficients called *unstandardized regression coefficients*. You can get the values for the equation from the output (see table 10.1).

Table 10.1. Regression equation and Stata output

Name in the regression equation	Stata name	Stata results: unstandardized coefficient
b_0	`_cons`	3.726
b_1	`educat`	-0.001
b_2	`inc`	-5.51e-08
b_3	`com3`	0.050
b_4	`hlthprob`	-0.297
b_5	`epht3`	-0.402

You can write this as an equation, calculating the estimated value of the dependent variable:

$$\widehat{\texttt{env\_con}} = 3.726 - 0.001(\texttt{educat}) - 0.000(\texttt{inc}) + 0.050(\texttt{com3}) - 0.297(\texttt{hlthprob}) - 0.402(\texttt{epht3})$$

The regression results next give a standard error, t value, and $P > |t|$ for each regression coefficient. If you divide each regression coefficient by its standard error, you obtain the t value that is used to test the significance of the coefficient. Each of the t ratios has $N - k - 1$ degrees of freedom. Subtract one degree of freedom for each of the $k = 5$ predictors and another degree for the constant. From the upper right corner of the results, you can see that there are 3,769 observations, and from the analysis of variance table, you can see that there are 3,763 degrees of freedom for the residual. If you forget the $N - k - 1$ rule for degrees of freedom, you can use the number for the residual in the analysis of variance table. You might want to include the degrees of freedom in your report, but you do not need to look up the probability in a table of t values because Stata gives you a two-tailed probability. For example, you might write that $b_5 = -0.402$, $p < 0.001$. In some fields, you would write $b_5 = -0.402$, $t(3763) = -31.69$, $p < 0.001$.

The unstandardized regression coefficients have a simple interpretation. They tell you how much the dependent variable changes for a unit change in the independent variable. For example, a 1-unit change in `com3`, identification with the community, produces a 0.05-unit change in `env_con`, environmental concern, holding all other variables constant. Without studying the range and distribution of each of the variables, it is hard to compare the unstandardized coefficients to see which variable is more or less important. The problem is that each variable can be measured on a different scale. Here we measured income using dollars, and this makes it hard to interpret the b value as a measure of the effect of income. After all, a \$1.00 change in your annual income is something you might not even notice, and this change surely would not have much of an effect on anything. This is reflected in $b_{\texttt{inc}} = -5.51\text{e}^{-08}$, which is a tiny value. If you have not seen this format for numbers, it is a part of scientific notation, which is used for tiny numbers. The e^{-08} tells us to move the decimal place eight places to the left. Thus this number is -0.0000000551. If we had used the natural logarithm of income or if we had measured income in 1,000s of dollars or even 10,000s of dollars, we would have obtained a different value. When you use income as a predictor, it is common to use a natural logarithm or measure income in 10,000s of dollars, where an income of 47,312 would be represented as 4.7312.

Here you included the option to get beta weights (β). These are more easily compared because they are based on standardizing all variables to have a mean of 0 and a standard deviation of 1. These beta weights are interpreted similarly to how you interpret correlations in that $\beta < 0.20$ is considered a weak effect, β between 0.2 and 0.5 is considered a moderate effect, and $\beta > 0.5$ is considered a strong effect. The beta weights tell you that education, income, and community identity all have a weak effect, and of these, only community identity is statistically significant, $p < .001$. Having a health problem in the family, `hlthprob`, has a beta weight that is still weak, $\beta = -0.173$, but it is statistically significant $p < 0.001$. Having environmental health concerns, `epht3`, has $\beta = -0.458$, $p < 0.001$, which is a moderate-to-strong effect.

You might summarize this regression analysis as follows: A model including education, income, community identity, environmental health concerns, and health problems in the family explains 29.8% of the variance in environmental concerns of adults in Oregon, $F(5,3763) = 320.15$, $p < 0.001$. Neither education nor income has a statistically significant effect. Community identity has a weak but statistically significant effect, $\beta = 0.074$, $p < 0.001$. The two health variables are the strongest predictors in this group: having a health problem in the family has $\beta = -0.173$, $p < 0.001$ and having environmental health concerns has $\beta = -0.458$.

If you do not want beta weights, you would simply rerun the regression without the option. When you do this you get a confidence interval for each unstandardized regression coefficient. If you have an analysis where your primary interest is in the unstandardized coefficients, this is a useful extension on the simple tests of significance.

10.4 Increment in R-squared: semipartial correlations

Beta weights are widely used for measuring the effect sizes of different variables. Because it is based on standardized variables, a beta weight tells you how much of a standard deviation the dependent variable changes for each standard deviation change in the independent variable. Except for special circumstances, beta weights range from -1 to $+1$, with a zero meaning that there is no relationship. Sometimes the interpretation of beta weights is problematic. If you have a categorical variable like gender, the interpretation is unclear. What does it mean to say that as you go up one standard deviation on gender, you go up β standard deviations on the dependent variable? Also, when you are comparing groups, the beta weights can be misleading. If you compare regression models predicting income for women and for men, the beta weights would be confounded with real differences in variance between women and men in the standardization of all variables.

Another approach to comparing variables is to see how much each variable increases R^2 if the variable is entered last. Effectively, this approach fits the model with all the variables except for the one you are interested in. You obtain an R^2 value for this model. Then you fit the model a second time, including the variable in which you are interested. This will have a larger R^2, and the difference between the two R^2 values will tell you how much the variable increases R^2. This increment is how much of the variance is uniquely explained by the independent variable, controlling for all the other independent variables. This is given different names in different fields of study. It is often called an *increment* in R^2, which is a very descriptive name. Some people call it a *part-correlation square*, as it measures the part that is uniquely explained by the variable. Others call it a *semipartial* R^2. Other people compare variables using what is called a *partial correlation*. We will not cover that here because it has less-desirable properties for comparing the relative importance of each variable.

Estimating the increment in R^2 would be a tedious process since you would have to fit the model twice for every independent variable and compute each of these differences. In this example, we would have to run five regression models. Fortunately, there is a user-contributed program by Richard Williams that automates this process. The command is pcorr2. Because it is a user-written command, you need to install it (type ssc install pcorr2); it is not available through a Stata dialog box. The command is simple. As with the regress command, you list the dependent variable first, and then follow it with the independent variables. Here are the command and results where we use inclog, the log of the income:

```
. pcorr2 env_con educat inclog com3 hlthprob epht3
(obs=3769)
Partial and Semipartial correlations of env_con with

    Variable |    Partial      SemiP   Partial^2     SemiP^2        Sig.
-------------+------------------------------------------------------------
      educat |    -0.0118    -0.0099      0.0001      0.0001       0.469
      inclog |     0.0145     0.0121      0.0002      0.0001       0.374
        com3 |     0.0875     0.0735      0.0076      0.0054       0.000
    hlthprob |    -0.1916    -0.1635      0.0367      0.0267       0.000
       epht3 |    -0.4584    -0.4320      0.2101      0.1866       0.000
```

Here you are interested only in the column labeled SemiP^2, semipartial R^2, which shows how much each variable contributes uniquely. The concern for environmental health problems has an increment to R^2 of 0.1866, $p < 0.001$. Although two other variables have statistically significant effects, this is the only variable that has a substantial increment to R^2. The semipartial R^2 is a conservative estimate of the effect of each variable because it measures only how much the R^2 increases when that variable is entered after all of the other variables are already in the model. You might notice that the sum of the semipartial R^2s is just 0.219 even though the model R^2 is 0.298. This happens because some of the predictors are explaining the same part of the variance in the dependent variable, and whatever is shared with other predictors is not unique. The semipartial R^2 estimates only the unique effect of each predictor.

10.5 Is the dependent variable normally distributed?

Now let's examine the distribution of the dependent variable. You would like this to be normal, but this can often be a problem. Let's create a histogram on env_con:

```
. histogram env_con, frequency normal kdensity
```

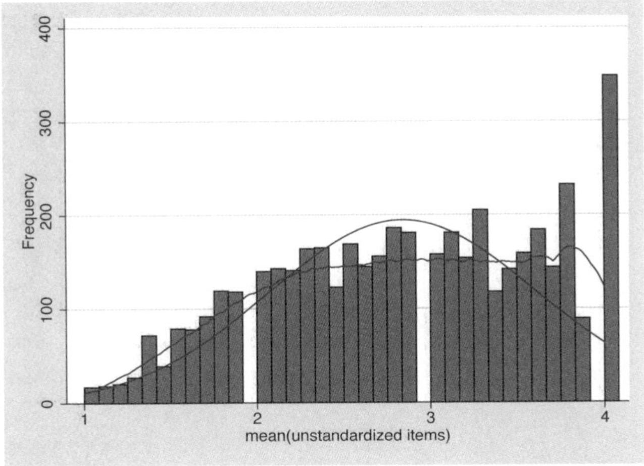

Figure 10.2. Histogram of dependent variable, `env_con`

Three distributions are represented in figure 10.2. The bars represent the actual distribution, which does not look too bad except for the big bunch of high scores (there are 348 people who have the maximum possible score of 4.0 on the environment concerns scale). Apparently, there is a sizable group of people in Oregon who share a serious concern about the environment. The smooth, bell-shaped curve represents how the data would be distributed if they were normal.

The other curve, called a kdensity curve, is a bit hard to see in figure 10.2. The kdensity curve is an estimation of how the population data would look, given our sample data. The actual data from the histogram, the normal curve, and the kdensity curve are similar up to a value of about 3.2–3.5. At the center of the distribution, the data and the kdensity curve are a little flatter than a normal curve. We have too few cases in the middle of the distribution to call it normal. The real problem, however, is on the right side of the distribution, where we have too many people who have a mean score of over about `3.5`. Remember that the scale ranged from a `1` to `4`, and a mean score of `3.5` or more means that the person has a strong environmental concern.

An alternative plot is the hanging rootogram (`ssc install hangroot`). Once the `hangroot` command is installed, we simply enter `hangroot env_con, bar`. This produces the graph in figure 10.3.

(*Continued on next page*)

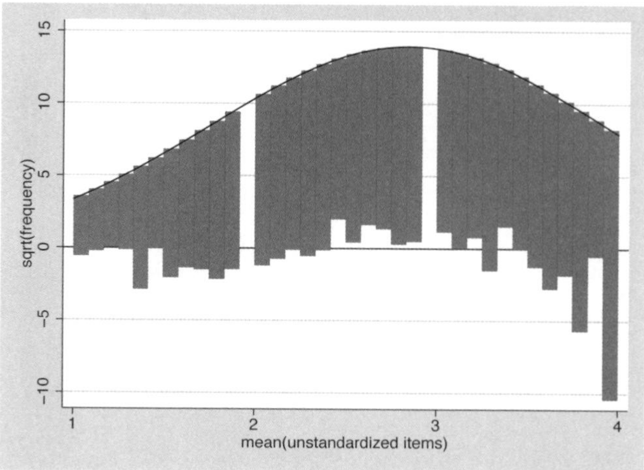

Figure 10.3. Hanging rootogram of dependent variable, `env_con`

This hanging rootogram has a smooth curve representing how the data would be distributed if it were normal. When a bar descends below the horizontal line, there are too many observations at this value compared to a normal distribution. At a value of 4, the bar drops far below the horizontal line and the bars are all below the line from a value of about 3.5 or higher. There are a few values toward the middle of the distribution that do not drop down to the horizontal line. There are not enough values in this area of the distribution.

The hanging rootogram is sometimes easier to read, in part, because the vertical axis is the square root of the frequency rather than the actual frequency. This makes it much easier to see deviations from normality in the tails where there are usually relatively few observations. This type of graph was suggested by John Tukey and the command for it was written by Maarten Buis.

We can evaluate the normality of our dependent variable by using two additional statistics that are known as skewness and kurtosis. We learned earlier about the `detail` option to the `summarize` command; this option provides these statistics. Just repeat the command now.

```
. summarize env_con, detail
                    mean(unstandardized items)
-------------------------------------------------------------
      Percentiles      Smallest
 1%     1.181818              1
 5%     1.545455              1
10%          1.8              1       Obs                4506
25%     2.272727              1       Sum of Wgt.        4506

50%          2.9                      Mean           2.842405
                        Largest       Std. Dev.      .7705153
75%          3.5              4
90%     3.818182              4       Variance       .5936939
95%            4              4       Skewness      -.2302043
99%            4              4       Kurtosis       2.059664
```

In the lower right column, you find skewness of -0.23 and kurtosis of 2.06. Skewness is a measure of whether a distribution trails off in one direction or another. For example, income is positively skewed because there are a lot of people with relatively low income, but there are just a few people who have extremely high incomes. By contrast, `env_con` is negatively skewed because there are a lot of people who have a high level of environmental concern, but there are relatively fewer who have a low level of concern. A normal distribution has skewness $= 0$. If the skewness is greater than this, the distribution is positively skewed, but if it is less than this (as with `env_con`), the distribution is negatively skewed.

Kurtosis measures how thick the tails of a distribution are. If you look at our histogram, you see that the tail to the left of the mean is a little too thick and the tail to the right of the mean is way too thick to be normally distributed. When a distribution has a problem with kurtosis indicated by thick tails, it will also have too few cases in the middle of the distribution. By contrast, when a distribution has a problem with kurtosis indicated by thin tails, it will also have too many cases in the middle of the distribution (peaked) for it to be normally distributed. This would happen for a variable in which most people were very close to the mean value.

A normal distribution will have a kurtosis of 3.00. A value of less than 3.00 means that the tails are too thick (hence, too flat in the middle), and a value of greater than 3.00 means that the tails are too thin (hence, too peaked in the middle). The kurtosis is 2.06 for `env_con`, meaning that the tails are too thick; you could tell that by looking at the histogram. Some statistical software, such as SAS, reports a value for kurtosis that is the actual value of kurtosis minus three so that a normal distribution would have a value of zero. Stata does not do this, so the correct value for a normal distribution is 3.00. You need to be careful when writing a report that will be read by people who rely on other programs that may have kurtosis $= 0$ for a normal distribution.

The `summarize` command does not give you the significance of the skewness or kurtosis coefficients. To get these, you need to run a command that you enter directly:

```
. sktest env_con

                    Skewness/Kurtosis tests for Normality
                                                         ------ joint ------
    Variable | Pr(Skewness)   Pr(Kurtosis)  adj chi2(2)    Prob>chi2
-------------+-------------------------------------------------------
     env_con |      0.000         0.000             .         0.0000
```

This does not report the test statistics Stata uses (the chi-squared is too big to fit in the space Stata provides for that column), but it does give us the probabilities. Combining this result with the results from the `summarize` command, we would say that skewness $= -0.23$, $p < 0.001$ and kurtosis $= 2.06$, $p < 0.001$. Both of these tested jointly have a $p < 0.001$. Hence, the distribution is not normal, and the skewness tells us the distribution has a negative skew, and the kurtosis tells us the tails are too thick.

When we are concerned about normality, we can try a robust regression estimate. The command involves adding an option after the comma: `regress env_con educat inc com3 hlthprob epht3, beta vce(robust)`. This robust regression produces the same R^2, b's, and βs but has standard errors that do not assume normality. Usually the t-values will be slightly smaller.

If you want yet another alternative, you can get bootstrap estimates of the standard errors by adding an option to the regression command: `regress env_con educat inc com3 hlthprob epht3, vce(bootstrap, reps(1000))`. Here Stata draws repeated random samples with replacement from your dataset and estimates the regression for each of these datasets. The standard errors are simply the standard deviations of the distribution of each parameter estimate across all of these samples. For example, the first sample might estimate $b_1 = 0.5$, the second sample might estimate $b_1 = 0.6$, and the 1,000th sample might estimate $b_1 = 0.4$. If these estimates are highly consistent, the standard deviation of these 1,000 estimates of b_1 will be very small. This standard deviation is the standard error of the sampling distribution of b_1 and is used to calculate the t test. One limitation of the bootstrap solution is that it does not report standardized β coefficients. You would need to get these from the regular ordinary least squares regression and then use the bootstrap solution to get the t tests.

10.6 Are the residuals normally distributed?

An important assumption of regression is that the residuals are normally distributed. At a risk of oversimplification, this means that we are as likely to overestimate a person's score as we are to underestimate their score, and we should have relatively few cases that are extremely overestimated or underestimated. This assumption becomes more difficult when we add the fact that this distribution should be about the same for any value or combination of values for the predictors. Imagine predicting income from a person's education. With a low education, we would predict a low income. As a person goes up the scale on education, we would predict higher income. However, for those with 16 or more years of education, there is a huge range in income. Some may be in low-paying service positions, and others may be in extremely high-paying positions. If we plotted a scattergram with the regression line drawn through it, we might expect the variance around the line to increase as education increased. This effect is illustrated in figure 10.4.

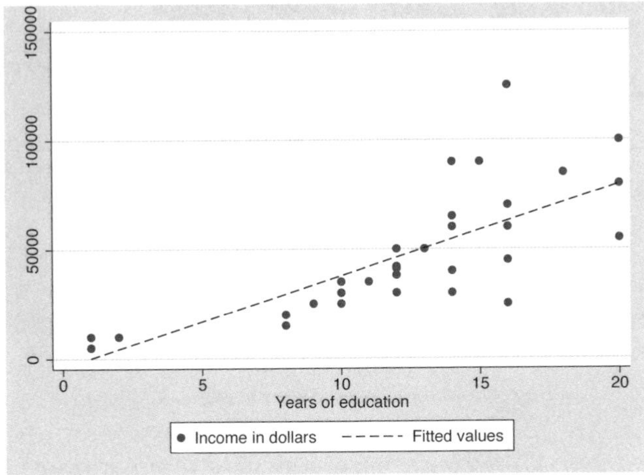

Figure 10.4. Heteroskedasticity of residuals

There is little error (we use the terms *error* and *residual*, interchangeably) up to about 12 years of education. The observations are very close to the predicted value. Beyond 12 years of education, however, the errors increase dramatically. If you locate 14 years on the horizontal axis and go straight up, we see that the predicted income is a bit more than \$50,000 per year. However, one person makes about \$90,000 per year, and another makes about \$25,000 a year. We cannot accurately predict income for those with a lot of education.

When we have multiple regression, it is difficult to do a graph like this because we have several predictors. The solution is to look at the distribution of residuals for different predicted values. That is, when we predict a small score on environmental concerns, are the residuals distributed about the same as they are when we predict a high score?

Stata has several options. Selecting Statistics ▷ Linear models and related ▷ Regression diagnostics ▷ Residual-versus-fitted plot opens a dialog box for the residual-versus-fitted value plot. On the Main tab, we see that we only need to click on OK. We could use the other tabs to make the graph more appealing (we used the Y axis tab to add a reference line at a value of zero), or we could just click on OK. Figure 10.5 shows the results.

(*Continued on next page*)

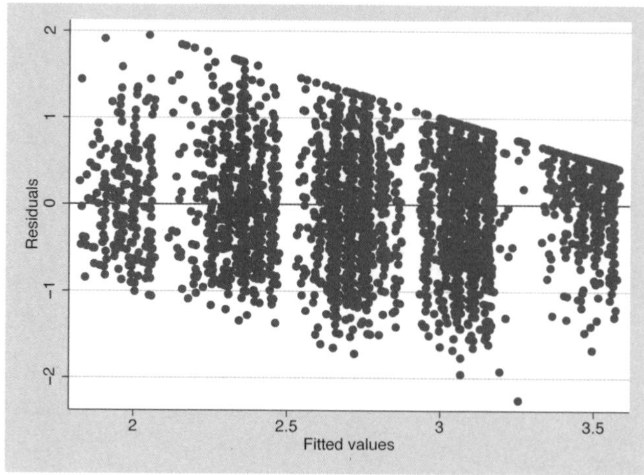

Figure 10.5. Residual-versus-fitted plot

This graph is a bit hard to read. Ideally, the observations would be normally distributed about the reference line. There would be about as many observations that have positive residuals as have negative residuals and these would be nicely distributed for all fitted values. Do you see the problem with this graph? A normal distribution has a potential range from minus infinity to plus infinity. Our dependent variable is on a scale that has a lower limit of 1 and an upper limit of 4. If you predict a person will have a score of 4, meaning they are highly concerned about environmental issues, we cannot underestimate their score because it is not possible to have a value bigger than 4. If the estimated value is 1 (no concern about environmental issues), then we cannot overestimate their score because it is not possible to have a value smaller than 1. You can see this in the graph with a drift from more positive residuals for low estimated scores to more negative residuals for high scores.

A residual-versus-predicted plot, similar to the plot in figure 10.5, makes the most sense when the dependent variable takes on a larger range of values. In those situations, it may show that there is increasing or decreasing error variance as the predicted value gets larger. With our current example, where the dependent variables are between 1 and 4, it makes the problem clear but does not suggest any solution.

Sometimes it is easier to visualize how the residuals are distributed by doing a graph of the actual score of env_con on a predicted value for this score. It is easiest to show how to do this using the commands. Before doing that, however, we might want to draw a sample of the data. Because this is a very large dataset, there will be so many dots in a graph that it will be hard to read and many of the dots will be right on top of each other. Because of this, we will include a step to sample 100 observations. Here are the commands:

```
. regress env_con educat inclog com3 hlthprob epht3, beta
. predict envhat
```

```
. preserve
. set seed 111
. sample 100, count
. twoway (scatter env_con envhat) (lfit env_con envhat)
. restore
```

This series of commands does the regression first. The next line, `predict envhat`, will predict a score on `env_con` for each person based on the regression equation. Here the new variable is called `envhat`, but we can use any name because the default is to predict the score based on the regression equation. Before we take a sample of 100 observations, we enter the command `preserve`, which allows us to `restore` the dataset back to its original size. The fourth line, `set seed 111` provides a fixed start for the random sampling, so we can repeat what we did and get the same results. The command `sample 100, count` will draw a random sample of 100 cases. We are doing this to make the graph easier to read. We might want to save our data before doing this because once we drop all but 100 people, they are gone. If you save the sample of 100 observations, make sure you give the file a different name. (The use of the `preserve` and `restore` commands is a simpler way to preserve the full dataset.) The next line is a `twoway` graph, as we have done before. The first section does a scattergram of `env_con` on `envhat`, and the second section does a linear regression line, regressing `env_con` on `envhat`. The final line has the command `restore`, which returns the complete dataset. The resulting graph is shown in figure 10.6, where we have used the Graph Editor to relabel some parts of the figure.

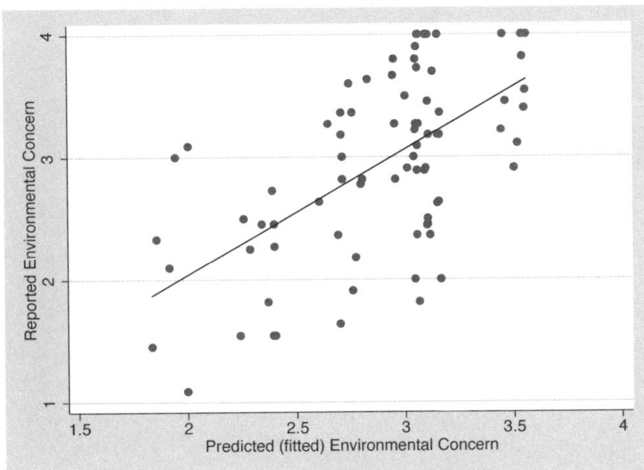

Figure 10.6. Actual value of environmental concern regressed on the predicted value

Examining this graph, we can see that in the middle there is not much of a problem. The problem is really only at the ends, where we are predicting a very high or very low score. That is, for predictions of a high score, there are more people who actually report being lower than what we predict, and for low scores it is just the opposite.

10.7 Regression diagnostic statistics

Stata has a strong set of regression diagnostic tools. We will cover only a few of them. Enter the command `help regress postestimation`, and you will see a basic list of what we cover as well as several other possibilities that are beyond the scope of this book.

10.7.1 Outliers and influential cases

When we are doing regression diagnostics, one major interest is in finding outliers. These are cases with extreme values, and they can have an extreme effect on an analysis. One way to think of outliers is to identify those cases the regression equation has the most trouble predicting; these are the cases that are farthest from the predicted values. These cases are of special interest for several reasons. You might want to examine them to see if there was a coding error, comparing your data to the original questionnaires. If you can contact these people, you might want to do a qualitative interview to better understand why the variables you thought would predict their score failed. You may discover factors that you had not anticipated that are important to them, and this could guide future research.

To find the cases we cannot predict, use postestimation commands. Select Statistics ▷ Postestimation ▷ Predictions, residuals, etc. to open the dialog box to predict the estimated value, `yhat`; the residual, `resid`; and the standardized residual, `rstandard`. We can also enter the following series of commands directly:

```
. regress env_con educat inclog com3 hlthprob epht3, beta
. predict yhat
. predict residual, resid
. predict rstandard, rstandard
. list respnum env_con yhat resid rstandard
> if abs(rstandard) > 2.58 & rstandard < .
```

The first prediction command, `predict yhat`, will predict the estimated score based on the regression. This has no option after a comma because the estimated score is the default prediction and the variable name, `yhat`, could be any name we choose. The next command, `predict residual, resid`, will predict the residual, or $Y - \widehat{Y}$, giving us a raw or unstandardized measure of how far the estimated value, `yhat`, is from the person's actual score on the environmental-concern outcome. The `resid` after the comma is an option that tells Stata to predict this residual value. The last prediction command, `predict rstandard, rstandard`, tells Stata to estimate the standardized residual for each observation and to create a new variable called `rstandard`. This is a z score that we can use to test how bad our prediction is for each case. We usually will be interested in our "bad" predictions, as these are the residual outliers. The `list` command lists all cases that have a standardized residual, i.e., z score, whose size is greater than 2.58 (we include the restriction that `rstandard` is also less than "`.`"). We picked 2.58 because this corresponds to the two-tailed 0.01 level of significance. In other words, we would expect residuals this large in either direction less than 1% of the time by chance. Here are the results of this string of commands:

```
. regress env_con educat inclog com3 hlthprob epht3, beta

      Source |       SS       df       MS              Number of obs =    3769
-------------+------------------------------           F(  5,  3763) =  320.37
       Model |  647.988857     5  129.597771           Prob > F      =  0.0000
    Residual |  1522.24781  3763  .404530376           R-squared     =  0.2986
-------------+------------------------------           Adj R-squared =  0.2976
       Total |  2170.23666  3768  .575965144           Root MSE      =  .63603

------------------------------------------------------------------------------
     env_con |      Coef.   Std. Err.      t    P>|t|                     Beta
-------------+----------------------------------------------------------------
      educat |  -.0028858   .0039829    -0.72   0.469                -.0108658
      inclog |   .0111218   .0124984     0.89   0.374                 .0132953
        com3 |    .049928   .0092714     5.39   0.000                 .0737784
    hlthprob |  -.2970166   .0248074   -11.97   0.000                -.1727021
       epht3 |  -.4015034   .0126902   -31.64   0.000                -.4569504
       _cons |   3.631318   .1260653    28.81   0.000                        .
------------------------------------------------------------------------------

. predict yhat
(option xb assumed; fitted values)
(738 missing values generated)

. predict residual, resid
(739 missing values generated)

. predict rstandard, rstandard
(739 missing values generated)
```

(Continued on next page)

```
. list respnum env_con yhat resid rstandard
> if abs(rstandard) > 2.58 & rstandard < .
```

	respnum	env_con	yhat	residual	rstandard
65.	100072	4	2.051867	1.948133	3.066239
170.	100189	4	2.347755	1.652245	2.602065
323.	100370	4	2.337523	1.662477	2.617791
539.	100626	4	2.334638	1.665362	2.622386
800.	100928	4	2.358169	1.641831	2.584356
803.	100931	1.454545	3.107825	-1.653279	-2.600516
1056.	101237	4	2.155571	1.844429	2.904261
1657.	101958	4	2.327333	1.672667	2.635602
1690.	101996	4	2.340227	1.659773	2.612668
2247.	102656	1.272727	3.195089	-1.922362	-3.027591
2418.	102866	4	2.196842	1.803158	2.838216
2608.	103091	1.818182	3.494809	-1.676627	-2.638001
3386.	200221	1.363636	3.040674	-1.677038	-2.638506
3463.	200325	1	2.650622	-1.650622	-2.596241
3655.	200587	1.111111	3.066463	-1.955352	-3.076585
3662.	200595	3.818182	1.906173	1.912009	3.009197
3679.	200621	4	2.358169	1.641831	2.584356
3736.	200704	4	2.344151	1.655849	2.607023
3743.	200712	4	2.169958	1.830042	2.883007
3745.	200716	1.3	3.066463	-1.766464	-2.779384
3762.	200743	1	2.652066	-1.652066	-2.603171
3795.	200785	1	3.255249	-2.255249	-3.553156
4147.	201423	1	2.719743	-1.719743	-2.706931
4172.	201472	4	2.239736	1.760264	2.77161
4348.	201767	4	2.351135	1.648865	2.595331

In this dataset, `respnum` is the identification number of the participant. Participant 100072 had an actual mean score of 4 for the scale. This is the highest possible score, yet we predicted this person would have a score of just 2.05. The residual is 1.95, indicating how much more environmentally concerned this person is than we predicted. The z score is 3.07. Toward the bottom of the listing, participant 200785 had a score of 1, indicating that he or she was at the low point on our scale of environmental concern. However, we predicted this person would have a score of 3.26, and the z score is -3.55. We might look at the questionnaires for these two people to see if there was a miscoding. If the coding was correct, we might try to contact the participant to find out other variables we need to add to the regression equation. Unstructured interviews with participants that have extreme outliers can be helpful. We may find that those who have positive outliers are politically liberal and those who have negative outliers are politically conservative. Then we could add a variable measuring liberalism as a predictor. Correlating `rstandard` with other possible predictors would also be helpful. A variable that is highly correlated with `rstandard` helps explain the variance in our outcome that is not already explained by the predictors in our model.

10.7.2 Influential observations: DFbeta

Stata offers several measures of the influence each observation has. dfbeta is the most direct of these. It indicates the difference between each of the regression coefficients when an observation is included and when the observation is excluded. You could think of this as redoing the regression model, omitting just one observation at a time and seeing how much difference omitting each observation makes. A value of DFbeta $> 2/\sqrt{N}$ indicates that an observation has a large influence. To open the dialog box for this command, select Statistics ▷ Linear models and related ▷ Regression diagnostics ▷ DFBETAs, but it is easier to type the one-word command, dfbeta, in the Command window:

```
. dfbeta
(739 missing values generated)
                         DFeducat:  DFbeta(educat)
(739 missing values generated)
                         DFinclog:  DFbeta(inclog)
(739 missing values generated)
                           DFcom3:  DFbeta(com3)
(739 missing values generated)
                       DFhlthprob:  DFbeta(hlthprob)
(739 missing values generated)
                          DFepht3:  DFbeta(epht3)
```

These results show that Stata created five new variables. DFeducat has a DFbeta score for each person on the education variable, DFinclog does this for the income variable, DFcom3 does this for the community identity variable, DFhlthprob does this for health problems, and DFepht3 does this for environmental-specific health concerns.

We could then do a listing of cases that have relatively large values of DFbeta, using the formula DFbeta $= 2/\sqrt{3769} = 0.03$ as our cutoff because we have $N = 3769$ observations. For example, if we wanted to know problematic observations for the educat variable, the list command would be

```
. list respnum rstandard DFeducat if abs(DFeducat) > 2/sqrt(3769) & DFeducat < .
```

We will not show the results of these problematic observations here. This list command, however, has a few features you might note. We use the absolute value function, abs(DFeducat). We also include a simple formula, 2/sqrt(3769), where 3,769 is the sample size in our regression. This listing is just for the educat variable. dfbeta gives us information on how much each observation influences each parameter estimate. This is more specific information than that provided by the standardized residual. If there are a few observations that have a large influence on the results, we might see if they were miscoded. If possible, we might interview the individuals to see why they picked the answer they did and how they interpreted the item.

10.7.3 Combinations of variables may cause problems

Collinearity and multicollinearity can be a problem in multiple regression. If two independent variables are highly correlated implicitly, it is difficult to know how important each of them is as a predictor. Multicollinearity happens when a combination of variables makes one or more of the variables largely or completely redundant. Figure 10.7 shows how this can happen with just two predictors, `X1` and `X2`, which are trying to predict a dependent variable, `Y`. The figure on the left shows that areas `a` and `c` represent the portion of `Y` that is explained by `X1`, and areas `b` and `c` represent the portion of `Y` that is explained by `X2`. Because `X1` and `X2` overlap a little bit (are correlated), the area `c` cannot be distributed to either `X1` or `X2`. Still, there is a lot of variance in both predictors that is not overlapping, and we can get a good estimate of the unique effect of `X1`, namely, area `a`, and the unique effect of `X2`, namely, area `b`. Earlier in this chapter, we referred to the areas represented by `a` and `b` as the semipartial R^2s.

The figure on the right shows what happens when the two predictors are more correlated and, hence, overlap more. Together, `X1` and `X2` are explaining much of the variance of `Y`, but the unique effects represented by areas `a` and `b` are relatively small. You can imagine that, as `X1` and `X2` become almost perfectly correlated, the areas `a` and `b` all but disappear. This example illustrates what could happen with just two predictors. You can imagine what could happen in multiple regression when there are many predictors that are correlated with each other. The more correlated the predictors, the more they overlap and, hence, the more difficult it is to identify their independent effects. In such situations, you can have multicollinearity in which one or more of the predictors are virtually redundant.

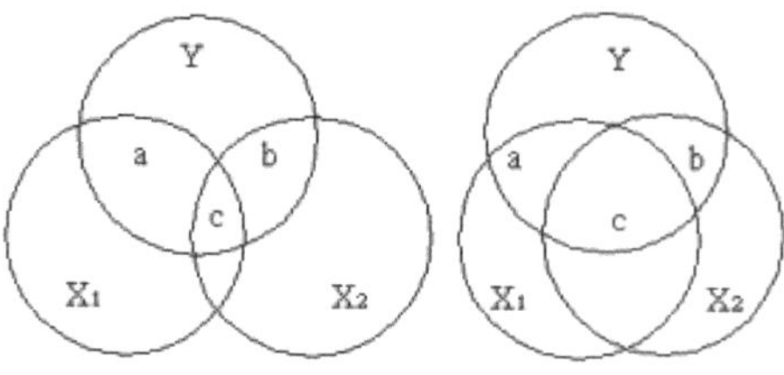

Figure 10.7. Collinearity

Stata can compute a variance inflation factor to assess the extent to which multicollinearity is a problem for each independent variable. This is computed by running the command `estat vif` after the regression. Here are the results for our example:

```
. estat vif

    Variable |       VIF       1/VIF
-------------+----------------------
      educat |      1.21    0.828780
      inclog |      1.20    0.835001
       epht3 |      1.12    0.893603
    hlthprob |      1.12    0.895876
        com3 |      1.01    0.993069
-------------+----------------------
    Mean VIF |      1.13
```

This command computes both the variance inflation factor (VIF) and its reciprocal (1/VIF). Statisticians usually look at the VIF. If this is more than 10 for any variable, there may be a multicollinearity problem, and you may need to consider making an adjustment to your model. None of the variables has a problem using this criterion. If the average VIF is substantially greater than 1.00, there still could be a problem. Here the mean VIF is 1.13, and this is not a problem.

The value of 1/VIF may have a more intuitive interpretation than the VIF. If we regressed `educat` on `inclog`, `epht3`, `hlthprob`, and `com3`, we get an R^2 of 0.17. This means that there is an overlap of education with all the other predictors of 17% of the variance in education. Alternatively, $1 - 0.17 = 0.83$ or 83% of the variance in education is not overlapping, i.e., is not explained by the other predictors. Thus 83% of the variance in `educat` is available to give us an estimate of the independent effect of education on environmental concern, controlling for the other predictors. Some people call 1/VIF the tolerance. It is 1.00 minus the R^2 you obtain if you regress an independent variable on the set of other independent variables. It tells how much of the variance in the independent variable is available to predict the outcome variable independently. When VIF = 10, this means that only 10% of the variance in an independent variable is really available after you adjust for the other predictors. When VIF $>$ 10 or 1/VIF $<$ 0.10, there may be a multicollinearity problem.

When you have a problem with multicollinearity, consider these solutions. Dropping a variable often helps. If there are two variables that both have a high VIF value, try dropping one of them and repeating the analysis. Dropping a variable that has a high VIF value is not as much of a problem as you might suspect because a variable that has more than 90% of its variance confounded with the other predictors probably does not explain much uniquely because it has little unique variance itself. Sometimes there are several closely related variables. For example, you might have a series of items that involve concerns people have about global warming. Rather than trying to include all of these individual items, you might create a scale that combines them into a variable.

Multicollinearity sometimes shows up in what may seem like strange parameter estimates. Normally, the standardized βs should not be outside the range of -1 to $+1$. Yet, you might have one variable that has $\beta = 2.14$ and a closely related variable that has $\beta = -1.56$. Sometimes these βs will not only be excessively large, but neither of them will be statistically significant. You almost certainly have a multicollinearity problem when this happens. If you drop either variable, the remaining variable may have a β that is within the normal range and is statistically significant.

10.8 Weighted data

Regression allows us to weight our cases. Many large datasets will use what is known as a weighted sample. They want to have an adequate number of observations of groups that otherwise would be a small subsample. One survey of a community that is mostly non-Hispanic whites may oversample Hispanics and African Americans. Another survey might oversample people who are cohabiting or people who have a disability. If you want to take a subsample of people who are disabled, cohabiting, Hispanic, or African American, this oversampling is critically important, as it means that your subsample will be large enough for meaningful analysis. However, if you want to generalize to the entire population, you will need to adjust for this oversampling.

Many surveys provide a weight that tells you how many people each respondent represents, based on the sample design. This sampling fraction is simply the population N for the group, say, the number of African Americans in the United States, divided by the sample n for African Americans in your sample. Other surveys provide a proportional weight so that if you oversampled African Americans by a factor of two (i.e., an African American was twice as likely to be sampled as other people), then each African American would have a weight value of 0.5 when you tried to generalize to the entire population.

When using a weighted sample, you need to consult the documentation on the survey to know what weight variable to use. The Oregon sample has both types of weights. We can list these for the first five observations.

```
. list finalwt finalwt2 in 1/5

     +----------------------+
     |  finalwt    finalwt2 |
     |----------------------|
  1. | 969.7892    1.602893 |
  2. | 384.8077    .6360203 |
  3. | 91.07467    .1505306 |
  4. |     5000    8.264132 |
  5. |  27.0839     .044765 |
     +----------------------+
```

With complex weighting, we can have different weights for each person. The first observation in the Oregon sample represents 969.79 Oregonians, and the fifth observation represents just 27.08 Oregonians, as indicated by the scores they have for the variable `finalwt`. The sum of all the `finalwt` scores will be close to the population of Oregon.

The probability weights represented by `finalwt2` accomplish the same thing for the sample. A person who is in a highly oversampled group (say, Native Americans) will have a `finalwt2` score of less than 1.00. A person who is not oversampled because we know we will have plenty of them in our sample (say, a white male) will have a `finalwt2` score of more than 1.00. The sum of all the `finalwt2` scores will approximate the total sample size.

Weighting is complicated and beyond the scope of this book. Consult the documentation for the dataset you are using to find out what is the right weight variable for you to use. Stata can adjust for different kinds of weights, but here we will just use what StataCorp calls `pweights`. In the dialog box for multiple regression, you will find a tab called Weights. Click on this tab, and click that you want to use *Sampling weights*; make sure you click on this (this is the name Stata's dialog system uses for what Stata calls `pweights`). Then all you need to do is enter the name of the weight variable; here use `finalwt`. The command and results are as follows:

```
. regress env_con educat inclog com3 hlthprob epht3 [pweight=finalwt], beta
(sum of wgt is   2.3167e+06)
Linear regression                                      Number of obs =    3769
                                                       F(  5,  3763) =  120.45
                                                       Prob > F      =  0.0000
                                                       R-squared     =  0.3080
                                                       Root MSE      =  .60742

------------------------------------------------------------------------------
             |               Robust
     env_con |      Coef.   Std. Err.      t    P>|t|                     Beta
-------------+----------------------------------------------------------------
      educat |  -.0116873   .0069538    -1.68   0.093                -.0454832
      inclog |   .0235766   .0244483     0.96   0.335                 .0292171
        com3 |   .0451292   .0180871     2.50   0.013                  .067008
    hlthprob |  -.2693781   .0447719    -6.02   0.000                -.1624003
       epht3 |  -.4071392   .0225652   -18.04   0.000                -.4670926
       _cons |   3.593785   .2535868    14.17   0.000                        .
------------------------------------------------------------------------------
```

When you do a weighted regression this way, Stata automatically uses the robust regression—whether you ask for it or not—because weighted data require robust standard errors. The results of this robust regression have important differences from the original regression. Because it is robust regression, we do not get the ANOVA table and cannot get an adjusted R^2. Because we used a weighted sample, we get a line saying that the `sum of wgt is 2.3167e+06`. When Stata encounters a number that is extremely small or extremely large, it uses this notation. The e+06 means to move the decimal place six places to the right. If we do this, the sum of the weights is 2,316,700. This is the approximate total adult population of Oregon at the time of this survey. It is good to check this value, which should be the total population size if you use this type of weight or the total sample size if you use the proportional weight.

The number of observations is 3,769, and this is the actual size of the sample used for tests of significance. We should check this to make sure that we did the weighting correctly. If we had used the other weight, `finalwt2`, the line would read `sum of wgt is 3.8291e+03`. Moving the decimal place over three spaces, this is 3,829, which is close to the actual sample size we have of 3,769. The Oregon Survey includes 4,508 observations. Because regression uses casewise deletion, we have only 3,769 observations with valid scores on all the variables used in the regression. Because these cases with complete data are not missing completely at random, the sum of the weight variable is not identical to 3,769.

You may have observed that many of the results are different from the unweighted results. For example, the F value is 120.45 compared with 320.37 for the unweighted sample. This is partly due to the weighting and partly due to using robust regression. The parameter estimates of the coefficients and the βs are different, as are the standard errors and t tests. When the weighted sample gives different results, this means that the weighting is important. Imagine that the authors of the study decided they needed many Native Americans, so they oversampled Native Americans by a factor of 10. This means that each Native American would count 10 times as much in the unweighted sample as they would if you wanted to generalize the overall population. If the relationship between the variables were somehow different for Native Americans than for other groups, these differences would be overrepresented by a factor of 10. Weighting makes every case represent its proportion of the population.

10.9 Categorical predictors and hierarchical regression

Smoking by adolescents is a major health problem. We will examine this problem in this section by using the `nlsy97_selected_variables.dta` dataset as a way of illustrating what can be done with categorical predictors. Let's pick a fairly simple model to illustrate the use of categorical variables. Say that we think smoking behavior depends on the adolescent's age (a continuous variable), gender (a dichotomous variable), peer influence (a continuous variable), and race/ethnicity (a multicategory nominal variable). Let's also say we think of these variables as being hierarchical. Start with just age and gender, and see how much they explain. Then add peer influence, and see how much it adds. Finally, add race/ethnicity to see if it has a unique effect above and beyond what is explained by age, gender, and peer influence.

When we are working with categorical predictors, it is important to distinguish between dichotomous predictors, such as gender, and multicategorical predictors, such as race, for which there are more than two categories. When we have a dichotomous predictor, such as gender, we can create an indicator variable coded `0` for one gender, say, females, and `1` for the other gender, males. We can then enter this as a predictor in a regression model. We can do this by using the `recode` command; select Data ▷ Create or change variables ▷ Other variable transformation commands ▷ Recode categorical variable. In the *Variables* box, type `gender97` because this is the name of the variable for gender in our dataset. Because males are coded as `1` and females as `2`, we need to do some recoding. An indicator or dummy variable is a 0/1 variable where a `1` indicates the presence of the characteristic and a `0` indicates the absence. Click on the Options tab, and check *Generate new variables*. Name the variable `male` because that is the category we will code as `1`. If we had named the variable `female`, we would code females as `1`. The choice is arbitrary and only changes the sign of the regression coefficient. Returning to the Main tab, enter the *Required* rule: `(1 = 1 Male) (2 = 0 Female)`. If you want to run the command directly from the command line, you enter

```
. recode gender97 (1 = 1 Male) (2 = 0 Female), generate(male)
```

Names for categorical variables

The conventions for coding categorical variables precede the widespread use of multiple regression. Conventions are slow to change, and we usually need to recode categorical variables before doing regressions. Most surveys code dichotomous response items using codes of `1` and `2` for `yes/no`, `male/female`, and `agree/disagree` items. We need to convert these to codes of `0` and `1`. With more than two category response options for categorical variables (e.g., religion, marital status), we need to recode the variable into a series of dichotomous variables, each of which is coded as `0` or `1`. We explain how to do this in the text, but there is some inconsistency in names used for these recoded variables.

A common name for a "0,1" variable is *dummy variable*. Others call these variables *indicator variables*, and still others call them *binary variables*. You should use whatever naming convention is standard for your content area. Here we will use them interchangeably.

When we pick a name for one of these variables, it is useful to pick a name representing the category coded as `1`. If we code women with `1` and men with `0`, we would call the variable `female`. Calling the variable `gender` would be less clear when we interpret results. If whites are coded `1` and nonwhites are coded `0`, it would make sense to call the variable `white`. Which category is coded `1` and which is coded `0` is arbitrary and affects only the sign of the coefficient. A positive sign signifies that the category coded `1` is higher on the dependent variable, and a negative sign signifies that it is lower on the dependent variable.

It is always important to check our changes, so enter a cross-tabulation, `tab2 gender97 male, missing`, to verify that we did not make a mistake. Once we have an indicator variable, we can enter it in the regression as a predictor just like any other variable. The unstandardized regression coefficient tells us the change in the outcome for a one-unit change in the predictor. Because a 1-unit change in the predictor, `male`, constitutes being a male rather than being a female, the regression coefficient is simply the difference in means on the dependent variable for males and females, controlling for other variables. Although this regression coefficient is easy to interpret, many researchers also report the βs. These make a lot of sense for a continuous variable; i.e., a one-standard-deviation change in the independent variable produces a β-standard-deviation change in the dependent variable. However, the βs are of dubious value for an indicator variable. It does not make sense for an indicator variable to vary by degree. Going from `0` for female to `1` for male makes sense, but going up or down one standard deviation on gender does not make much sense.

Using dummy variables is more complex when there are more than two possible scores. Consider race: We might want a variable to represent differences between whites, African Americans, Hispanic Americans, and others. We might code a variable `race` as 1 for non-Hispanic white, 2 for African American, 3 for Hispanic American, and 4 for other. This variable, race, is a nominal-level categorical variable, and it makes no sense to think of a code of 4 being higher or lower than a code of 1 or 2 or 3. These are simply nominal categories, and the order is meaningless. To represent these four racial or ethnic groups, we need three indicator variables. In general, when there are k categories, we need $k - 1$ indicator variables.

How does this work? First, choose one category that serves as a reference group. It makes sense to pick the group we want to compare with other groups. In this example, it would make sense to pick non-Hispanic white as our reference group. This is a large enough group to give us a good estimate of its value, and we will probably often want to compare other groups with whites. It would make little sense to pick the "other" group as the reference category because this is a combination of several different ethnicities and races (e.g., Pacific Islander, Native American). Also there are relatively few observations in the "other" category.

Next generate three dummy or indicator variables (aa, hispanic, and other), allowing us to uniquely identify each person's race/ethnicity. A score of 1 on other means that the person is in the other category. A score of 1 on hispanic means that the person is Hispanic. A score of 1 on aa means that the person is African American. You may be wondering how we know a person is white. A white non-Hispanic person will have a score of 0 on aa, 0 on hispanic, and 0 on other.

The dataset used two questions to measure race/ethnicity. This is done to more accurately represent Hispanics who may be of any race. The first item asks respondents if they are Hispanic. If they say no, they are then asked their race. The coding is fairly complicated, and here are the Stata statements used.

```
. gen race=race97
. replace race=1 if race97==1 & ethnic97==0
. replace race=2 if race97==2 & ethnic97==0
. replace race=3 if ethnic97==1
. replace race=4 if race97>3 & ethnic97==0
. tab2 race race97 ethnic97
. recode race (2 = 1 African_American) (1 3/4 = 0 Other), generate(aa)
. recode race (3 = 1 Hispanic) (1/2 4 = 0 Other), generate(hispanic)
. recode race (4 = 1 Other_race) (1/3 = 0 W_AA_H), generate(other)
. tab1 aa hispanic other
```

The first line creates a new variable, race, which is equal to the old variable, race97. We never want to change the original variable. We then change the code of race based on the code the person had on two variables, their original race97 variable and their ethnicity, ethnic97. You can run the command codebook race97 ethnic97 to see how these were coded. For example, we make race=1 if the person had a code of 1 on race97 and a code of 0 on ethnic97. We use the command tab2 race race97 ethnic97 to check that we have done this correctly. We then use the recode command to generate the three dummy variables. Finally, we do frequency tabulations for the three dummy variables. Here only African Americans will have a code of 1 on aa, only Hispanics will have a code of 1 on hispanic, only others will have a code of 1 on other, and only non-Hispanics whites will have a code of 0 on all three indicator variables.

Now we are ready to do the multiple regression. In this example, we will enter the variables in blocks. The first block is estimated by using

```
. regress smday97 age97 male
> if !missing(smday97, age97, male, psmoke97, aa, hispanic, other), beta

      Source |       SS       df       MS              Number of obs =    3469
-------------+------------------------------           F(  2,  3466) =   94.60
       Model |  22839.1255     2  11419.5628           Prob > F      =  0.0000
    Residual |  418378.464  3466  120.709309           R-squared     =  0.0518
-------------+------------------------------           Adj R-squared =  0.0512
       Total |   441217.59  3468  127.225372           Root MSE      =  10.987

------------------------------------------------------------------------------
     smday97 |      Coef.   Std. Err.      t    P>|t|                     Beta
-------------+----------------------------------------------------------------
       age97 |   1.837199    .133713    13.74   0.000                 .2272644
        male |   .2676277   .3734257     0.72   0.474                 .0118543
       _cons |  -20.31917   1.994527   -10.19   0.000                        .
------------------------------------------------------------------------------
```

We have a special qualification on who is included in this regression with the `if` qualifier. By inserting `if !missing(smday97, age97, male, psmoke97, aa, hispanic, other)`, we exclude people who did not answer all of the items. If we did not do this, the number of observations for each regression might differ. When you put this qualification with a command, there are a couple of things to remember: (1) `!missing` means "not missing"; programmers like to use the exclamation mark to signify "not". (2) You must insert the commas between variable names in this particular command; this is inconsistent with how Stata normally lists variables without the commas.

The $R^2 = 0.05$, $F(2, 3466) = 94.60$, $p < 0.001$. Although age and gender explain just a little bit of the variance in adolescent smoking behavior, their joint effect is statistically significant. How important is age as a predictor? For each year an adolescent gets older, he or she smokes an expected 1.84 more days per month. This is highly significant: $t(3468) = 13.74$, $p < 0.001$. This seems like a substantial increase in smoking behavior. $\beta = 0.23$ suggests that this is not a strong effect, however. What about gender—is it important? Because we coded males as `1` and females as `0`, the coefficient 0.27 for `male` indicates that males smoke an average of 0.27 more days per month than females do, controlling for age. This difference does not seem great, and it is not significant. The β is very weak, $\beta = 0.01$, but we will not try to interpret β because gender is a categorical variable.

The next regression equation adds peer influence. Repeat the command, but add the variable `psmoke97`, which represents the percentage of pairs who smoke. The command and results are

```
. regress smday97 age97 male psmoke97
> if !missing(smday97, age97, male, psmoke97, aa, hispanic, other), beta

      Source |       SS       df       MS              Number of obs =    3469
-------------+------------------------------           F(  3,  3465) =  134.03
       Model |  45876.8692     3  15292.2897           Prob > F      =  0.0000
    Residual |   395340.72  3465  114.095446           R-squared     =  0.1040
-------------+------------------------------           Adj R-squared =  0.1032
       Total |   441217.59  3468  127.225372           Root MSE      =  10.682

------------------------------------------------------------------------------
     smday97 |      Coef.   Std. Err.      t    P>|t|                     Beta
-------------+----------------------------------------------------------------
       age97 |   1.286052   .1356611     9.48   0.000                 .1590866
        male |    .948632   .3662008     2.59   0.010                 .0420187
    psmoke97 |   2.247035   .1581336    14.21   0.000                 .2403992
       _cons |  -19.46011   1.940058   -10.03   0.000                        .
------------------------------------------------------------------------------
```

This new model explains 10% of the variance in smoking behavior: $R^2 = 0.10$, $F(3, 3465) = 134.03$, $p < 0.001$. Does psmoke97 make a unique contribution? There are three ways of getting an answer. First, we can see that the regression coefficient of 2.25 is significant: $t(3468) = 14.21$, $p < 0.001$ and $\beta = 0.24$.

A second way of testing the unique effect of adding psmoke97 is to use the pcorr2 command that we installed earlier to estimate how much variance is explained by each predictor uniquely, as well as what was already explained by age and gender. Once you have the command installed (run ssc install pcorr2 to install pcorr2), enter pcorr2 smday97 age97 male psmoke97 if !missing(smday97, age, male, psmoke97, aa, hispanic, other), which results in the following:

```
. pcorr2 smday97 age97 male psmoke97
> if !missing(smday97, age97, male, psmoke97, aa, hispanic, other)
(obs=3469)

Partial and Semipartial correlations of smday97 with

    Variable |     Partial      SemiP    Partial^2     SemiP^2       Sig.
-------------+-------------------------------------------------------------
       age97 |      0.1590     0.1524       0.0253      0.0232      0.000
        male |      0.0440     0.0417       0.0019      0.0017      0.010
    psmoke97 |      0.2347     0.2285       0.0551      0.0522      0.000
```

So we can say that peer smoking influence adds 5% to the explained variance, semi-partial $R^2 = 0.05$, $p < 0.001$. If we did not have access to this user-written command, we could subtract the two R^2 values for the two models: $0.10 - 0.05 = 0.05$.

If you have not installed the pcorr2 command, you could compute the significance of the R^2 change a third way by hand, using a formula in a standard statistics textbook. However, this will give you the same result as the t test for psmoke97 when there is a variable added in the block. It is important to see that the R^2 change for a variable and the semipartial R^2 are the same thing. Both tell us how much more variance is explained by peer influence.

The next model we want to estimate includes race/ethnicity. We can estimate this by simply adding the variables aa, hispanic, and other to the regression command. Here are the command and its results:

```
. regress smday97 age97 male psmoke97 aa hispanic other
> if !missing(smday97, age97, male, psmoke97, aa, hispanic, other), beta

      Source |       SS       df       MS              Number of obs =    3469
-------------+------------------------------           F(  6,  3462) =   90.35
       Model |  59732.584     6  9955.43067           Prob > F      =  0.0000
    Residual | 381485.005  3462  110.192087           R-squared     =  0.1354
-------------+------------------------------           Adj R-squared =  0.1339
       Total |  441217.59  3468  127.225372           Root MSE      =  10.497

------------------------------------------------------------------------------
     smday97 |      Coef.   Std. Err.      t    P>|t|                     Beta
-------------+----------------------------------------------------------------
       age97 |   1.307936   .1333508     9.81   0.000                 .1617937
        male |   1.054237   .3600216     2.93   0.003                 .0466964
    psmoke97 |   2.169816   .1556779    13.94   0.000                  .232138
          aa |  -4.676537   .4592872   -10.18   0.000                -.1674766
    hispanic |  -3.223828   .4638049    -6.95   0.000                -.1144782
       other |   .1552411    1.13163     0.14   0.891                 .0021882
       _cons |  -17.99777   1.911617    -9.41   0.000                        .
------------------------------------------------------------------------------
```

When we add the set of three indicator variables that collectively represent race and ethnicity, we increase our R^2 to 0.14, $F(6, 3462) = 90.35$, $p < 0.001$. This represents a $0.1354 - 0.104 = .0314$, or 3.1%, increase in R^2 when we add the set of three indicator variables that represent race/ethnicity.

The t tests for `aa` and `hispanic` are statistically significant, and the t test for `other` is not. This shows that both African Americans and Hispanics smoke fewer days per week than do white non-Hispanics. If we want to know whether the set of three indicators is statistically significant (this refers to the three variables simultaneously and tests the significance of the combined race/ethnicity variable), we can run the following `test` command immediately after the regression and before doing another multiple regression because it uses the most recent results:

```
. test aa hispanic other

 ( 1)  aa = 0
 ( 2)  hispanic = 0
 ( 3)  other = 0

       F(  3,  3462) =   41.91
            Prob > F =    0.0000
```

This `test` command is both simple and powerful. It does a test that three null hypotheses are all true. It is testing that the effect of `aa` is zero, `hispanic` is zero, and `other` is zero. It gives us $F(3, 3462) = 41.91$, $p < 0.001$. Thus we can say that the effect of race/ethnicity is weak because it adds just 3.1% to the explained variance, but it is statistically significant. If you paid careful attention to the unstandardized regression coefficients, you may object to the statement that the effect is weak. Uniquely explaining 3.1% of the variance sounds weak, which is why we described it this way. However, $b = -4.68$ for African Americans, and this means that they are expected to smoke almost five fewer days per month than white non-Hispanics. If you think this is a substantial difference, you might want to focus on this difference rather than the explained variance.

More on testing a set of parameter estimates

We have seen one important use for the `test` command, namely, testing a set of indicator variables (`aa`, `hispanic`, `other`) that collectively define another variable (race/ethnicity). There are many other uses for this command. You may want to test whether a set of control variables is significant. When you have interaction terms, you may want to test a set of them. The test is always done in the context of the multiple regression most recently estimated. In this example, if we ran the command `test age97 male`, we would get a different result from when we entered these two variables by themselves. This is because we are testing whether these two variables are simultaneously significant, controlling for all the other variables that are in the model.

There is a useful command called `nestreg`. The procedures we have been running here involve nested regressions. The regressions are nested in the sense that the first regression is nested in the second because all the predictors in the first regression are included in the second. Likewise, the second regression is nested in the third regression, and so on. Some people call this hierarchical regression. If you call it hierarchical regression, you should not confuse it with hierarchical linear modeling, which is the name of a program that does multilevel analysis or mixed regression, which is related to the program in Stata called `xtmixed`. Nested regression is used where we have blocks of variables we want to enter in a sequence, each step adding another block. Our method, using a series of regressions followed by the `test` command, is extremely powerful and flexible. However, the `nestreg` command was created to automate this process.

To run the `nestreg` command, you add the `nestreg:` prefix (the colon must be included) to the regression command. You then write a `regress` command, but put each block of predictors in parentheses. Here we have three blocks of predictors `(age97 male)`, `(psmoke97)`, and `(aa hispanic other)`. The usual options for the `regress` command are available. We selected the option to produce the beta weights.

```
. nestreg: regress smday97 (age97 male) (psmoke97) (aa hispanic other), beta
Block  1: age97 male

      Source |       SS       df       MS              Number of obs =    3469
-------------+------------------------------           F(  2,  3466) =   94.60
       Model |  22839.1255     2  11419.5628           Prob > F      =  0.0000
    Residual |  418378.464  3466  120.709309           R-squared     =  0.0518
-------------+------------------------------           Adj R-squared =  0.0512
       Total |   441217.59  3468  127.225372           Root MSE      =  10.987

------------------------------------------------------------------------------
     smday97 |      Coef.   Std. Err.      t    P>|t|                     Beta
-------------+----------------------------------------------------------------
       age97 |   1.837199    .133713    13.74   0.000                 .2272644
        male |   .2676277   .3734257     0.72   0.474                 .0118543
       _cons |  -20.31917   1.994527   -10.19   0.000                        .
------------------------------------------------------------------------------
```

```
Block  2: psmoke97
      Source |       SS       df       MS              Number of obs =    3469
-------------+------------------------------           F(  3,  3465) =  134.03
       Model |  45876.8692     3  15292.2897           Prob > F      =  0.0000
    Residual |   395340.72  3465  114.095446           R-squared     =  0.1040
-------------+------------------------------           Adj R-squared =  0.1032
       Total |   441217.59  3468  127.225372           Root MSE      =  10.682

------------------------------------------------------------------------------
     smday97 |      Coef.   Std. Err.      t    P>|t|                     Beta
-------------+----------------------------------------------------------------
       age97 |   1.286052   .1356611     9.48   0.000                 .1590866
        male |    .948632   .3662008     2.59   0.010                 .0420187
    psmoke97 |   2.247035   .1581336    14.21   0.000                 .2403992
       _cons |  -19.46011   1.940058   -10.03   0.000                        .
------------------------------------------------------------------------------
Block  3: aa hispanic other
      Source |       SS       df       MS              Number of obs =    3469
-------------+------------------------------           F(  6,  3462) =   90.35
       Model |  59732.584      6  9955.43067           Prob > F      =  0.0000
    Residual | 381485.005   3462  110.192087           R-squared     =  0.1354
-------------+------------------------------           Adj R-squared =  0.1339
       Total |   441217.59  3468  127.225372           Root MSE      =  10.497

------------------------------------------------------------------------------
     smday97 |      Coef.   Std. Err.      t    P>|t|                     Beta
-------------+----------------------------------------------------------------
       age97 |   1.307936   .1333508     9.81   0.000                 .1617937
        male |   1.054237   .3600216     2.93   0.003                 .0466964
    psmoke97 |   2.169816   .1556779    13.94   0.000                  .232138
          aa |  -4.676537   .4592872   -10.18   0.000                -.1674766
    hispanic |  -3.223828   .4638049    -6.95   0.000                -.1144782
       other |   .1552411    1.13163     0.14   0.891                 .0021882
       _cons |  -17.99777   1.911617    -9.41   0.000                        .
------------------------------------------------------------------------------
```

Block	F	Block df	Residual df	Pr > F	R2	Change in R2
1	94.60	2	3466	0.0000	0.0518	
2	201.92	1	3465	0.0000	0.1040	0.0522
3	41.91	3	3462	0.0000	0.1354	0.0314

This command will run the three regressions and report increments in R^2 and whether these are statistically significant. A special strength of this command is that it works with many types of nested models, as you can see by typing `help nestreg`. The `nestreg` command is hard to find under the menu system. You find it at Statistics ▷ Other ▷ Nested model statistics. In addition to working with our multiple regression example, the dialog box provides a way of evaluating nested models by using 18 specialized extensions of multiple regression such as logistic regression.

The results show the three regressions we ran earlier—calling them `Block 1`, `Block 2`, and `Block 3`—and listing the variables added with each block. After the last block is entered, there is a summary table showing the change in R^2 for the block along with its significance. For example, the second block added the variable `psmoke97`. This block increased the R^2 by 0.052, from 0.052 to 0.104. This increment in R^2 is significant, $F(1,3465) = 201.92$, $p < 0.001$. The significance of $R^2 = 0.104$ at this stage is taken from the `Block 2` regression, $F(3,3465) = 134.03$, $p < 0.001$. Avoid confusing the F test for the increment in R^2 from the summary table and the F test for the R^2 itself from the regression for the block. Adding the race/ethnicity variables to the model in `Block 3` increased R^2 by 0.031, and this is significant, $F(3,3462) = 41.91$, $p < 0.001$. So, too, is the final $R^2 = 0.135$, $F(6,3462) = 90.35$, $p < 0.001$.

Tabular presentation of hierarchical regression models

You may find it helpful to have a table showing the results of your nested (hierarchical) regressions. The table should show b's, standard errors, and standardized betas for each model you fit. You should also include R^2 for each model and the F test of the change in the R^2. In many fields, it is conventional to show the significance of regression coefficients by using asterisks. One asterisk is used for parameter estimates that are significant at the 0.05 level, two asterisks for parameters significant at the 0.01 level, and three asterisks for parameters significant at the 0.001 level. Some fields, e.g., psychology through the American Psychological Association, have recommended tables that only report the coefficients for the variables that are added at each step. This approach ignores the fact that the coefficients for the variables from previous blocks will change when the new block of variables is added. This can become misleading because a variable in block 1 may be significant at that step, but it may become insignificant when subsequent blocks are added. A detailed example of a summary table format for nested regression is shown at http://oregonstate.edu/~acock/tables/regression.pdf.

Nested regression is commonly used to estimate the effect of personality characteristics or individual beliefs after you have controlled for background variables. If you wanted to know whether motivation was important to career advancement, you would do a multiple regression entering background variables first and then adding motivation as a final step. For example, you might enter gender, race, and parents' education in the first step as background variables over which the individual has no control; then enter achievement variables, such as education, in a second step; and finally, enter motivation. If motivation makes a significant increase in R^2 in the final step, then you have a much better test of the importance of motivation than if you had just done a correlation of career achievement and motivation without entering the control variables first.

10.10 Fundamentals of interaction

In many situations, the effect of one variable depends on where you are on another variable. For example, there is a well-established relationship between education and income. The more education you have, the more income you can expect, on average. There is also an established relationship between gender and income. Men, on average, make more money than women. With this in mind, we could do a regression of income on both education and gender to see how much of a unique effect each predictor has. There is, however, an important relationship that we could miss by doing this. Not only may men make more money than women, but the payoff they get for each additional year of education may be greater than the payoff women receive. Thus, men are advantaged in two ways. First, they make more money on average. Second, because they are men, they get a greater return on their education. This second issue is what we mean by interaction. That is, the effect of education on income depends on your gender—stronger effect for men, weaker effect for women. How can we test this relationship using multiple regression?

As an example here, we will fit two models. The first model includes just education and gender as predictors, or "main effects", to distinguish them from interaction effects. The second model adds the interaction between education and gender. Let's use `c10interaction.dta`, a dataset with hypothetical data. The first model is fitted using `regress inc educ male`, where `inc` is income measured in thousands of dollars, `educ` is education measured in years, and `male` represents gender, with men coded `1` and women coded `0`. Here are the results:

```
. regress inc educ male, beta

      Source |       SS       df       MS              Number of obs =     120
-------------+------------------------------           F(  2,   117) =   37.19
       Model |  100464.105     2  50232.0527           Prob > F      =  0.0000
    Residual |  158015.895   117   1350.5632           R-squared     =  0.3887
-------------+------------------------------           Adj R-squared =  0.3782
       Total |      258480   119  2172.10084           Root MSE      =   36.75

------------------------------------------------------------------------------
         inc |      Coef.   Std. Err.      t    P>|t|                     Beta
-------------+----------------------------------------------------------------
        educ |   8.045694   1.008586     7.98   0.000                 .5775017
        male |   19.04991   6.719787     2.83   0.005                 .2052297
       _cons |  -42.54411    14.2919    -2.98   0.004                        .
------------------------------------------------------------------------------
```

We can explain 38.87% of the variance in income by using education and gender, $F(2, 117) = 37.19$, $p < 0.001$, and both education and gender are significant in the ways we anticipated. Each additional year of education yields an expected increase in income of $8,046, and men expect to make $19,050 more than women. (Remember, these are hypothetical data!) We can make a graph of this relationship. First, generate a predicted income for men as a new variable and a predicted income for women as a new variable. Just after the regression, run the following commands:

```
. predict incfnoi if male==0
. predict incmnoi if male==1
```

The first command generates a predicted value for income for women (`male==0`), and the second command does the same for men. Then do an overlaid two-way graph by using the dialog box we discussed earlier in the book. The graph command used to produce figure 10.8 is long:

```
. twoway (connected incmnoi educ if male == 1, lcolor(black) lpattern(dot)
> msymbol(diamond) msize(large)) (connected incfno educ if male == 0,
> lcolor(black) lpattern(solid) msymbol(circle) msize(large)),
> ytitle(Income in thousands) xtitle(Education) legend(order(1 "Men" 2 "Women"))
> scheme(s2manual)
```

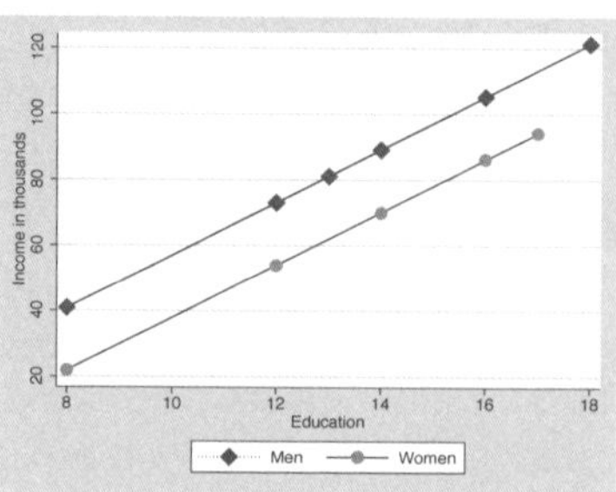

Figure 10.8. Education and gender predicting income, no interaction

There is one line for the education-to-income relationship for men (the higher line) and another line for women. This shows a strong positive connection between education and income, as well as a big gender gap. We use this model that includes just the main effects and has both lines parallel because it is the best model, as long as we assume that the gender gap is constant. If the size of the gender gap increases with increasing age, this model would not be able to show that.

Testing for interaction takes us a huge step further. When we test for interaction, we want to see if the slopes are different. In other words, we want to see if the slope between education and income is different and perhaps steeper for men than it is for women. You may be thinking that the graph shows the lines to be parallel, but this is because we set up the equation without an interaction term; without the interaction term, Stata gives us the best parallel lines. To see if the lines really should not have this restriction, we add an interaction term. We get the interaction term as the product of gender and education. The command to generate the interaction term is

```
. gen ed_male = educ*male
```

where `ed_male` is the product of education times gender. Then we refit the model, adding the interaction term:

```
. regress inc educ male ed_male, beta

      Source |       SS       df       MS              Number of obs =     120
-------------+------------------------------           F(  3,   116) =   34.89
       Model |  122604.719     3  40868.2397           Prob > F      =  0.0000
    Residual |  135875.281   116  1171.33863           R-squared     =  0.4743
-------------+------------------------------           Adj R-squared =  0.4607
       Total |      258480   119  2172.10084           Root MSE      =  34.225

------------------------------------------------------------------------------
         inc |      Coef.   Std. Err.      t    P>|t|                     Beta
-------------+----------------------------------------------------------------
        educ |   3.602369   1.388076     2.60   0.011                 .2585699
        male |  -91.88539   26.27242    -3.50   0.001                -.9899052
     ed_male |   8.196446   1.885263     4.35   0.000                 1.287508
       _cons |   16.84834   19.07279     0.88   0.379                        .
------------------------------------------------------------------------------
```

These results show that we can explain 47.43% of the variance, $F(3, 116) = 34.89$, $p < 0.001$, in income after we add the effect of interaction. This is an increase of 8.56%. We know this is significant because the t test for the interaction term is significant, $t(116) = 4.35$, $p < 0.001$. We could also get this increase by using `pcorr2` for the semipartial R^2 or by using `nestreg: regress inc (educ male) (ed_male), beta`.

Centering quantitative predictors before computing interaction terms

Although we have not done it here, some researchers choose to center quantitative independent variables, such as education, before computing the interaction terms. The interaction term is the product of the centered variables, but the interaction term will not be centered itself. Centering involves subtracting the mean from each observation. If the mean of education were 12 years, centered education would be `generate c_educ = educ - 12`. A better way to do this is to summarize the variable first. Stata then saves the mean to many decimal places in memory as a variable called `r(mean)`. We would use two commands: `summarize educ` and then `generate educ_c = educ - r(mean)`. If you plan to center many variables, you may want to install the user-written command `center` (type `findit center` or `ssc install center`). This command was written by Ben Jann. Once it is installed, you can enter `center educ` to automatically create a new variable named `c_educ`.

If you center these variables, the mean has a value of zero. Because the intercept is the estimated value when the predictors are zero, the intercept is the estimated value when the centered variable is at its mean. This often makes more sense, especially when a value of zero for the uncentered variable is a rare event, as it is in the example of years of education. The centered education variable becomes a measure of how far a person is above or below average on years of education rather than how many years of education they have had. This centering can help you interpret the intercept.

Interpreting the interaction term (`ed_male`) and the main effects (`educ`, `male`) is tricky when the interaction is significant. Just glancing at them seems to suggest ridiculous relationships. The −91.89 for the `male` variable might look like men make significantly less than women, but we cannot interpret it this way in the presence of interaction. We need to make a separate equation for each level of the categorical variable. Thus we need to make an equation for men and a separate equation for women. This is not hard to do; just remember that women have a score of zero on two variables: `male` and `ed_male` (`educ` $\times 0 = 0$). For men, by substituting a value of `1` for the `male` variable, the equation simplifies to

$$\widehat{\texttt{inc}} = (16.85 - 91.89) + (3.60 + 8.20)\texttt{educ}$$

$$\widehat{\texttt{inc}} = -75.04 + 11.80(\texttt{educ})$$

For women, by substituting a value of `0` for the `male` variable, the equation simplifies to

$$\widehat{\texttt{inc}} = 16.85 + 3.60(\texttt{educ})$$

Here the payoff of one additional year of education for men is \$11,800 compared with just \$3,600 for women. The adjusted constant or intercept for men of $-\$75,040$ does not make a lot of sense. In the data, the lowest education is 8 years, and the constant refers to a person who has no education. Do not interpret an intercept that is out of the range of the data.

A nice way to see how the interaction works is with a graph. Construct an overlaid two-way graph, as before. After doing the regression with the interaction term, use the `predict` command to estimate income for women (`if male==0`) and for men (`if male==1`) and then produce a graph:

```
. predict incf if male==0

. predict incm if male==1

. twoway (connected incm educ if male==1, lcolor(black) lpattern(dot)
> msymbol(diamond) msize(large)) (connected incf educ if male==0, lcolor(black)
> lpattern(solid) msymbol(circle) msize(large)), ytitle(Income in thousands)
> xtitle(Education) legend(order(1 "Men" 2 "Women")) scheme(s2manual)
```

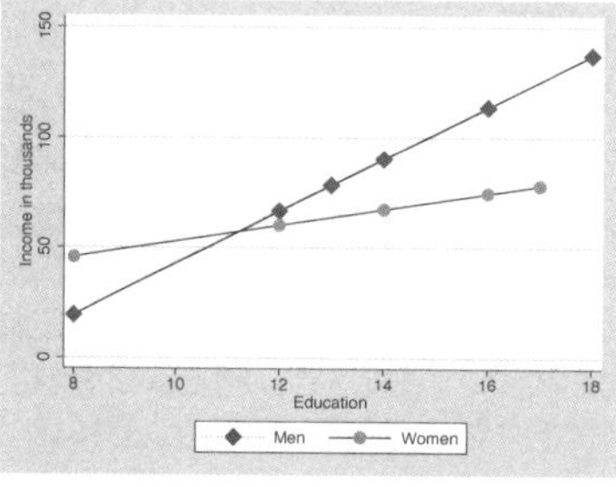

Figure 10.9. Education and gender predicting income, with interaction

Figure 10.9 shows that, over the range of the data, the payoff of an additional year of education is much steeper for men than it is for women. There is also a point at which the lines cross. Men with substantially less than a high school diploma make less money than women with comparable education. However, for people with a high school diploma or more, men not only make more money but make increasingly more money as education increases.

These are hypothetical data, but if they were actual data, here is what we could say: At the lowest level of education, women have some advantage over men. For those with a high school diploma or more money, men not only make more, but the gap increases with each additional year of education. A year of education increases the estimated earnings of men by $11,800, compared with just $3,600 for women. This difference is highly statistically significant.

Do not compare correlations across populations

We have found a significant interaction showing that women are relatively disadvantaged in the relationship between education and income. We did this by examining the unstandardized slopes, which showed that the payoff for an additional year of education was $11,800 for men, compared with $3,600 for women. If these were actual data, this would be compelling evidence of a gross inequity. Some researchers make the mistake of comparing standardized beta weights or correlations to make this argument, but this is a serious mistake. The standardized beta weights depend on the form of the relationship (shown with the unstandardized slopes in figure 10.9) and the differences in variance. Unless the men and women have identical variances, comparing the correlations or beta weights can yield misleading results. Correlation is a measure of fit or the clustering of the observations around the regression line. You can have a steep slope like that for men, but the observations may be widely distributed around this, leading to a small correlation. Similarly, you can have a much flatter slope, like that for women, but the observations may be closely packed around the slope, leading to a high correlation. Whether the correlation between education and income is higher or lower for women than it is for men measures how closely the observations are to the estimated values. How steep the slopes are, measured by the unstandardized regression coefficients as we have done, measures the form of the relationship.

10.11 Summary

Multiple regression is an extremely powerful and general strategy for analyzing data. We have covered several applications of multiple regression and diagnostic strategies. These provide you with a strong base, and Stata has almost unlimited ways of extending what we have done. Here are the main topics we have covered:

- What multiple regression can do and how it is an extension of the bivariate correlation and regression we covered in chapter 8
- The basic command for doing multiple regression, including the option for standardized beta weights
- How we measure the unique effects of a variable, controlling for a set of other variables. This included a discussion of semipartial correlation and the increment in R^2
- Diagnostics of the dependent variable and how we test for normality of the dependent variable
- Diagnostics for the distribution of residuals or errors in prediction
- Diagnostics to find outlier observations that have too much influence and may involve a coding error
- Collinearity and multicollinearity, along with how to evaluate the seriousness of the problem and what to do about it
- How to use weighted samples where different observations have different probabilities of being included, such as when there is oversampling of certain groups and we want to generalize to the entire population
- How to work with categorical predictors, both with just two categories and with more than two categories
- How to work with nested (hierarchical) regression in which we enter blocks of variables at a time
- How to work with interaction between a continuous variable and a categorical variable

If you are a beginner, you may be amazed by the number of things you can do with regression and by what we have covered in this chapter. If you are an experienced user, we have only whetted your appetite. The next chapter on logistic regression includes several references that you should pursue if you want to know more about Stata's capabilities for multiple regression. Stata has special techniques for working with different types of outcome variables, working with panel data, and working with multilevel data where subgroups of observations share common values on some variables.

The next chapter is the first extension of multiple regression. I will cover the case in which there is a binary outcome; this happens often. You will have data with which you want to predict whether a couple will get divorced, whether a business will survive its first year or not, whether a patient recovers, whether an adolescent reports having considered suicide, or whether a bill will become a law. In each case, the outcome will be either a success or a failure. Then we will learn how to predict a success or a failure.

10.12 Exercises

1. Use the gss2002_chapter10.dta dataset. You are interested in how many hours a person works per week. You think that men work more hours, older people work fewer hours, and people who are self-employed work the most hours. The GSS has the needed measures including sex, age, wrkslf, and hrs1 (hours worked last week). Recode sex and wrkslf into dummy variables, and then do a regression of hours worked on the predictors. Write out a prediction equation showing the β. What are the predicted hours worked for a female who is 20 years old and works for herself? How much variance is explained, and is this statistically significant?

2. Use the census.dta dataset. The variable tworace is the percentage of people in each state and the District of Columbia that report being a combination of two or more races. Predict this by using the percentage of the state that is white (white), the median household income (hhinc), and the percentage with a B.A. degree or more education (ba). Predict the estimated value, calling it yhat. Predict the standardized residual, calling it rstandard. Plot a scattergram like the one in figure 10.4. Do a listing of states that have a standardized residual of more than 1.96. Interpret the graph and the standardized residual list.

3. Use the census.dta dataset. Repeat the regression in the previous analysis. Compute the DFbeta score for each predictor, and list the states that have relatively large DFbeta values on any of the predictors. Explain why the problematic states are problematic.

4. Use the gss2002_chapter10.dta dataset. Suppose that you want to know if a man's socioeconomic status depends more on his father's or his mother's socioeconomic status. Run a regression (use sei, pasei, and masei), and control for how many hours a week he works (hrs1). Do a test to see if pasei and masei are equal. Create a graph similar to that in figure 10.6, and carefully interpret it for the distribution of residuals.

5. Use the gss2002_chapter10.dta dataset. You are interested in predicting the socioeconomic status of adults. You are interested in whether conservative political views predict higher socioeconomic status uniquely, after you have included background and achievement variables. The socioeconomic variable is sei. Do a nested regression. In the first block, enter gender (you will need to recode the variable, sex, into a dummy variable, male), mother's education (maeduc), father's education (paeduc), mother's socioeconomic index (masei), and father's socioeconomic index (pasei). In the second block, enter education (educ). In the third block, add conservative political views (polviews). Carefully interpret the results. Create a table summarizing the results.

6. Use the c10interaction.dta dataset. Repeat the interaction analysis presented in the text, but first center education and generate the interaction term to be male times the centered score on education. Centering involves subtracting the mean from each observation. You can either use the summarize command to get

the mean for education and then subtract this from each score, or run `findit center` and install the program called `center`. Run `center educ` to generate a new variable called `c_educ`. Then generate the interaction term as `generate educc_male = c_educ*male`. After running the regression (`regress inc male c_educ educc_male`) and two-way overlay graph, compare the results with those in the text. Why is the intercept so different? Hint: The intercept is when the predictors are zero—think about what a value of zero on a centered variable is.

11 Logistic regression

11.1 Introduction to logistic regression

The regression models that we covered in chapter 10 focused on quantitative outcome variables that had an underlying continuum. There are important applications of regression in which the outcome is binary—something either does or does not happen—and we want to know why. Here are a few examples:

- A woman is diagnosed as having breast cancer
- A person is hired
- A married couple get divorced
- A new faculty member earns tenure
- A participant in a study drops out of the program
- A candidate is elected

We could quickly generate many more examples where something either does or does not happen. Binary outcomes were fine as predictors in chapter 10, but now we are interested in them as dependent variables. Logistic regression is a special type of regression that is used for binary-outcome variables.

11.2 An example

One of the best predictors of marital stability is the amount of positive feedback a spouse gives her or his partner.

- You observe 20 couples in a decision-making task for 10 minutes.
- You do this 1 week prior to their marriage and record the number of positive responses each of them makes about the other. An example would be "you have a good idea" or "you really understand my goals".
- You look just at the positive comments made by the husband and call these `positives`.
- You wait 5 years and see whether they get divorced. You call this `divorce`.
- You assign a code of `1` to those couples who get divorced and a code of `0` to those who do not.
- The resulting dataset has just two variables and is called `divorce`.

The following is a list of the data in the `divorce.dta` dataset:

```
. list divorce positives

     +--------------------+
     | divorce   positi~s |
     |--------------------|
  1. |       0         10 |
  2. |       0          8 |
  3. |       0          9 |
  4. |       0          7 |
  5. |       0          8 |
     |--------------------|
  6. |       0          5 |
  7. |       0          9 |
  8. |       0          6 |
  9. |       0          8 |
 10. |       0          7 |
     |--------------------|
 11. |       1          1 |
 12. |       1          1 |
 13. |       1          3 |
 14. |       1          1 |
 15. |       1          4 |
     |--------------------|
 16. |       1          5 |
 17. |       1          6 |
 18. |       1          3 |
 19. |       1          2 |
 20. |       1          0 |
     +--------------------+
```

If we graph this relationship, it looks strange because the outcome variable, `divorce`, has just two possible values. (The graph has been enhanced by using the Graph Editor, but the basic command is `scatter divorce positives`.)

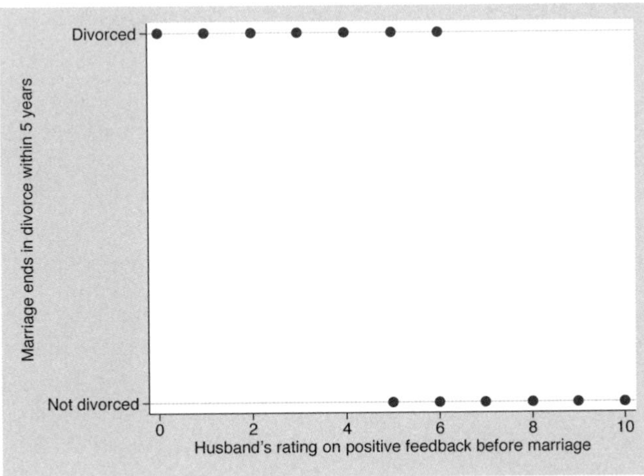

Figure 11.1. Positive feedback and divorce

A couple with a very low rating, say, 0–4 on positives, is almost certain to have a score of 1 on divorce. With a score of 7–10, a couple is almost certain to not get divorced, i.e., to score 0 on divorce. A couple somewhere in the middle on positives, with a score of 5 or 6, is hard to predict. This is the zone of transition. It would not make sense to use a straight line as a prediction rule like we did with bivariate regression.

If we use logistic regression to predict the probability of a divorce based on the husband's rating on giving positive feedback prior to the marriage (we will soon learn how to do this), we get a predicted probability of divorce based on the score a husband has for positives. This conforms to what is sometimes called an S-curve (figure 11.2).

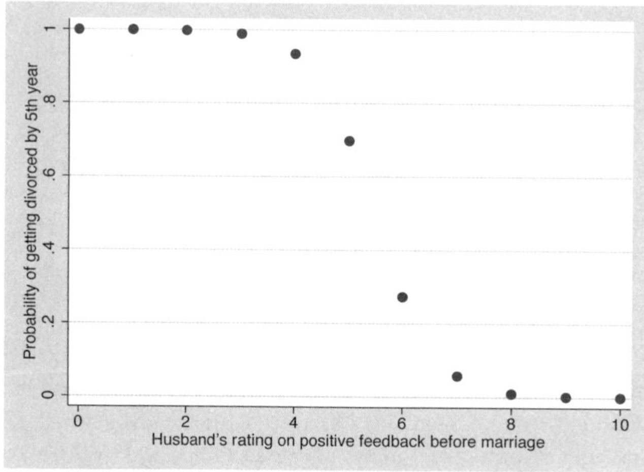

Figure 11.2. Predicted probability of positive feedback and divorce

Now this makes sense; it is a nonlinear relationship. The probability of divorce is nonlinearly related to the number of positive responses of the husband. If he gives relatively few positive responses, the probability of a divorce is very high; if he gives a lot of positive feedback, the probability of divorce is very low. Those husbands who are somewhere in the middle on positive feedback are the hardest to predict. This is exactly what we would expect and what our figure illustrates.

To fit a model like this, logistic regression does not estimate the probability directly. Instead, we estimate something called a logit. When the logit is linearly related to the predictor, the probability conforms to an S-curve like the one in figure 11.2. The reason we go through the trouble of computing a logit for each observation and using that in the logistic regression as the dependent variable is that the logit can be predicted by using a linear model. Notice that the probability of getting a divorce varies from 0 to 1, but the logit ranges from about -10 to about 10 in this example. Although the probability of divorce is not linearly related to positives, the logit is. The relationship between the positive feedback the husband gives and the logit of the `divorce` variable appears in figure 11.3. From a statistical point of view, figure 11.3 predicting the logit of `divorce` is equivalent to figure 11.2 predicting the probability of `divorce`.

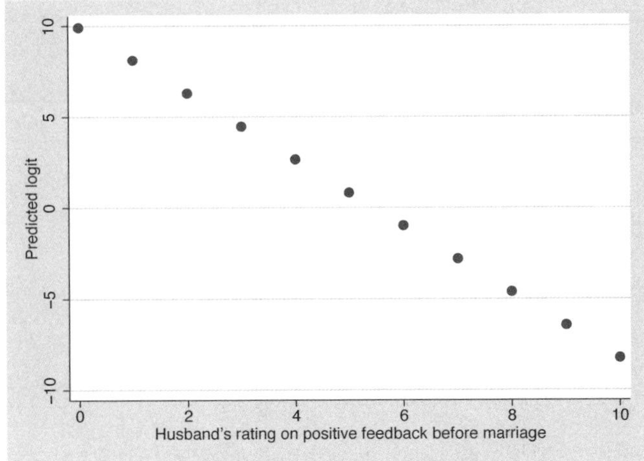

Figure 11.3. Predicted probability of positive feedback and logit of divorce

Before logistic regression was available, people did ordinary least squares (OLS) regression, as described in chapter 10, when there was a binary-outcome variable. One problem with this is that the probability of the outcome being a 1 is much more likely to follow an S curve than it is to follow a straight line. A second problem is that there is an absolute lower limit of zero and an absolute upper limit of one for a binary variable. If you predict 1 using OLS regression, you are saying that the probability is 1.0. If you predict 0.5, you are saying that the probability is 0.5. What if you are predicting −0.2 or 1.5? These are impossible values, but they can happen by using OLS regression. Figure 11.4 shows what would happen using OLS regression because OLS regression has no way of knowing that your outcome cannot be below 0 or above 1.

```
. regress divorce positives

      Source |       SS       df       MS              Number of obs =      20
-------------+------------------------------           F(  1,    18) =   42.95
       Model |  3.52343538     1  3.52343538           Prob > F      =  0.0000
    Residual |  1.47656462    18  .082031368           R-squared     =  0.7047
-------------+------------------------------           Adj R-squared =  0.6883
       Total |           5    19  .263157895           Root MSE      =  .28641

------------------------------------------------------------------------------
     divorce |      Coef.   Std. Err.      t    P>|t|     [95% Conf. Interval]
-------------+----------------------------------------------------------------
   positives |  -.1381739    .021083    -6.55   0.000    -.1824677   -.0938801
       _cons |   1.211596   .1260582     9.61   0.000     .9467574    1.476434
------------------------------------------------------------------------------

. predict divols
(option xb assumed; fitted values)

. twoway (scatter divorce positives) (lfit divols positives)
```

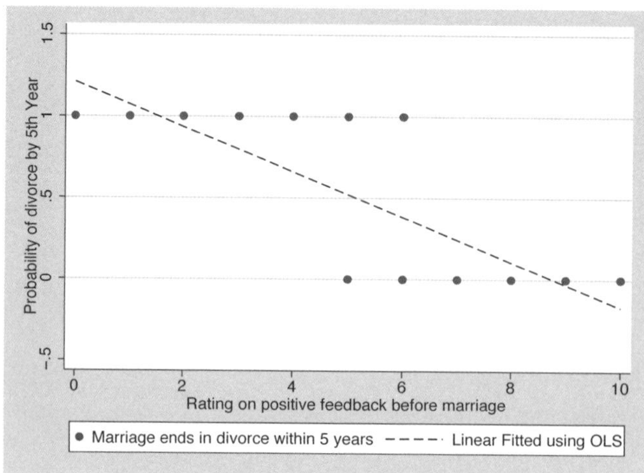

Figure 11.4. Positive feedback and divorce using OLS regression

You can see the problem with using a straight line with OLS regression. A couple with a score of `10` on `positives` would have a probability of getting divorced within 5 years of less than zero. A couple with a score of `0` on `positives` would have a probability of about 1.25. Both of these are impossible values because a probability must not be greater than 1.0 nor less than 0.0. Comparing figure 11.2 and figure 11.4, you can see that a logistic regression approach makes sense for binary-outcome variables.

11.3 What is an odds ratio and a logit?

We will explain odds ratios and logits by using a simple example with hypothetical data, `environ.dta`. Suppose that you are interested in the relationship between environmental support and support for a liberal candidate in a community election. Your survey asks items that you can use to divide your sample into two categories of environmental concern, namely, high or low environmental concern. You also ask them if they support a particular liberal candidate, and they either support this candidate or they do not. We have two variables we will call `environ` (`1` means high environmental concern, `0` means low environmental concern) and `libcand` (`1` means support this liberal candidate, `0` means that they do not). Here is how the relationship between these variables looks:

```
. tab2 environ libcand, row
-> tabulation of environ by libcand
```

Key
frequency *row percentage*

Environmental concern: 1 high, 0 low	support liberal candiate: 1 yes, 0 no 0	1	Total
0	6 60.00	4 40.00	10 100.00
1	3 30.00	7 70.00	10 100.00
Total	9 45.00	11 55.00	20 100.00

- The probability of supporting a liberal candidate (1 on support liberal) is 11/20, or 0.55.
- The probability of supporting a liberal candidate if you have low environmental concern (1 on support liberal if 0 on Environmental concern) is 4/10, or 0.4.
- The probability of supporting a liberal candidate if you have high environmental concern (1 on support liberal if 1 on Environmental concern) is 7/10, or 0.7.

From these two probabilities, we can see that there is a relationship. With only 4 of the 10 people who have a low environmental concern supporting the liberal candidate compared with 7 of the 10 people who have high environmental concern, this looks like a fairly strong difference.

What are the odds of supporting a liberal candidate? This is the ratio of those who support the candidate to those who do not.

- The odds of supporting a liberal candidate are $11/9 = 1.22$. This means that there are 1.22 people supporting the liberal candidate for each person opposing the liberal candidate.
- The odds of supporting a liberal candidate if you have low environmental concern are $4/6 = 0.67$. This also means that among those with low environmental concern there are just .67 people supporting the liberal candidate for each person opposing the liberal candidate.
- The odds of supporting a liberal candidate if you have high environmental concern are $7/3 = 2.33$. This indicates that, among those with high environmental concern,

there are 2.33 people supporting the liberal candidate for each person who opposes the liberal candidate.

11.3.1 The odds ratio

We need a way to combine these two odds, 2.33 and 0.67, into one number. We can take the ratio of the odds. The odds ratio of supporting the liberal candidate is $2.33/0.67 = 3.48$. The odds of supporting a liberal candidate are 3.48 times as great if you have high environmental concern as they are if you have low environmental concern.

The odds ratio gives us useful information for understanding the relationship between environmental concern and support for the liberal candidate. Environmental concern is a strong predictor of this support because the odds of a person high on environmental concern supporting the liberal are 3.48 times as great as the odds of a person with low environmental concern supporting the liberal.

11.3.2 The logit transformation

Odds ratios make a lot of sense for interpreting data, but for our purposes they have some problems as a score on the dependent variable. The distribution of the odds ratio is far from normal.

An odds ratio of 1.0 means that the odds are equally likely and that the predictor makes no difference. Thus an odds ratio of 1.0 is equivalent to a beta weight of 0.0. If those with and without environmental concerns were equally likely to support the liberal candidate, the odds ratio would be 1.0.

An odds ratio can go from 1.0 to infinity for situations where the odds are greater than 1.0. By contrast, the odds ratio can go from 1.0 to just 0.0 where the odds are less than 1.0. This makes the distribution extremely asymmetrical, which could create estimation problems. On the other hand, if we take the natural logarithm of the odds ratio, it will not have this distributional problem. This logarithm of the odds ratio is the logit:

$$\text{Logit} = \ln(\text{odds ratio})$$

Although the probability of something happening makes a lot of sense to most people and the odds ratio also makes sense to most of us, a lot of people have trouble understanding the score for a logit. Because the logistic regression is predicting a logit score, the values of the parameter estimates are difficult to interpret. To get around this problem, we can reverse the transformation process and get coefficients for the odds ratio that are easier to interpret.

11.4 Data used in rest of chapter

The National Longitudinal Survey of Youth, 1997 (nlsy97_chapter11.dta) asked a series of questions about drinking behavior among adolescents between the ages of 12 and 16. One question asked how many days, if any, each youth drank alcohol in the past month, drday97. We have dichotomized this variable into a binary variable called drank30 that is coded 1 if they reported having had a drink in the past month or 0 if they reported not having had a drink in the past month.

Say that we want to see if peer behavior is important and if having regular meals together with the family is important. Our contention is that the greater the percentage of your peers who drink, the more likely you are to drink. By contrast, the more often you have meals together with your family, the less likely you are to drink. Admittedly, having meals together with your family is a limited indicator of parental influence. Because older youth are more likely to drink, we will control for age. We will also control for gender.

Predicting a count variable

We have a variable that is a count of the number of days the youth reports drinking in the last month. We have dichotomized this as none or some. This makes sense if we are interested in predicting whether a youth drinks or not. What if we were interested in predicting how often a youth drinks rather than simply whether they did? Certainly, there is a difference between an adolescent who drank once in the last month and an adolescent who drank 30 days in the last month.

Stata has commands for estimating a dependent variable that is a count, although these commands are beyond the scope of this book. Using OLS regression with a count variable can be problematic when the count is of a rare event such as the number of days a youth drank in the past month. The distribution of rare events is skewed and sometimes has a substantial percentage of participants reporting a count of zero.

Here are two things we could do. One is to use what is called *Poisson regression of the count*. Poisson regression is useful for counts that are skewed, with most people doing the behavior rarely or just a few times. A second thing Stata can do is called *zero-inflated Poisson regression* in which there is not only a skewed distribution but there are more zeros than you would expect with a Poisson distribution. If you asked how many times a female adolescent had a fist fight in the past 30 days, most would say zero times. Zero-inflated Poisson regression simultaneously estimates two regressions, one for the binary result of whether the count is zero and the other for how often the outcome occurred. Different factors may predict the onset of the outcome than would predict the count of the outcome. Long and Freese (2006) have an excellent book that discusses these extensions for what we cover in this chapter.

Interpreting a logistic regression depends on understanding something of the distribution of the variables. First, do a summary of the variables so that we know their means and distributions. If we need to eliminate any observations that did not answer all the items, we drop the cases that have missing values on any of the variables. It would also be important to do a tabulation, although those results are not shown here. The summary of the variables is

```
. summarize drank30 age97 pdrink97 dinner97 male
> if !missing(drank30, age97, pdrink97, dinner97, male)

    Variable |       Obs        Mean    Std. Dev.       Min        Max
-------------+--------------------------------------------------------
     drank30 |      1654     .382104    .4860487          0          1
       age97 |      1654    13.67352    .9371347         12         16
    pdrink97 |      1654    2.108222    1.214858          1          5
    dinner97 |      1654    4.699516    2.349352          0          7
        male |      1654    .5405079    .4985071          0          1
```

We see that the mean for drank30 is 0.382. This means that 38.2% of the students reported drinking at least once a month. Logistic regression also works well even when the outcome is rare, such as whether a person dies after surgery. When the mean is around 0.50, the OLS regression and logistic regression produce consistent results, but when the probability is close to 0 or 1, the logistic regression is especially important.

11.5 Logistic regression

There are two commands for logistic regression: logit and logistic. The logit command gives the regression coefficients to estimate the logit score. The logistic command gives us the odds ratios we need to interpret the effect size of the predictors. Select Statistics ▷ Binary outcomes. From here, we can select Logistic regression or Logistic regression (reporting odds ratios). Selecting the latter produces the dialog box in figure 11.5.

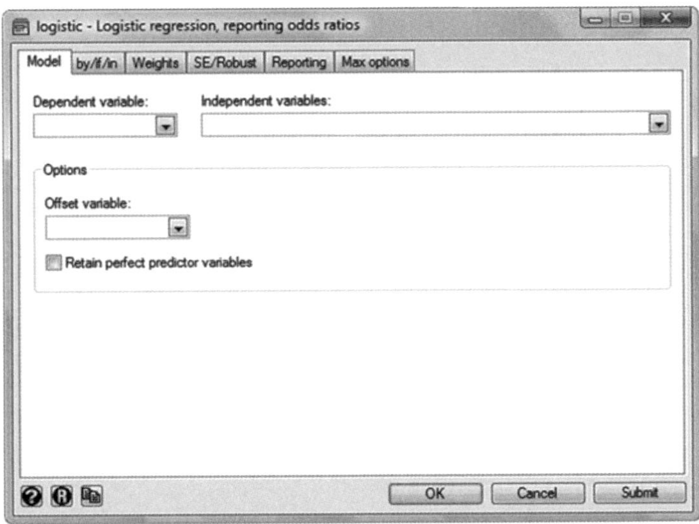

Figure 11.5. Dialog box for doing logistic regression

This dialog box is pretty straightforward. Type `drank30` as the *Dependent variable* and `age97 male pdrink97 dinner97` as the *Independent variables*. The *Offset variable* is for special purposes and is not discussed here. The option to *Retain perfect predictor variables* should not be checked in most applications. In a rare case where there is a category that has no variation on the outcome variable, Stata drops the predictor. Checking this box forces Stata to include the variable, which can lead to instabilities in the estimates. The other tabs have the usual set of options to restrict our sample, weight the observations, or use different estimators.

The only difference between the dialog boxes for logistic regression and for logistic regression with odds ratios is that the default for logistic regression is the estimated regression coefficients and the default for the logistic regression with odds ratios is the odds ratios. Regardless of which dialog box you choose, on the Reporting tab, we can specify whether we want the regression coefficients or the odds ratios. The following commands and output show the results of both dialogs. The first results are for the command `logistic`, where we get the odds ratios, and the second results are for the command `logit`, where we get the regression coefficients.

(*Continued on next page*)

```
. logistic drank30 age97 male pdrink97 dinner97
Logistic regression                               Number of obs   =       1654
                                                  LR chi2(4)      =      78.01
                                                  Prob > chi2     =     0.0000
Log likelihood = -1061.0474                       Pseudo R2       =     0.0355

------------------------------------------------------------------------------
     drank30 | Odds Ratio   Std. Err.      z    P>|z|     [95% Conf. Interval]
-------------+----------------------------------------------------------------
       age97 |   1.169241    .0684191     2.67   0.008     1.042546    1.311332
        male |   .9794922    .1046935    -0.19   0.846     .7943646    1.207764
    pdrink97 |   1.329275    .0598174     6.33   0.000     1.217056    1.451841
    dinner97 |    .942086    .0208682    -2.69   0.007     .9020603    .9838878
------------------------------------------------------------------------------

. logit drank30 age97 male pdrink97 dinner97
Iteration 0:   log likelihood = -1100.0502
Iteration 1:   log likelihood =  -1061.142
Iteration 2:   log likelihood = -1061.0474
Iteration 3:   log likelihood = -1061.0474
Logistic regression                               Number of obs   =       1654
                                                  LR chi2(4)      =      78.01
                                                  Prob > chi2     =     0.0000
Log likelihood = -1061.0474                       Pseudo R2       =     0.0355

------------------------------------------------------------------------------
     drank30 |      Coef.   Std. Err.      z    P>|z|     [95% Conf. Interval]
-------------+----------------------------------------------------------------
       age97 |   .1563548    .0585158     2.67   0.008     .0416659    .2710437
        male |   -.020721    .1068855    -0.19   0.846    -.2302128    .1887708
    pdrink97 |   .2846336    .0450001     6.33   0.000     .1964351    .3728321
    dinner97 |  -.0596587     .022151    -2.69   0.007    -.1030739   -.0162434
       _cons |  -2.947557    .7927494    -3.72   0.000    -4.501317   -1.393797
------------------------------------------------------------------------------
```

Both commands give the same results, except that `logit` gives the coefficients for estimating the logit score, and `logistic` gives the coefficients for estimating the odds ratios. The results of the `logit` command show the iterations Stata went through in obtaining its results. Logistic regression relies on maximum likelihood estimation rather than ordinary least squares; this is an iterative approach where various solutions are estimated until the best solution of having the maximum likelihood is found. The results show the number of observations followed by a likelihood-ratio (LR) chi-squared test and something called the pseudo-R^2. In this example, the likelihood-ratio chi-squared(4) $= 78.01$, $p < 0.001$. There are several coefficients that are called pseudo-R^2. The one reported by Stata is the McFadden pseudo-R^2. This is often a small value, and it should not be confused with R^2 for OLS regression. Many researchers do not report this measure.

It is difficult to interpret the regression coefficients for the `logit` command. We can see that both the percentage of peers who drink and the number of dinners a youth has with his or her parents are statistically significant. The peer variable has a positive coefficient, meaning that the more of the youth's peers who drink, the higher the logit for the youth's own drinking. The `dinner97` variable has a negative coefficient, and this is as we expected; that is, having more dinners with your family lowers the logit for drinking. We can also see that age is significant and in the expected direction. Sex, by contrast, is not statistically significant, so there is no statistically significant evidence that male versus female adolescents are more likely to drink.

The first output is for the `logistic` command, which gives us the odds ratios. We need to spend a bit of time interpreting these odds ratios. The variable `age97` has an odds ratio of 1.17, $p < 0.01$. This means that the odds of drinking are multiplied by 1.17 for each additional year of age. When the odds ratio is more than 1, the interpretation can be simplified by subtracting 1 and then multiplying by 100: $(1.17 - 1.00) \times 100 = 17\%$. This means that for each increase of 1 year, there is a 17% increase in the odds of drinking. This is an intuitive interpretation that can be understood by a lay audience. You have probably seen this in the papers without realizing the way it was estimated. For example, you may have seen that there is a 50% increase in the risk of getting lung cancer if you smoke. People sometimes use the word *risk* instead of saying there is a 50% increase in the odds of getting cancer. What would you say to a lay audience? Each year older an adolescent gets, the odds ratio of drinking increases by 17%. You need to be careful when you interpret the odds ratio. There is a separate coefficient called the relative-risk ratio, and this is discussed in the box about the odds ratio versus the relative-risk ratio. Lay audiences often act as if both of these were the same thing, but they are not and they can be different when the outcome is not a rare event. Odds ratios tell us what happens to the odds of an outcome whereas risk ratios tell us what happens to their probability.

What happens if you compare a 12-year-old with a 15-year-old? The 15-year-old is 3 years older, so you might be tempted to say the risk factor is $3 \times 17\% = 51\%$ greater. However, this underestimates the effect. If you are familiar with compound interest, you have already guessed the problem. Each additional year builds on the last year's compounded rate. To compare a 15-year-old with a 12-year-old, you would first cube the odds ratio, $1.17^3 = 1.60$ (to compare a 16-year-old with a 12-year-old, you would compute $1.17^4 = 1.87$). Thus the odds ratio of drinking for a 15-year-old is $(1.60 - 1.00) \times 100 = 60\%$ greater than that for a 12-year-old. The odds ratio for a 16-year-old is $(1.87 - 1.00) \times 100 = 87\%$ greater than that for a 12-year-old.

When an odds ratio is less than 1.00, you need to change the calculation just a bit. We like to talk of the decrease in the odds ratio, so we need to subtract the odds ratio from 1.00 to find this decrease. Thus, for each extra day per week an adolescent has dinner with his or her family, the odds of drinking are reduced by $(1.00 - 0.94) \times 100 = 6\%$. Consider two adolescents. One has no family dinners, and the other has them every day of the week. Once again, you could be tempted to say that the odds of drinking are now $(7 \times 6\%) = 42\%$ lower. This is just as wrong for odds ratios less than one as it is for odds ratios larger than one. Just as before, we must raise the original odds ratio to a power: the odds of drinking for the adolescent who eats at home is $0.94^7 = 0.65$ times as large as the odds of drinking for the adolescent who never eats with his or her family. Again we want to express this as a percentage change. Because the multiplier is less than 1.00, we compute the change as $(1.00 - 0.65) \times 100 = 35\%$, meaning that a youth who has dinner with the family every night of the week has 35% lower odds of drinking than a youth who has no meals with his or her family.

Using Stata as a calculator

When doing logistic regression, we often need to calculate something to help us interpret the results. We just said that $1.17^4 = 1.87$. To get this result, in the Stata Command window, enter `display 1.17^4`, which results in 1.87.

The odds ratios are a transformation of the regression coefficients. The coefficient for `pdrink97` is 0.2846. The odds ratio is defined as $\exp(b)$, which for this example is $\exp(.2846)$. This means that we exponentiate the coefficient. More simply, we raise the mathematical constant e to the b power, e^b or $e^{0.2846}$ for the `pdrink97` variable. This would be a hard task without a calculator. To do this within Stata, in the Command window, type `display exp(0.2846)`, and 1.33 is displayed as the odds ratio.

It is hard to compare the odds ratio for one variable with the odds ratio for another variable when they are measured on different scales. The `male` variable is binary, going from `0` to `1`, and the variable `dinner` goes from `0` to `7`. For binary predictor variables, you can interpret the odds ratios and percentages directly. For variables that are not binary, you need to have some other standard. One solution is to compare specific examples, such as having no dinners with the family versus having seven dinners with them each week. Another solution is to evaluate the effect of a 1-standard-deviation change for variables that are not binary. This way, you could compare a 1-standard-deviation change in dinners per week with a 1-standard-deviation change in peers who drink. When both variables are on the same scale, the comparison makes more sense, and this is exactly what we can do when we use the standard of a 1-standard-deviation change.

Using the 1-standard-deviation change as the basis for interpreting odds ratios and percentage change is a bit tedious if you need to do the calculations by hand. There is a command, `listcoef`, that makes this easy. Although this command is not part of standard Stata, you can do a `findit spost` command and install the most recent version of `spost`, which will include `listcoef`. There are several items that appear when you enter `findit spost`. Toward the end is a section called `packages found`. Under this heading, go to `spost9_ado`, and click on the link to install it. When you do this installation, you will install a series of commands that are useful for interpreting logistic regression. Be sure to install the latest version of this package of commands because versions for earlier Stata releases, e.g., Stata 7, are still posted. To run this command, after doing the logistic regression, enter the command `listcoef, help`. The `help` option gives a brief description of the various values the command estimates. Here are the results:

```
. listcoef, help
logistic (N=1654): Factor Change in Odds
  Odds of: 1 vs 0

------------------------------------------------------------------------------
     drank30 |        b         z     P>|z|      e^b     e^bStdX      SDofX
-------------+----------------------------------------------------------------
       age97 |   0.15635    2.672   0.008    1.1692    1.1578      0.9371
        male |  -0.02072   -0.194   0.846    0.9795    0.9897      0.4985
    pdrink97 |   0.28463    6.325   0.000    1.3293    1.4131      1.2149
    dinner97 |  -0.05966   -2.693   0.007    0.9421    0.8692      2.3494
------------------------------------------------------------------------------
       b = raw coefficient
       z = z-score for test of b=0
   P>|z| = p-value for z-test
     e^b = exp(b) = factor change in odds for unit increase in X
 e^bStdX = exp(b*SD of X) = change in odds for SD increase in X
   SDofX = standard deviation of X
```

This gives the values of the coefficients (called b), their z scores, and the probabilities. These match what we obtained by using the `logit` command. The next column is labeled `e^b` and contains the odds ratios for each predictor. The label may look strange if you are not used to working with logarithms. If you raise the mathematical constant e, which happens to be about 2.718, to the power of the coefficient b, you get the odds ratio. Thus $e^{0.15635} = 1.1692$.

We can interpret the odds ratio for `male` directly because `male` is a binary variable. The other variables are not binary, so we use the next column, which is labeled `e^bStdX`. This column displays the odds ratio for a 1-standard-deviation change in the predictor. For example, the odds ratio for a 1-standard-deviation change in age is 1.16, and this is substantially smaller than the odds ratio for a 1-standard-deviation change in the percentage of peers who drink, 1.41.

If you prefer to use percentages, you can use the command

```
. listcoef, help percent
logistic (N=1654): Percentage Change in Odds
  Odds of: 1 vs 0

------------------------------------------------------------------------------
     drank30 |        b         z     P>|z|        %      %StdX      SDofX
-------------+----------------------------------------------------------------
       age97 |   0.15635    2.672   0.008      16.9      15.8      0.9371
        male |  -0.02072   -0.194   0.846      -2.1      -1.0      0.4985
    pdrink97 |   0.28463    6.325   0.000      32.9      41.3      1.2149
    dinner97 |  -0.05966   -2.693   0.007      -5.8     -13.1      2.3494
------------------------------------------------------------------------------
       b = raw coefficient
       z = z-score for test of b=0
   P>|z| = p-value for z-test
       % = percent change in odds for unit increase in X
   %StdX = percent change in odds for SD increase in X
   SDofX = standard deviation of X
```

For communicating with a lay audience, this last result is remarkably useful. The odds of drinking are just 2.1% lower for males than for females, and this is not statistically significant. Having a 1-standard-deviation-higher percentage of peers who drink increases a youth's odds of drinking by 41.3%, and having dinner with his or her family 1 standard deviation more often reduces the odds by 13.1%. Both of these influences are statistically significant.

It is often helpful to create a graph showing the percentage change in the odds ratio associated with each of the predictors. This could be done with any graphics package, such as PowerPoint. To do this using Stata, we need to create a new dataset, `c11barchart.dta`. This dataset has just four variables, which we will name `Age`, `male`, `peers`, and `dinners`. There is just one observation, and this is the appropriate percentage change in the odds ratio. For `Age`, `peers`, and `dinners`, we enter 15.8, 41.3, and −13.1, respectively, because we are interested in the effects of a 1-standard-deviation change in each of these variables. For `male`, we enter −2.1 because we are interested in a 1-unit change for this dichotomous variable (female versus male). Then we construct a bar chart. You can use Graphics ▷ Bar chart to create this graph. The command is

```
. graph bar (asis) Age male peers dinner, bargap(10)
> blabel(name, position(outside)) ytitle(Percentage Change in Risk)
> title(Percentage Change in Risk of Drinking by)
> subtitle("Age, Gender, Percent of Peers Drinking, Meals with Family")
> legend(off) scheme(s2manual)
```

Note that if your title or subtitle includes commas, as we have shown in the `subtitle()` option above, you must enclose the title in quotes. We had Stata put the variable labels just above or below each bar. We must have very short labels for this to work. Also, on the Bar tab, we put a *Bar gap* of 10 so the bars would be separated a little bit. The resulting bar chart is shown in figure 11.6.

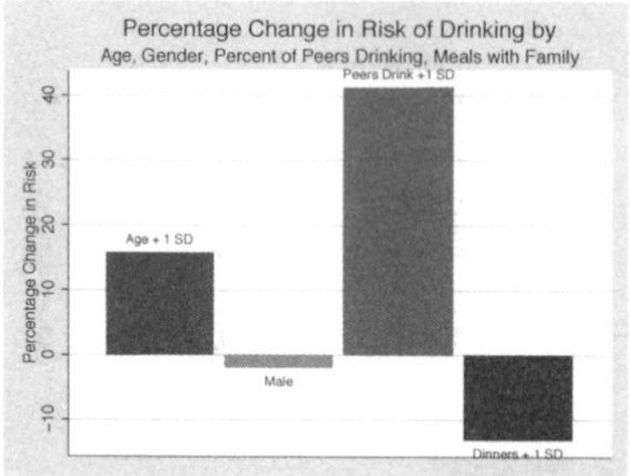

Figure 11.6. Risk factors associated with teen drinking

This bar graph shows that as adolescents get older there is a substantial increase in the odds of their drinking, that drinking risks are roughly comparable for males and females, that having peers who drink is a substantial risk factor, and that the more the adolescent shares dinners with his or her family the lower the odds of drinking.

Odds ratio versus relative-risk ratio

In section 11.3, we showed how the odds ratio is computed. There we noted that, among the 10 people who favored environmental protection, 7 voted liberal and 3 voted conservative, creating odds of voting liberal of $7/3 = 2.33$. By contrast, the 10 people who did not favor environmental protection had odds of voting liberal of $4/6 = 0.67$. The resulting odds ratio was $2.33/0.67 = 3.48$.

Relative-risk ratio approaches this for probabilities. There are 7 of 10 (70% or 0.70) pro-environmentalists supporting a liberal candidate. There are 4 of 10 (40% or 0.40) people who are not pro-environmental supporting a liberal candidate. The relative risk of supporting a liberal candidate for people who are pro-environment is $0.70/0.40 = 1.75$. People who are not used to odds ratios often misinterpret the odds ratio as if it were the relative risk. Generally, the odds ratio will be larger than the relative-risk ratio. The difference is small with a rare outcome, but with an outcome like this there is a huge difference. Being pro-environment makes the odds of voting for a liberal 3.48 times as great. Being pro-environment increases the risk of voting for a liberal candidate 1.75 times. Popular media such as newspapers often misinterpret an odds ratio as if it were a risk ratio. In this example, there is a 348% increase in the odds of voting liberal if you are pro-environment, but there is only a 75% increase in the risk of voting liberal.

Stata's `logit` and `logistic` commands do not estimate the relative-risk ratio (sometimes called incidence relative risk). However, Joseph Hilbe wrote a command, `oddsrisk`, that converts odds ratios to risk ratios. You can install this command by entering `ssc install oddsrisk`.

The relative risk is appealing, but it should not be used in a study that controls the number of people in each category. For example, you should not use this if you are comparing 100 stroke victims to 100 people who did not have a stroke on their participation in a rigorous physical activity. By controlling the number of people in each group, you have artificially (nonrandomly) inflated the number of stroke victims because 50% of your sample have had a stroke—a far greater percentage than would be obtained in a representative sample. For more on relative risk, visit http://www.childrensmercy.org/stats/journal/oddsratio.asp, which was written by Steve Simon.

11.6 Hypothesis testing

There are many types of hypothesis testing that can be done with logistic regression. I will cover some of them here. As with multiple regression, we have an overall test of the model, tests of each parameter estimate, and the ability to test hypotheses involving combinations of parameters. As noted above, with logistic regression there is a chi-squared test that has k degrees of freedom, where k is the number of predictors. For our current example, the $\chi^2(4) = 78.01$, $p < 0.001$. This tells us only that the overall model has at least one significant predictor.

11.6.1 Testing individual coefficients

There are two tests we can use to test individual parameters, and they often yield slightly different probabilities. The most common is a Wald test, commonly reported in journal articles. The z test in the Stata output is actually the square root of the Wald chi-squared test. Some other statistical packages report this by using the Wald chi-squared test (evaluated with 1 degree of freedom), but these are identical to the square of the z test in Stata.

Many statisticians prefer the likelihood-ratio chi-squared test to the Wald chi-squared test or the equivalent z test reported by Stata. The likelihood-ratio chi-squared test for each parameter estimate is based on comparing two logistic models, one with the individual variable we want to test included and one without it. The likelihood-ratio test is the difference in the likelihood-ratio chi-squared values for these two models (this appears as `LR chi2(1)` near the upper right corner of the output). The difference between the two likelihood-ratio chi-squared values is 1 degree of freedom. In standard Stata, the likelihood-ratio test is rather tedious to get. If you wanted the best possible test of the significance of the effect of age on drinking, with the `nlsy97_chapter11.dta` dataset, you would need to run a model, `logistic drank30 male dinner97 pdrink97`, and record the chi-squared. Then you would run a model, `logistic drank30 age97 male dinner97 pdrink97` and record the chi-squared. The difference between the chi-squared values with 1 degree of freedom would be the best test of the effect of age on drinking.

Stata lets us simplify this process a bit with the following set of commands:

```
. logistic drank30 male dinner97 pdrink97
. estimates store a
. logistic drank30 age97 male dinner97 pdrink97
. lrtest a
```

The first line runs the logistic regression without including `age97`. The second line saves all the estimates and stores them in a vector labeled `a`. This vector includes the `LR chi2(1)` value. The third line runs a new logistic regression, but this time it includes `age97`. The last line, `lrtest a`, subtracts the chi-squared values and estimates the probability of the chi-squared difference. Although this is not too difficult, doing this for each of the four predictors would be a bit tedious. An automated approach is available with a command written by user Zhiqiang Wang, `lrdrop1`, which gives you the likelihood-ratio test for each parameter estimate, along with some other information we will not discuss (`lrdrop1` works with the `logit`, `logistic`, and `poisson` regression commands). Type the command `ssc install lrdrop1` to install this command. Immediately after running the full model (with all four predictors), simply type the following command:

```
. lrdrop1
Likelihood Ratio Tests: drop 1 term
logistic regression
number of obs = 1654
------------------------------------------------------------------------
 drank30      Df       Chi2      P>Chi2     -2*log ll    Res. Df   AIC
------------------------------------------------------------------------
Original Model                              2122.09      1649     2132.09
  -age97       1        7.18     0.0074     2129.27      1648     2137.27
    -male      1        0.04     0.8463     2122.13      1648     2130.13
-pdrink97      1       40.62     0.0000     2162.72      1648     2170.72
-dinner97      1        7.23     0.0072     2129.33      1648     2137.33
------------------------------------------------------------------------
Terms dropped one at a time in turn.
```

Here we can see that the likelihood-ratio chi-squared for the age97 variable is $\chi^2(1) = 7.18$, $p < 0.01$. If we take the square root of 7.18, we obtain 2.68, which is close to the square root of the Wald chi-squared, i.e., the z test provided by Stata of 2.67. The likelihood-ratio chi-squared and the Wald statistic are not always this close, but the two approaches rarely produce different conclusions. Although the likelihood-ratio test has some technical advantages over the Wald test, the Wald test is usually what people report—perhaps because it is a bit tedious to estimate the likelihood-ratio chi-squared in other statistical packages. If the Wald test and the likelihood-ratio tests differ, the likelihood-ratio test is preferred. These results are identical whether you are testing the regression coefficients by using the `logit` command or the odds ratios by using the `logistic` command. Because this `lrdrop1` command is so easy to use and because the likelihood-ratio test is invariant to nonlinear transforms, Stata users may want to report the likelihood-ratio test routinely. When they do so, it is important to note that it is the likelihood-ratio chi-squared test rather than the Wald approximation.

11.6.2 Testing sets of coefficients

There are three situations for which we would need to test a set of coefficients rather than a single coefficient. These parallel the tests we used for multiple regression in chapter 10.

- We have a categorical variable that requires several dummies. Suppose that we want to know if marital status has an effect, and we have three dummies representing different marital statuses (divorced or separated, widowed, and never married, with married serving as the reference category). We want to know if marital status is significant, so we need to test the set of coefficients rather than just one coefficient.
- We want to know if a set of variables is significant. The set might be whether peer and family variables are significant when age and gender are already included in a model to predict drinking behavior. We need to test the combination of coefficients for peer and family rather than testing these individually.

- We have two variables measured on the same scale, and we want to know if their parameter estimates are equal. Suppose that we want to know if mother's education is more important than father's education as a predictor of whether an adolescent plans to go to college. Both predictors are measured on the same scale (years of education), so we can test if the mother's education or the father's education have significantly different coefficients.

The first two of these are tested by the hypothesis that the set of coefficients are all zero. For the marital status set, the three simultaneous null hypotheses would be

H_0: Divorce effect = 0

H_0: Widow effect = 0

H_0: Never-married effect = 0

For the second example, the set of two simultaneous null hypotheses would be

H_0: pdrink97 effect = 0

H_0: dinner97 effect = 0

We will illustrate how to test for the first two sets with the data we have been using. The full model is

```
. logistic drank30 age97 male pdrink97 dinner97
```

We want to know if `pdrink97` and `dinner97`, as a set, are significant.

There are several ways of testing this. We could run the model twice:

(a) `logistic drank30 age97 male`

(b) `logistic drank30 age97 male pdrink97 dinner97`

We would then see if the difference of chi-squared values was significant. Because the second model added two variables, we would test the chi-squared with 2 degrees of freedom.

We can also use the `test` command. We run this after we run the full logistic regression. In other words, we run the command `logistic drank30 age97 male pdrink97 dinner97`, and then we run the command `test pdrink97 dinner97`. Here are the results:

```
. test pdrink97 dinner97
 ( 1)  pdrink97 = 0
 ( 2)  dinner97 = 0
           chi2(  2) =   48.85
         Prob > chi2 =    0.0000
```

The variables **pdrink97** and **dinner97**, as a set, are statistically significant, chi-squared(2) = 48.85, $p < 0.001$. We can say that one or both of these variables are statistically significant. When the set of variables is significant, we should always look at the tests for the individual coefficients because this overall test only tells us that at least one of them is significant.

The third reason for testing a set of parameter estimates is to test the equality of parameter estimates for two variables (assuming that both variables are measured on the same scale). We might do a logistic regression and find that both mother's education and father's education are significant predictors of whether an adolescent intends to go to college. However, we may want to know if the mother's education is more or less important than the father's education. Say that we want to test whether the parameter estimates are equal. Our null hypothesis is

H_0: mother's education effect = father's education effect

We will not illustrate this test because it uses a different dataset, but the command is simple. Let's assume that mother's education is represented by the variable `maeduc` and father's education is represented by `faeduc`. The command is `test maeduc==faeduc`. It is optional to include both equal-signs.

11.7 Nested logistic regressions

In chapter 10, we discussed nested regression (sometimes called hierarchical regression), where blocks of variables are entered in a planned sequence. The `nestreg` command is extremely general, applying across a variety of regression models including logistic, negative binomial, Poisson, probit, ordered logistic, tobit, and others. It also works with the complex sample designs for many regression models. Although most of these models are beyond the scope of this book, we can illustrate how `nestreg` can be generalized with logistic regression.

We have been interested in whether the percentage of an adolescent's peers who drink and the number of days a week the adolescent eats with his or her family influence drinking behavior. In doing this, we controlled for gender and age. We might decide to enter the four predictors in three blocks. In the first block, we enter gender; in the second block, we enter age; and in the third block, we enter dinners with the family and peer drinking. We now apply the format used in chapter 10 for logistic regression to get the following results:

(Continued on next page)

```
. nestreg: logistic drank30 (male) (age97) (dinner97 pdrink97)
Block  1: male
Logistic regression                               Number of obs   =       1654
                                                  LR chi2(1)      =       4.37
                                                  Prob > chi2     =     0.0365
Log likelihood =  -1097.864                       Pseudo R2       =     0.0020

------------------------------------------------------------------------------
     drank30 | Odds Ratio   Std. Err.      z    P>|z|     [95% Conf. Interval]
-------------+----------------------------------------------------------------
        male |   .8087911   .0820956    -2.09   0.037     .6628816    .9868173
------------------------------------------------------------------------------
Block  2: age97
Logistic regression                               Number of obs   =       1654
                                                  LR chi2(2)      =      27.87
                                                  Prob > chi2     =     0.0000
Log likelihood = -1086.1173                       Pseudo R2       =     0.0127

------------------------------------------------------------------------------
     drank30 | Odds Ratio   Std. Err.      z    P>|z|     [95% Conf. Interval]
-------------+----------------------------------------------------------------
        male |   .8304131   .0849835    -1.82   0.069     .6794902    1.014858
       age97 |   1.304566    .072308     4.80   0.000     1.170272    1.454272
------------------------------------------------------------------------------
Block  3: dinner97 pdrink97
Logistic regression                               Number of obs   =       1654
                                                  LR chi2(4)      =      78.01
                                                  Prob > chi2     =     0.0000
Log likelihood = -1061.0474                       Pseudo R2       =     0.0355

------------------------------------------------------------------------------
     drank30 | Odds Ratio   Std. Err.      z    P>|z|     [95% Conf. Interval]
-------------+----------------------------------------------------------------
        male |   .9794922   .1046935    -0.19   0.846     .7943646    1.207764
       age97 |   1.169241   .0684191     2.67   0.008     1.042546    1.311332
    dinner97 |    .942086   .0208682    -2.69   0.007     .9020603    .9838878
    pdrink97 |   1.329275   .0598174     6.33   0.000     1.217056    1.451841
------------------------------------------------------------------------------

  +-------------------------------+
  | Block |   chi2    df   Pr > F |
  |-------+-----------------------|
  |     1 |   4.37     1   0.0366 |
  |     2 |  23.01     1   0.0000 |
  |     3 |  48.85     2   0.0000 |
  +-------------------------------+
```

These results show that each block is statistically significant. Our main interest is in the third block, and we see that adding dinners with your family and peer drinking has $\chi^2(2) = 48.85$, $p < 0.001$. These results duplicate those obtained with the `lrtest` command. The results do not duplicate those for the first two blocks (age and gender) using the `lrtest`. Here we are testing if each additional block makes a statistically significant improvement above what was done by the preceding blocks. We would say that "Block 1", `male`, is significant. Gender was not significant in the overall model, controlling for the other three variables, but "Block 1" contains only the variable `male`. "Block 2", `age97`, increases the significance, but it does not control for "Block 3" variables because they have yet to be added to the model.

11.8 Summary

Logistic regression is a powerful data analysis procedure that on one hand is fairly complicated but on the other hand can be presented to a lay audience in a clear and compelling fashion. Policy makers are likely to understand what odds mean, and they want to know which factors raise the odds of a success and reduce the odds of a failure. Although it is easy to let the language surrounding logistic regression intimidate you, the results are easy to understand and explain. To continue with the example from this chapter, policy makers are definitely interested in the factors that increase or decrease the odds that an adolescent will drink. With logistic regression, this is exactly the information you can provide.

Some inexperienced researchers, upon learning logistic regression, may use it for too many applications. Logistic regression makes sense whenever you have an outcome that is clearly binary. If you have an outcome that is continuous or a count of how often something happens, you may want to use other procedures. For example, if you have a scale measuring political conservatism and this is what you want to explain, it is probably not wise to dichotomize people into conservative versus liberal categories just so you can do logistic regression. When you have a continuous variable and you dichotomize it, you lose a lot of variance. The difference between a person who is just a little right of center and a radical conservative is lost if they both go into the same category. In our example, we lost the difference between an adolescent who drank only occasionally and one who drank 30 days a month. This was probably okay because we wanted to find out what explained whether the child drank or did not drink. Therefore, the variance in how often an adolescent drinks was not part of our question.

This chapter covered the following topics that prepare you to do logistic regression:

- Examples of when the technique is appropriate
- Key concepts, including odds, odds ratios, and logit
- How a linear estimation of a logit value corresponds to a nonlinear relationship for the probability of a success
- Two Stata commands, `logit` and `logistic`, that are used for doing logistic regression
- The regression coefficients and the odds ratios
- How to interpret the results for categorical and continuous predictors
- How to associate a percentage change in the odds ratio with each variable and how to summarize this in a bar chart
- Alternatives for hypothesis testing, including two ways to test the significance of each variable's effect and how to test the significance of sets of predictors

Chapters 10 and 11 provide the core techniques used by many social scientists. Although these are both complex procedures, they can provide informative answers to many research and policy questions. I have covered many useful and powerful methods of data analysis using Stata, and you are prepared for the next steps. Many researchers can do everything they need to do with the techniques we have covered, but we have tapped only the core capabilities of Stata. An exciting thing about working with Stata is that it is a constantly evolving statistical package, and one that few of us will outgrow.

11.9 Exercises

1. Use the `severity.dta` dataset, which has three variables. The `severity` variable is whether the person sees prison sentences as too severe (coded `1`) or not too severe (coded `0`). The `liberal` variable is how liberal the person is, varying from `1` for conservative to `5` for liberal. The variable `female` is coded `1` for women and `0` for men. Do a logistic regression analysis with `severity` as the dependent variable and `liberal` and `female` as the independent variables. Carefully interpret your results.

2. Use the `gss2002_chapter11.dta` dataset. What predicts who will support abortion for any reason? Recode `abort12` to be a dummy variable called `abort`. Use the following variables as predictors: `reliten`, `polviews`, `premarsx`, and `sei`. You may want to create new variables for these so that a higher score goes with the name. For example, `polviews` might be renamed `conservative` because a higher score means more conservative. Do a logistic regression followed by the `listcoef` command. Carefully interpret the results.

3. Using the results from the last exercise, create a bar graph showing the percentage change in odds of supporting abortion for any reason associated with each predictor. Justify using a 1-unit change or a 1-standard-deviation change in the score on each predictor.

4. Using the results from the logistic regression, compute likelihood-ratio chi-squared tests for each predictor, and compare them with the standard z tests in the logistic regression results.

12 Measurement, reliability, and validity

12.1 Overview of reliability and validity

The quality of our measurement is the foundation for all statistical analysis. We must always keep two facts in mind: First, our analysis can be no better than our measurement. If we have poor measurement of our variables, then the relationship we find will be inaccurate and will often underestimate the true strength of the relationship. Second, virtually every variable we measure has some error, and often this error is substantial.

Imagine doing a study of the relationship between postpartum depression and body fat among a group of women two months after childbirth. Suppose we find a weak relationship, say, $r = 0.10$. Is this because postpartum depression is not strongly related to body fat percentage? Perhaps, but it could be due to poor measurement. Our depression scale might have an alpha (α) reliability of 0.90 and our measure of body fat percentage might have a test–retest reliability of 0.80. We will explain these measures of reliability later, but both of them indicate that at least some of the variance in our two measures is not related to what we call them, but is simply measurement error. Reliability means that our measurement process will produce consistent results.

We also need to worry about validity. Many measures of body fat percentage are not only prone to error, i.e., low reliability, but they are also biased. For example, a strong athlete may have a completely wrong estimated fat percentage. An NFL running back may be 5'11" and weigh 235 pounds. An estimator of his body fat such as the BMI that uses just these two values might grossly overestimate his percentage of body fat. In his case, his weight is coming from musculature rather than from fat. Just adding his waist size of 32 inches would greatly improve the accuracy of the estimate. This is a problem of validity. The answer we get is off the mark. It may or may not be reliable, but it is not valid.

This chapter discusses what we can do to evaluate the reliability and validity of our measures. Our discussion is largely limited to what is known as classical measurement theory.

12.2 Constructing a scale

The most typical situation is having a set of k items that measure a concept. We might have 10 items measuring depression, 5 items measuring attitude toward abortion, or 15 items measuring perceived risk of sexually transmitted HIV. How does Stata generate a scale score? Some researchers like to generate the sum of the items. If each of 10 items is on a scale of 1 to 4 (*strongly agree*, *agree*, *disagree*, and *strongly disagree*), the sum (total score) would range from a minimum of 10 to a maximum of 40. Other researchers like to generate the mean of the items, and in this example the mean would be between 1 and 4. One advantage of generating the mean is that it is on the same scale as the individual items and hence can be interpreted more easily. For example, a sum of 25 might not have an obvious meaning, but a mean of 2.5 would indicate the participant was halfway between *agree* and *disagree* on the 4-point scale.

The 2006 General Social Survey (`gss2006_chapter12.dta`) asked people a series of 7 items, `empathy1`–`empathy7`, designed to measure empathy. Each of these was scored from `1` for *Does not describe me very well* to `5` for *Describes me very well*. There was also a category of `8` for *Don't know* that we treat as a missing value.

If you run a `tabulate` on the empathy items, you will see that some of the items, `empathy2`, `empathy4`, and `empathy5`, are worded so a higher score goes with less empathy, and the others, `empathy1`, `empathy3`, `empathy6`, and `empathy7`, are worded so a higher score goes with more empathy. When we cover factor analysis, we will see how such an arrangement introduces a methods factor. For now, we will recode the three negative items so each item has a higher score indicating greater empathy. Always coding a scale so a higher score represents more of whatever is being measured makes it easier for readers to understand what you are doing. We discussed how to recode variables by using the `recode` command. The command is

```
. recode empathy2 empathy4 empathy5 (1=5 "Does not describe very well")
> (2=4) (3=3) (4=2) (5=1 "Describes very well"), pre(rev) label(empathy)
```

Because we are recoding three variables with one command, this is a bit more complicated. We are applying value labels to two values, the new value 5 and the new value 1, and we did not bother to make labels for the other values. When we have value labels in the command, we need to give a nickname for this set of labels. This could be any value label name we have not used already in a dataset. I chose the `label(empathy)` option for the nickname. We never overwrite variables but generate new variables, and to name those variables, we used the option to add a prefix of `rev` to each variable by using the `pre(rev)` option. Thus, our seven variables are now named `empathy1`, `revempathy2`, `empathy3`, `revempathy4`, `revempathy5`, `empathy6`, and `empathy7`.

12.2.1 Generating a mean score for each person

Stata has a convenient command for generating the mean score for each person. Because each participant's data is in a row, we want the row mean for the selected items. The command name is `egen`, which stands for extended generation. We name the new variable we want to generate `empathy`, and we make this equal to the `rowmean` of the listed variables. The variables are listed inside parentheses and have no commas separating them. Here is the command:

```
. egen empathy = rowmean(empathy1 revempathy2 empathy3 revempathy4 revempathy5
> empathy6 empathy7)
```

(*Continued on next page*)

Requiring a 75% completion rate

The `rowmean` calculates the mean of the items from the set that a participant answers. If Sara answered all but the last empathy item, she would get a mean for the six items she answered. If George answered only the first three items, he would get the mean for those three items. Some researchers will want to exclude people who did not answer at least 75% of the items (they will want to exclude people who did not answer at least 5 of the 7 items). One way to do this is to count how many items a person answers and then exclude people who did not answer at least that many items. Here we would count the number of missing values by using `egen miss = rowmiss(empathy1 revempathy2 empathy3 revempathy4 revempathy5 empathy6 empathy7)`. This creates a new variable, `miss`, that is the count of the number of items that have a missing value. Because we require 5 of the 7 questions be answered, we would only compute the empathy score for those who have a count of less than 3 (they answered 5, 6, or all 7 of the items). We can use this variable to restrict our sample; our new `rowmean` command will be: `egen empathya = rowmean(empathy1 revempathy2 empathy3 revempathy4 revempathy5 empathy6 empathy7) if miss < 3`.

12.3 Reliability

There are four types of reliability, and these are illustrated in table 12.1 (see Shultz and Whitney 2005).

Table 12.1. Four kinds of reliability and the appropriate statistical measure

Type of reliability	Reliability focus	Name of procedure	Statistical measure
Stability	Changes in participants	Test–retest correlation Test–retest agreement	correlation, r intraclass correlation, ρ_I
Equivalence	Sampling of content domain	Alternate forms	correlation, $\mathrm{r}_{xx'}$
Internal consistency	Sampling of content domain	Split-half Alpha reliability	correlation, $\mathrm{r}_{X_a X_b}$ alpha, α
Rater consistency	Interrater agreement	Agreement	kappa, κ

Stability means that if you measure a variable today using a particular scale and then measure it again tomorrow using the same scale, your results will be consistent. For depression, we would expect new mothers who are higher than average on depression today will still be higher than average tomorrow.

A problem generating a total scale score

Many researchers generate a total score rather than a mean. We mentioned that this is often hard to interpret because it is not on the same scale as the original items. Another problem is that Stata, as well as other software such as SPSS, assigns a value of zero to missing data when generating the total score. If you have items that range from 1 to 5, assigning a zero to missing values makes no sense. If a person answers four of the items and picks a 5 for each of them, the total score will be 20. If another person answers seven items and picks a 3 for each of them, the score will be 21. It does not make sense that a person who picked a neutral answer to every item has more empathy than the person who answered the most empathetic response to each of the items. We could restrict scores to those who answered all of the items, but this would be throwing out a lot of information. In the first box, Sara answered six of the seven items, and we would be acting like we had no information on her level of empathy. There is a simple solution if a person insists on having a total score. Generate the row mean and then multiply it by the number of items. You could restrict your sample to those who answered at least 75% of the items.

Equivalence means that you have two measures of the same variable and they produce consistent results. For example, you might have the mothers write an essay on motherhood, and then your coders would rate the essay from 1 to 10 based on indications in the essay that the mother is depressed. Then you might give the new mothers a questionnaire that includes a 20-item scale measuring depression. The alternate forms or alternate measures could be evaluated for reliability by correlating the score on one measure with the score on the other measure.

Internal consistency is widely used with scales that involve several items in the same format. Likert-type items that have responses varying from strongly agree to strongly disagree are one example. A reliability assessment might be done by generating the mean score on the first half of the items and the mean score on the second half and then correlating these two scores. A reliable test would be internally consistent if the score for the first half of the items was highly correlated with the score for the second half of the items. Below we will discuss the coefficient alpha, α, which is a better measure of internal consistency.

Rater consistency is important when you have observers rating a video, observed behavior, essay, or something else where two or more people are rating the same information. Here reliability means that each pair of raters are giving consistent results. If a person who wrote an essay is rated high on depression by one rater and low on depression by a second rater, then we have reason to say there is low reliability measured for rater consistency.

12.3.1 Stability and test–retest reliability

A researcher uses the identical measure twice and simply correlates the results. The data (`retest.dta`) consists of the score at time one, called `score1`, and the score at time two, called `score2`.

The test–retest correlation is computed by using the command

```
. pwcorr score1 score2, obs sig

             |   score1   score2
-------------+------------------
      score1 |   1.0000
             |
             |       10
             |
      score2 |   0.7702   1.0000
             |   0.0091
             |       10       10
             |
```

The option `obs` reports the number of observations, and the option `sig` reports the significance level. From the results, we see that there is a high test–retest reliability, $r = 0.77, p < 0.01$. We would expect the correlation to be substantial, say, $r > 0.5$. If the correlation is not substantial, then the measure is not giving reliable results.

Test–retest reliability has some limitations as a measure of reliability. First, it is difficult to distinguish between a lack of a reliable measure and real change. If the length of time between the two measures is very long, or if some event related to the concept has occurred between the measurements, there may be real change. In that case, the correlation would be low regardless of the reliability of the true measure. If you have a measure of health risks associated with smoking tobacco, there might be a television special on the topic between the first and second measurements. The measure might be reliable, but the correlation could be low if those who viewed the television program changed their beliefs. If you try to minimize the risk of this happening by having the two measurements close together, say, one in the morning and the other in the afternoon, participants may recall their responses and try to be consistent.

A second limitation of using this correlation to measure reliability is that correlation does not measure agreement in a strict sense. If the scores at time one had been 2, 3, 4, 5 and the corresponding scores at time two had been 4, 6, 8, 10, then the regression equation

$$\widehat{Y} = 0 + 2X$$

would have a correlation of $r = 1.0$. This is because each score at the second measurement is exactly twice the time one score. There is no strict agreement, only relative agreement, with the higher the score at time one being matched to the higher the score at time two.

12.3.2 Equivalence

An alternative approach to reliability uses a different form of the measure for the second measurement. This alternative form has different items and, hence, the issue of participants trying to remember how they answered items the first time is eliminated. The alternative forms can be administered at separate times or one right after the other. The main problem with this approach is that it is extremely difficult to develop two versions of a measure that we can assume are truly equivalent. With this approach, the correlation, r, between the scores on the two forms is all that is needed to measure reliability. A low correlation means either that the measure is not reliable or that the measures are not truly equivalent.

12.3.3 Split-half and alpha reliability—internal consistency

Instead of developing alternative forms of a measure, one approach is known as split-half reliability. This is easier to use because you only need to have one measurement time. Suppose you have 10 items. You compute a score for the first five items and a separate score for the last five items. Then you correlate these two scores. There are various ways of splitting the items. A random selection would probably be best. One problem with this approach is that we are using only 5-item measures to estimate the reliability of a 10-item measure. A 10-item measure will be more reliable than a 5-item measure, so our correlation for the two 5-item measures will underestimate the correlation for the 10-item measure. To correct this, we can use the Spearman–Brown prophecy formula (developed by Spearman and Brown independently over a century ago).

The major limitation of this approach is that we could get a different result with each split we make. Using the first half and the second half of the items would yield a different reliability estimate than using the even versus the odd items or randomly selecting items. Also, if the ordering of items is important, this method could be misleading.

Given these concerns with the split-half correlation, the standard for measuring reliability within classical measurement-theory framework is to use a coefficient called Cronbach's alpha (α). In general, an $\alpha > .80$ is considered good reliability, and many researchers feel an $\alpha > 0.70$ is adequate reliability.

There are two widely used interpretations of α. The simplest is that it is the expected value of all corrected split-half correlations. This approach is a simple interpretation. The second interpretation of α is the proportion of the observed variance that represents true variance. If we believe our items are an unbiased sample of items from a domain of possible items and hence that it is valid, we can interpret α as the ratio of the true variance to the total variance (the total variance contains both the true variance and error variance): $\alpha = \sigma^2_{\text{true}}/(\sigma^2_{\text{true}} + \sigma^2_{\text{error}})$. This approach is elegant in that an $\alpha = 0.80$ means that 80% of the variance in the scale represents the true score on the variable, and 20% is random error. However, for this interpretation to be used, we need to assume that the scale is valid. We discuss validity assessment techniques later in this chapter.

There are two formulas for computing α. One of these uses the covariance matrix, and the other uses the correlation matrix. You can think of the correlation matrix as a special transformation of the covariance matrix, where all of the variances are rescaled to be equal and fixed at a value of 1.0. Analyzing the covariance matrix is called the unstandardized approach, and analyzing the correlation matrix is called the standardized approach. If items have similar variances, the two approaches will yield similar results. The unstandardized alpha is more general because it does not require us to assume that all items have the same variance. The unstandardized alpha is the default in Stata and is generally recommended.

We now return to using the `gss2006_chapter12.dta` dataset. We will use the dialog box to generate the `alpha` command because there are several options that might be hard to remember if you were writing the command directly. Open the dialog box by selecting Statistics ▷ Multivariate analysis ▷ Cronbach's alpha. On the Main tab, simply enter the variables, `empathy1 revempathy2 empathy3 revempathy4 revempathy5 empathy6 empathy7`. Under the Options tab, there are several powerful options. The first one takes the sign of each item as is. This makes sense to select because we already reverse coded the items that were worded negatively. If we do not check this option, Stata will decide if any items should have their signs reversed. The second option will delete cases with missing values on any item. We will not check this option for now. Next we have an option to list individual interitem correlations and covariances. We will not check this now, but it is useful when combined with selecting a standardized solution to see if there are problematic items, i.e., items that have low or negative correlations with the other items.

Next is an option that will generate the scale for us. I already showed you how to generate this score earlier by using the `egen rowmean` command. It is reasonable to check this if you want Stata to generate the mean score for you. But only do this after you are satisfied with your scale and its reliability. If you check this option, there is a box highlighted in which you enter the name of the computed scale score. We check the next option to display item-test and item-rest correlations. We do not check the box to include variable labels, but with big scales this can be useful. We do check the box to require a certain number of items and enter `5`. This means that people who answer 5, 6, or 7 of the items will be included, and people who answer fewer than 5 items will be dropped. At the bottom of the dialog box is the option to select a standardized solution. This can be useful when combined with asking for correlations if your reliability is low to see where there is a problem. We will not check this for now. The command and results are as follows:

(Continued on next page)

```
. alpha empathy1 revempathy2 empathy3 revempathy4 revempathy5 empathy6 empathy7,
> asis item min(5)
Test scale = mean(unstandardized items)
```

Item	Obs	Sign	item-test correlation	item-rest correlation	average inter-item covariance	alpha
empathy1	1349	+	0.6701	0.5151	.3722749	0.7034
revempathy2	1346	+	0.5957	0.3933	.3954405	0.7332
empathy3	1348	+	0.5733	0.4027	.4118086	0.7281
revempathy4	1349	+	0.6303	0.4474	.3827332	0.7191
revempathy5	1342	+	0.6088	0.4277	.3932707	0.7233
empathy6	1347	+	0.6762	0.5358	.3764256	0.7004
empathy7	1349	+	0.6701	0.5204	.3748609	0.7026
Test scale					.3866902	0.7462

We have selected three options. The `asis` (as is) option means that we do not want Stata to change the signs of any of our variables. We want a display of the item analysis, and we require that there are a minimum of 5 answered items, `min(5)`. The bottom row of the output table, `Test scale`, reports the α for the scale (0.7462). Above this value is the α we would obtain if we dropped each item, one at a time. For example, if we dropped `empathy7`, our α would be 0.7026. Dropping this item would make our scale less reliable. Sometimes you will find an item that, if dropped, would improve α. Carefully used, this information can maximize your α. However, you need caution in doing this because you are capitalizing on chance. Imagine having 100 items and computing α. Then drop one item to increase α and redo the analysis using 99 items. You could continue until you had a very high α with, say, 20 items, but you would have not tested to see if you are capitalizing on chance.

The column labeled `Obs` reports the number of observations. This varies slightly because we used pairwise deletion for people who answered at least 5 of the 7 questions. The column labeled `Sign` contains all plus signs, meaning all the items are positively related to the scale. If any item had a minus sign in this column, you would need to reverse code the item or drop it. The `item-test correlation` column reports the correlation of each item with the total score of the 7 items. All of these are strong. Because the total score includes the item, the values in this column are artificially inflated. Therefore, Stata also reports the `item-rest correlation`. This is the correlation of each item with the total of the other items. For example, the correlation between `empathy1` and the total of the other 6 items is 0.5151. The correlations are useful, but the most valuable information is in the `alpha` column, where we can see whether each item is useful (dropping it would lower α) or unuseful (dropping it would increase α).

Alpha, average correlation, number of items

The unstandardized alpha is a function of the number of items, the mean covariance of the items, and the mean variance of the items. The standardized alpha is a function of just the number of items and their mean correlation. We will have a large α whenever we have either a large average correlation among the items, $\bar{r}$, or a large number of items, k. This can lead to misleading results when you have just a few items that are reasonably consistent or when you have a large number of items that are not consistent.

For example, we might not consider a set of items with an average correlation of $\bar{r} = 0.1$ as having much in common. With a 5-item scale, this results in an alpha of 0.36; with a 20-item scale, the alpha becomes 0.69, and with a 50-item scale, the alpha is 0.85. This shows an important qualification on how we view alpha. A 50-item scale with a reliability of just 0.85 contains items that do not have much in common! Some major scales are like this where they gain consistency from the large number of items and not from the items sharing much common variance.

The opposite happens when we have scales with just a few items. Many large-scale surveys attempt to measure many concepts and have just a few items to measure each of the concepts. If we have an average correlation among items of 0.3, a 3-item scale will yield an alpha of just 0.47, whereas a 10-item scale yields an alpha of 0.87. Adding just a few items greatly enhances the reliability of a scale, and 10-item scales should be adequate even if the average correlation is only moderate. At the other extreme, some researchers feel the need to have many items. Increasing our scale from 10 items to 30 items, and measuring one-third as many concepts in the survey as a result, raises the alpha from 0.87 to 0.90. For many applications, this small improvement will not offset the resulting increase in the length of the questionnaire.

12.3.4 Kuder–Richardson reliability for dichotomous items

The equivalent of alpha for items that are dichotomous is the Kuder–Richardson measure of reliability. The dichotomous items need to be coded as a 1 for "Yes" or "Correct" and a 0 for "No" or "Incorrect". This measure is sometimes used with multiple-choice tests where students get each item right or wrong. It is also used when there are simple yes/no response options to the items. To run this procedure, you need to first install a command called `kr20`. This command was written by Herve Caci. You can do this by entering `ssc install kr20`. The results follow for running `alpha` and `kr20` on a set of items about using the web for art that were asked on the 2002 General Social Survey (`kuder-richardson.dta`):

```
. alpha newartmus1 newartmus2 newartview newartinfo newartmus3, asis item
Test scale = mean(unstandardized items)

                                                                  average
                                 item-test     item-rest      inter-item
Item           |  Obs  Sign    correlation   correlation      covariance      alpha
---------------+------------------------------------------------------------------
newartmus1     |   90    +        0.5767        0.3432        .0578652       0.5626
newartmus2     |   90    +        0.6600        0.3966        .0485019       0.5327
newartview     |   90    +        0.5948        0.3123        .0563046       0.5784
newartinfo     |   90    +        0.6420        0.3831        .0506242       0.5404
newartmus3     |   90    +        0.6423        0.3749        .0506242       0.5448
---------------+------------------------------------------------------------------
Test scale     |                                               .052784       0.6066
----------------------------------------------------------------------------------

. kr20 newartmus1 newartmus2 newartview newartinfo newartmus3
Kuder-Richardson coefficient of reliability (KR-20)
Number of items in the scale = 5
Number of complete observations =    90

                       Item        Item      Item-rest
    Item     | Obs  difficulty  variance  correlation
-----------+----------------------------------------
 newartmus1|  90    0.7889      0.1665    0.3412
 newartmus2|  90    0.5889      0.2421    0.3944
 newartview|  90    0.6222      0.2351    0.3105
 newartinfo|  90    0.6556      0.2258    0.3809
 newartmus3|  90    0.6111      0.2377    0.3728
-----------+----------------------------------------
      Test |        0.6533                0.3600
KR20 coefficient is 0.6138
```

The `alpha` = 0.607 and the Kuder–Richardson reliability coefficient, `KR20` = 0.614, are similar. The `kr20` provides another set of information in the column labeled `Item difficulty`. This shows that the items were about equally difficult except for the first item. When constructing a test or measure with dichotomous items, we often want to have items of varying difficulty so that even the best participants are challenged by some of the items.

12.3.5 Rater agreement—kappa (κ)

Reliable coding of observational data is often extremely difficult. If you have two people trained to code videotapes of interactions for evidence of conflict, it is hard to have both of them code everything the same way. If your two coders do not agree on how they are coding the observational data, then it is difficult to have much confidence in the analysis based on their coding.

When collecting observational data, we often have two or more raters code their observations, and we measure the reliability of their agreement. We might have two physicians rating the health of the same patients on a 5-point scale. We might have two clinicians identify the mental health condition of patients. A widely used measure of interrater agreement is the coefficient kappa (κ).

Suppose that we were observing people from the press who were asking questions of a political leader. We might have several categories such as

1. Friendly questions that the political leader would like to have asked—Do you feel education is important?
2. Difficult questions that the political leader would like to have asked—Could you elaborate on the benefits of your plan to reform the Social Security system?
3. Unwanted questions that the political leader would not like to have asked—Why did your Press Secretary resign?
4. Extremely unwanted and difficult questions that the political leader would not like to have asked—How can you justify the $125,000 donation you received from the Committee to Promote Higher Pay for Men?

Perhaps at the beginning of the politician's tenure, there is a honeymoon period in which most of the questions are ones the leader wanted to answer because they are easy or because they allow the leader to clarify his or her plans. Perhaps later in their tenure, the questions become less desirable from the politician's perspective. You have two coders rate a press conference in which 20 questions were asked, and their hypothetical ratings are in `kappa1.dta`.

We can do a tabulation of this data:

```
. tabulate coder1 coder2

                 |                      coder2
          coder1 |      Easy  Hard want   Unwanted  Very unwa |     Total
-----------------+--------------------------------------------+----------
            Easy |         5          1          0          0 |         6
     Hard wanted |         0          3          0          0 |         3
        Unwanted |         1          2          2          0 |         5
   Very unwanted |         0          1          1          4 |         6
-----------------+--------------------------------------------+----------
           Total |         6          7          3          4 |        20
```

The coders seem to agree on the classification of the friendly questions and the unwanted questions. They seem to have a harder time distinguishing the difficult but wanted questions from the difficult unwanted types. Ideally, all of the observations should fall on the diagonal where coder one and coder two have the same classification of the question.

To estimate κ, we enter the command `kap coder1 coder2`. Note that we typed `kap` and not `kappa`. The latter is used only for data entered in a special way. The output is as follows:

```
. kap coder1 coder2

             Expected
Agreement   Agreement     Kappa   Std. Err.         Z      Prob>Z
-----------------------------------------------------------------
  70.00%      24.00%     0.6053     0.1221       4.96      0.0000
```

We would describe this by writing that the judges agree exactly on the classification 70% of the time, i.e., $(5+3+2+4)/20$. The coefficient of agreement is $\kappa = 0.61, z = 4.96, p < 0.001$, indicating a reasonable level of agreement.

Some people would use the percentage agreement instead of κ, but the percentage agreement exaggerates the amount of agreement because it ignores the agreement we would expect by chance. κ only gives us credit for the extent the agreement exceeds what we would have expected to get by chance alone.

What is a strong kappa?

We would like alpha to be 0.80 or higher for a scale to be considered reliable. The expectations for kappa are different. Because kappa adjusts the percentage agreement for what would be expected by chance, kappa tends to be lower than alpha. Landis and Koch (1977) suggested these guidelines; see also Altman (1991):

Alpha	Strength of agreement
Under 0.20	Poor
0.21 – 0.40	Fair
0.41 – 0.60	Moderate
0.61 – 0.80	Good
0.81 – 1.00	Very good

Thus our $\kappa = 0.605$ is between moderate and good.

If you conceptualize the classes as ordinal, it is possible to estimate a weighted kappa. This κ_{weighted} gives partial credit for being close. If the two judges are close, the weighted kappa gives us partial credit. Stata allows us to use default value for the weights or to enter our own criterion. This is explained in the *Stata Base Reference Manual*, along with a detailed example. Entering the command `help kap` provides a brief explanation. We will not cover weighted kappa here.

Some researchers are able to use three raters rather than two, and this offers much more information. With three raters, kappa not only reports the overall κ but can even give us a κ for each category. For example, we may find that the coders have a high level of reliability for coding items that are in the "Easy" or "Hard wanted" categories, but a very low reliability for the "Unwanted" or "Very unwanted" categories. Here we might find that combining the "Unwanted" and "Very unwanted" categories into one category is necessary to have interrater reliability. The extension of kappa to three raters is described in the *Stata Base Reference Manual*.

12.4 Validity

A good measure needs to be reliable, but it also needs to be valid. A valid measure is one that measures what it is supposed to be measuring. A measure of maternal depression may be reliable, meaning it gets consistent results, without being valid. In order to be valid, the measure should have a high score for mothers who actually are depressed and a low score for mothers who are not. A measure is not valid when it is measuring other phenomena. For example, asking a person if they have high anxiety is not an especially valid item to measure depression. A person may have high anxiety just before an examination, but this is different than being depressed. The gist of validity is simply that a measure needs to measure what you are trying to measure and not something else. Assessing validity involves comparing a measure to one or more standards. There are many possible standards, resulting in several types of validity checks.

12.4.1 Expert judgment

In making a scale, you first need to carefully define the content domain of what you are measuring. Once you have done this, you share your items and your definition of the domain with a small group of experts in the field and have them critique each item. This is a good measure of content validity if your definition of the content domain and the quality of your experts are both good.

Defining the content domain is often difficult. Suppose you are measuring self-reported health. What does this include? Does it include mental health? Does it include living a healthy lifestyle? Is it limited to physical health? What about ability to perform activities of daily living? What about flexibility, strength, and endurance? Are you interested in a global rating of health or more specific aspects?

Once you have carefully defined the domain of your concept, your panel of experts can judge if you have items that represent this domain. Judges may be academic scholars who have published in the area or others who have worked in the field. A psychiatric social worker, a physician, and an academic scholar on health might be good candidates for your panel of experts. In other cases, you might want to include people who are experiencing issues related to your content. A member of a youth gang would be an expert if you were measuring gang-related violence.

One measure of content validity is the content validity ratio (CVR) (Lawshe 1975; Shultz and Whitney 2005). Judges rate each item as *essential*, *useful*, or *not necessary*. For each item, you compute the CVR by using

$$\mathrm{CVR}_i = \frac{n_e - \frac{N}{2}}{\frac{N}{2}}$$

where n_e is the total number of judges rating item i as *essential*, and N is the number of judges. Assume you have a 10-item scale and four judges, with three of the four judges rating the first item as *essential*. The CVR_i would be $(3-2)/2 = 0.50$. You can keep the items that have a relatively high CVR and drop those that do not. However, a limitation of this is that it is no better than the quality of your definition of your domain and the quality of the judges.

Face validity is a type of content validity, but it does not require expert judgment. People like those who will be measured by your scale can often be most helpful for evaluating face validity. If there are subgroups that might have special understandings (gender, ethnicity, etc.), they should be asked to say whether the items make sense for measuring the variable you are trying to measure. Imagine you are developing a scale to measure the self-regulation in a group of 8-year-olds. It would be good to ask a small number of 8-year-olds to read each item aloud and explain what the item is trying to say or ask. Then ask each of these 8-year-olds to pick an answer and explain why they picked that answer. You will almost always find that a few of the items are interpreted differently by these 8-year-olds than the way you intended and that the reason they pick an answer is extraneous to what you are measuring.

12.4.2 Criterion-related validity

Criterion-related validity picks one or more criteria or standards for evaluating a scale. The standard may be a concurrent measure or a predictive measure. If you have a 10-item scale of self-reported health, you need to find a standard. This could be a more comprehensive 40-item measure that has been used in the literature but is too long for your purposes. You include both your 10-item measure and the 40-item measure in a pilot survey and correlate the two scale scores. You might also use as a standard a clinical evaluation of the person's health.

Predictive standards are criteria that your measure should predict, and these typically require a time interval. A measure of motivation to achieve given to high school juniors might use grade point averages during their senior year as a criterion. Where correlations are used, a value of 0.30 is considered moderate support, and a correlation of 0.50 is considered strong support.

If you are measuring intention of medical students to work in rural areas, you might wait until the group graduates and see how many of them actually do work in a rural area. This would involve predictive validity. Because working in a rural area is a binary 0/1 variable, you would use logistic regression with your scale score as the predictor. If your scale score is a substantively and statistically significant predictor of whether they work in a rural area or not, then this would be evidence of predictive criterion validity. If you are measuring the commitment to quit smoking, you might use the number of sessions a person attends as your criterion in which case you would compute a simple correlation.

Criterion-related validity can also be used to evaluate individual items. If you believe the overall scale is valid, you can correlate each item with the total score and drop those that have low correlations with the total score. Stata's command `alpha` with the option `item` provides this information, as was shown in the output on page 294 under the columns labeled `item-test correlation` and `item-rest correlation`.

It is also possible to divide the sample into quartiles based on the total score. Then you would compare the top quartile to the bottom quartile on each item by using a t test. Try this as an exercise.

12.4.3 Construct validity

The term construct validity is reserved for analyses that involve more than one approach to assessing validity. Some of the analyses that fall under the rubric of construct validity include the following:

- Comparing known groups. A measure of depression might be evaluated by comparing the mean score for a group of people being treated for post-traumatic stress disorder to a group who are not being treated. With two groups you would use a t test, and with more than two groups you would use ANOVA.

- Internal consistency. Although we presented alpha reliability as a measure of reliability, it is also reasonable to think of it as tapping validity. If you have a series of items measuring commitment to stop smoking, they should all be internally consistent, as indicated by item-rest correlations and alpha, as well as by comparing the top and bottom quartiles. These techniques were covered in the previous section. If the scale is valid, then $\alpha = 0.80$ signifies that 80% of the variance represents true variance in the variable.

- Correlational analysis. We would expect criterion-related validity to be demonstrated. Better yet, with construct validity we would use multiple criteria. Our motivation scale given in the junior year of high school should predict grade point average in the senior year, college plans, etc. These are known as *convergent* criteria-related validity because we expect these correlations to all be positive. *Divergent* criteria-related validity would identify additional criteria for which we would expect a negative relationship. We would expect our motivational scale score to be negatively related to such criteria as frequency of arrest, use of hard drugs, likelihood of dropping out of high school, and unprotected sexual behavior. With construct validity, we are looking at a pattern of correlations. An extension of this is to use factor analysis (see next section) and possibly confirmatory factor analysis.

- Qualitative analysis. We should interview people who complete our set of items and ask them to explain each answer they gave. Having a group of fifth-grade students answer questions about how much they like school by using a computer-assisted interview program that includes neat graphics may be measuring how much they like school, but it also may measure how much they like "foolin' around on the computer". Asked why they give a positive response to an item, we might learn that it is because they would rather play on the computer than sit in a class. A surprising number of scales have been developed by academics that have never been evaluated by the target population. We should always evaluate if the people in our target population are interpreting the items and the options the way we expected.

12.5 Factor analysis

Ideally, a scale measures just one concept. We sometimes ignore this, and the results can be misleading. Imagine that Juanita has a GRE score on the math part of the examination of 750 (very high), but her score on the verbal part of the examination is just 250 (very low). Her total score is 1,000. A second student, Erica, has a 750 on the verbal part and a 250 on the math part of the examination, so she also has a total score of 1,000. It would be a huge mistake to say that both students have equal ability because their total scores are the same. Using one number to represent different things (math and verbal abilities) can lead to scales that lack validity. Such scale scores are often poor predictors of outcomes. Predicting that Juanita and Erica would do equally well in a graduate statistics program or a graduate history program because both have the same combined GRE score of 1,000 would be questionable. Most likely, we would get a better prediction if we treated the quantitative score and the verbal score as two separate variables and did a multiple regression.

Some concepts are multidimensional. For example, health might have one dimension for physical health and a separate dimension for psychological health. We would expect these two dimensions of health to be correlated, but the correlation should be small enough that we recognize them as two distinct dimensions of health. In such a case, we might not want to average items. Instead, we might want to make two scales, one for physical health and one for psychological health. Some researchers would ignore our advice to have two scales and simply combine them. If you do this, it would make more sense to call this an index rather than to call it a scale. Ideally, a scale should represent one dimension.

How do we know a set of items represents one dimension? This is where we use factor analysis. If the items represent one dimension, then all of them should be moderately to strongly correlated with each other, and the pattern of correlations should be consistent. If we had a 6-item scale, it might look something like table 12.2.

Table 12.2. Correlations you might expect for one factor

	Item 1	Item 2	Item 3	Item 4	Item 5	Item 6
Item 1	1.00					
Item 2	0.60	1.00				
Item 3	0.45	0.40	1.00			
Item 4	0.50	0.55	0.45	1.00		
Item 5	0.70	0.55	0.45	0.40	1.00	
Item 6	0.35	0.40	0.45	0.40	0.35	1.00

What would happen if we had two dimensions with the first three items representing physical health and the last three items representing psychological health? This is illustrated in table 12.3, where the first three items are highly correlated with each other (top left triangle), the last three items are highly correlated with each other (bottom right triangle), but the two sets of items are only moderately correlated with each other (bottom left rectangle).

Table 12.3. Correlations you might expect for two factors

	Item 1	Item 2	Item 3	Item 4	Item 5	Item 6
Item 1	**1.00**					
Item 2	**0.60**	**1.00**				
Item 3	**0.80**	**0.70**	**1.00**			
Item 4	0.30	0.25	0.35	**1.00**		
Item 5	0.20	0.25	0.25	**0.70**	**1.00**	
Item 6	0.25	0.20	0.15	**0.75**	**0.65**	**1.00**

Factor analysis is a collection of techniques that does an exploratory analysis to see if there are clusters of items that go together. When constructing a scale, you would hope that all the items you include in your scale form one factor or principal component.

There are two types of analysis that are often lumped together under the label of factor analysis. The default is exploratory factor analysis, which Stata calls principal factor analysis. We will use PF to refer to this type of factor analysis. The second type is principal-component factor analysis (often called principal component analysis), which Stata calls principal-component factor analysis. We will use PCF to refer to this approach. Each of these procedures has distinct objectives.

PF analysis attempts to identify a small number of latent variables or dimensions that explain the shared variance of a set of measures. If you believe that an attitude toward use of methamphetamines includes a cognitive component (beliefs), an affective component (emotional response, feelings), and a behavioral component (predispositions), you might do a PF analysis of a set of 30 items to see if the items fall into these three dimensions and label each dimension as a latent variable. Here the latent variables explain how you respond to the items. In PF analysis, the variance is partitioned into the shared variance and unique or error variance. The shared variance is how much of the variance in any one item can be explained by the rest of the items. This shared variance is estimated using a multiple R^2. In practical terms, factor analysis analyzes a correlation matrix in which the 1.00s on the diagonal are replaced by the square of the multiple correlation of each item with the rest of the items. PF analysis is not trying to explain all of the variance, just the variance that is shared among the set of items.

PCF analysis has a different approach, and its strength is in the area of data reduction. PCF analysis tries to identify components that are composites of the items. It does not distinguish between the common or shared variance and the unique or error variance. In practical terms, PCF analysis analyzes a correlation matrix in which the 1.00s on the diagonal stay there. It is trying to explain all of the variance. The interpretation of components is sometimes more difficult than the interpretation of factors because the components include both common and error variance. PCF analysis is used when developing a scale where one dimension is identified to represent the core of a set of items. If we have 20 items measuring your satisfaction with a drug, and we find that 15 of the items load highly on the first dimension in a PCF analysis, then we may want to use a composite score of the 15 items as a measure of satisfaction and drop the other 5 items.

Factor analysis has a special vocabulary of its own. The following is a list of some of the most important terminology. You can read this now, but it may be most helpful to refer to it after you have read the section where each term is used.

- Factor or component. Depending on the type of analysis you are doing, you will identify factors or components. Factors are the focus of PF analysis, and components are the focus of PCF analysis. Although there are important technical differences, many researchers refer to both of these as factors.

- Extraction. There are different ways of extracting a solution. This involves analyzing the correlation matrix (sometimes the covariance matrix) to obtain an initial factor solution. The initial factor solution may subsequently be rotated to aid interpretation.

- Eigenvalues. Both PCF analysis and PF analysis produce a value for each factor or component that is known as the eigenvalue for that factor. In the case of PCF analysis, because all the variance is being explained, the sum of the eigenvalues for all possible components is the number of items. If there are 10 items, the sum of the eigenvalues will be 10. Therefore, a component that has an eigenvalue of less than 1 is not helpful for data reduction because it accounts for less than one item. The factors will be ordered from the most important, which has the largest eigenvalue, to the least important, which has the smallest eigenvalue. In PF analysis the sum of the eigenvalues will be less than the number of items, and their interpretation is complex.

- Commonality. This is how much of each item is explained by the rest of the items. It is estimated by doing a multiple regression of each item on the remainder of items. We refer to the resulting R^2 as the commonality or shared variance. PF analysis tries to explain this shared variance, whereas PCF analysis tries to explain all of the variance.

- Loadings. Both PCF analysis and PF analysis produce a matrix with the items in the rows and the factors in the columns. The elements of this matrix are called loadings, and they are used to assess how clusters of items are most related to one or another of the factors. If an item has a loading over 0.4 on a factor, it is considered a good indicator of that factor.

- Simple structure. This is a pattern of loadings where each item loads strongly on just one factor and a subset of items load strongly on each factor. When an item loads strongly on more than one factor, it is factorially confounded.

- Scree plot. This is a graph showing the eigenvalue for each factor. Its name comes from geology, where scree collects at the bottom of a cliff. It is used to decide how many factors are included by throwing out those that are considered scree. When doing a PCF analysis, we usually drop factors that have eigenvalues in the neighborhood of 1.0 or smaller.
- Rotation. Neither PF analysis nor PCF analysis have a unique solution without making a series of assumptions. It is often helpful to rotate the initial solution to obtain a solution that is more interpretable.
- Orthogonal versus oblique. Some rotations require that the factors be uncorrelated. These are orthogonal solutions. Other rotations allow the factors to be correlated and are called oblique solutions. There are many options for each type of rotation, and each type of rotation makes different assumptions to obtain identification.
- Factor score. This is a score on each factor. Rather than simply averaging or adding the items that load above some value such as 0.4 on the factor to get a scale score, a factor score weights each item based on how related it is to the factor. Also, the factor score is scaled to have a mean of 0.0 and a variance of 1.0.

Should you use PF analysis or PCF analysis? If you want to explain as much of the variance in a set of items as possible with one dimension, then PCF analysis is reasonable. This is reasonable when you have a set of items that you believe all measure one concept. In this situation, you would be interested in the first principal component. You would want to see if it explained a substantial part of the total variance for the entire set of items, and you would want most of the items to have a loading of 0.4 or above on this component.

If, on the other hand, you want to identify two or more latent variables that represent interpretable dimensions of some concept, then PF analysis is probably best. For example, you might have 20 items asking about a person's level of alienation. A PF analysis might find that there are subsets of items that fall into different dimensions such as value isolation, powerlessness, and normlessness. With PF analysis, the focus is on finding two or more interpretable dimensions or factors rather than maximizing the amount of variance that is explained.

In this chapter, we limit our coverage to PCF analysis because it relates to developing a measure of a concept. Using the Stata menu, you can select Statistics ▷ Multivariate analysis ▷ Factor and principal component analysis. Here you will find several options:

- Factor analysis
- Factor analysis of a correlation matrix
- Principal component analysis (PCA)
- PCA of a correlation or covariance matrix
- Postestimation

We will focus on the first of these, factor analysis. This option allows us to do either PF analysis or PCF analysis. (Be careful not to select Principal component analysis (PCA).) The terminology used in factor analysis is highly specialized, and the same terms can be used differently by different software packages.

The second menu option, Factor analysis of a correlation matrix, allows you to enter a correlation matrix and do the factor analysis of it. This is useful when reading an article that includes a correlation matrix because you do not need the raw data to do the factor analysis; you just need the correlation matrix.

The third and fourth options involve Stata's application of principal component analysis (PCA). We will not consider these options here. The last option, Postestimation, includes things we can do after we obtain the first solution to help us interpret the results. These include rotating the solution, creating scree plots, and estimating factor scores.

What's in a name

Because factor analysis is an area of long-standing specialization, there are many differences of opinion as to what is the best approach to it. Many of these distinctions are not of great importance when we use factor analysis to develop or evaluate a scale. There are many technical terms that you rarely see in other areas of statistics. To make matters especially confusing, different software may use the same name for different procedures. We will only cover the procedures for factor analysis as Stata implements them. But, do not be surprised if you hear the same names used differently with other packages.

In Stata, PCF analysis produces the same results as what SPSS calls principal component analysis. However, what Stata calls principal component analysis (PCA—the third and fourth options in the menu) has important differences. We will use the name PCF analysis to refer to what SPSS calls principal component analysis.

The default in Stata is to do a PF analysis; in SPSS, the default is to do a PCF analysis.

12.6 PCF analysis

Here is an example of PCF analysis using 14 items from the 2006 General Social Survey (gss2006_chapter12_selected.dta). There are four options for how Stata extracts factors from your data. Other packages offer more options not available in Stata, but the options in Stata should meet most needs for people developing or evaluating a scale.

We will examine the 14 items from the 2006 General Social Survey that measure desire to invest in different national programs. Our goal is to select items that we can use in a scale. Here is a compact codebook of the items:

```
. codebook natspac natenvir natheal natcity natcrime natdrug nateduc natrace
> natarms natfare natroad natsoc natchld natsci, compact
Variable     Obs Unique  Mean     Min  Max  Label
-------------------------------------------------------------------------------
natspac     1407   3  2.237385    1     3  SPACE EXPLORATION PROGRAM
natenvir    1446   3  1.375519    1     3  IMPROVING  PROTECTING ENVIRONMENT
natheal     1451   3  1.297726    1     3  IMPROVING  PROTECTING NATIONS HEALTH
natcity     1349   3  1.634544    1     3  SOLVING PROBLEMS OF BIG CITIES
natcrime    1448   3  1.446133    1     3  HALTING RISING CRIME RATE
natdrug     1428   3   1.45028    1     3  DEALING WITH DRUG ADDICTION
nateduc     1462   3  1.317373    1     3  IMPROVING NATIONS EDUCATION SYSTEM
natrace     1348   3  1.780415    1     3  IMPROVING THE CONDITIONS OF BLACKS
natarms     1442   3  2.144938    1     3  MILITARY, ARMAMENTS, AND DEFENSE
natfare     1434   3  2.110879    1     3  WELFARE
natroad     2899   3  1.756468    1     3  HIGHWAYS AND BRIDGES
natsoc      2864   3  1.407472    1     3  SOCIAL SECURITY
natchld     2758   3  1.517041    1     3  ASSISTANCE FOR CHILDCARE
natsci      2786   3  1.688442    1     3  SUPPORTING SCIENTIFIC RESEARCH
-------------------------------------------------------------------------------
```

Running tabulate with the codebook command or the fre command, if you have that command installed, we see that these items have a higher score, implying less support for spending. We can reverse these codes, reverse the corresponding labels, and add a prefix of "r" to each variable to remind us that we have reversed them:

```
. recode natspac natenvir natheal natcity natcrime natdrug nateduc natrace
> natarms natfare natroad natsoc natchld natsci
> (1=3 "Too little") (2=2 "About right") (3=1 "Too much"), prefix(r) label(revnat)
(756 differences between natspac and rnatspac)
(1081 differences between natenvir and rnatenvir)
(1169 differences between natheal and rnatheal)
(839 differences between natcity and rnatcity)
(992 differences between natcrime and rnatcrime)
(1027 differences between natdrug and rnatdrug)
(1152 differences between nateduc and rnateduc)
(746 differences between natrace and rnatrace)
(967 differences between natarms and rnatarms)
(919 differences between natfare and rnatfare)
(1392 differences between natroad and rnatroad)
(1973 differences between natsoc and rnatsoc)
(1756 differences between natchld and rnatchld)
(1570 differences between natsci and rnatsci)
```

Now, a higher score implies greater support. A researcher may think there is one dimension representing how liberal or conservative a person is regarding government spending. This researcher may feel that a liberal would score high on all of these items (big spenders) and a conservative would score low, meaning they want to minimize government spending initiatives. Another person may think there are two separate dimensions, one focused on investing in social welfare issues—e.g., education, health, welfare, environment—and one focused on investing in the national infrastructure. A third researcher may believe it is more complicated and want to see what dimensions are needed to explain this set of 14 items.

To run a PCF analysis, we select Statistics ▷ Multivariate analysis ▷ Factor and principal component analysis ▷ Factor Analysis. The first tab is labeled Model, and here we list our 14 recoded variables. The second tab is labeled Model 2. Here we select *Principal-component factor*. This generates the following results:

```
. factor rnatspac rnatenvir rnatheal rnatcity rnatcrime rnatdrug rnateduc
> rnatrace rnatarms rnatfare rnatroad rnatsoc rnatchld rnatsci, pcf
(obs=1082)

Factor analysis/correlation                    Number of obs    =      1082
    Method: principal-component factors        Retained factors =         4
    Rotation: (unrotated)                      Number of params =        50

    --------------------------------------------------------------------------
         Factor  |   Eigenvalue   Difference        Proportion   Cumulative
    -------------+------------------------------------------------------------
        Factor1  |      3.00301      1.62066            0.2145       0.2145
        Factor2  |      1.38235      0.12372            0.0987       0.3132
        Factor3  |      1.25863      0.22702            0.0899       0.4031
        Factor4  |      1.03161      0.06974            0.0737       0.4768
        Factor5  |      0.96187      0.05047            0.0687       0.5455
        Factor6  |      0.91140      0.09801            0.0651       0.6106
        Factor7  |      0.81339      0.02930            0.0581       0.6687
        Factor8  |      0.78409      0.05945            0.0560       0.7247
        Factor9  |      0.72464      0.01659            0.0518       0.7765
       Factor10  |      0.70804      0.04088            0.0506       0.8271
       Factor11  |      0.66717      0.06485            0.0477       0.8747
       Factor12  |      0.60231      0.01414            0.0430       0.9178
       Factor13  |      0.58817      0.02484            0.0420       0.9598
       Factor14  |      0.56333            .            0.0402       1.0000
    --------------------------------------------------------------------------
    LR test: independent vs. saturated:  chi2(91) = 1701.47 Prob>chi2 = 0.0000
```

```
Factor loadings (pattern matrix) and unique variances
```

Variable	Factor1	Factor2	Factor3	Factor4	Uniqueness
rnatspac	-0.0405	0.6150	-0.4468	-0.0108	0.4204
rnatenvir	0.5562	0.0698	-0.1925	0.1629	0.6221
rnatheal	0.6298	0.0420	0.0923	0.2638	0.5235
rnatcity	0.5264	-0.0580	-0.0677	-0.4598	0.5036
rnatcrime	0.4533	0.1160	0.5154	-0.4234	0.3361
rnatdrug	0.5381	0.1048	0.3346	-0.2872	0.5051
rnateduc	0.5997	0.0035	0.0425	0.3838	0.4912
rnatrace	0.5921	-0.2286	-0.2660	-0.1885	0.4909
rnatarms	-0.1087	0.4926	0.4682	0.1719	0.4968
rnatfare	0.4725	-0.1726	-0.3547	-0.0629	0.6172
rnatroad	0.1421	0.5022	0.0704	-0.1269	0.7065
rnatsoc	0.4196	-0.1209	0.3374	0.4575	0.4862
rnatchld	0.5369	-0.1049	-0.1130	0.1343	0.6699
rnatsci	0.3057	0.6059	-0.2865	0.0493	0.4549

If these 14 items all fall along one dimension, we would expect the first factor to have a dominant effect. We can assess this by using the `Eigenvalues`. Remembering that in PCF analysis the total of the eigenvalues for all factors is the number of items, 14 in this case, we can see that the first factor has an eigenvalue of 3.00, and this is 21% of the total possible of 14. By comparison, the second factor has an eigenvalue of just 1.38. It looks like there is a strong first factor.

If you look at the matrix showing the "Factor loadings", you can go down the column labeled `Factor1` to see the loadings for each item on the first factor (component). Notice how many of the items have a loading greater than 0.40. The exceptions are for the space programs, armaments, roads, and science. The rest of the items have strong loadings and include the environment, health, crime, drugs, education, race, welfare, social programs, and children. This seems to be a general liberal spending program, and we could make a scale using these 10 items, dropping the other 4 items.

Generally, we should consider any factor that has an eigenvalue of more than 1.0. A visual way to examine the eigenvalues is with a scree plot. This is a postestimation command that we run just after doing a factor analysis. Select Statistics ▷ Multivariate analysis ▷ Factor and principal component analysis ▷ Postestimation ▷ Scree plot of eigenvalues. This dialog box gives us a variety of options, but it takes less work to enter the command directly than it does to find the dialog box. The command is `screeplot`, and it produces the graph in figure 12.1.

(Continued on next page)

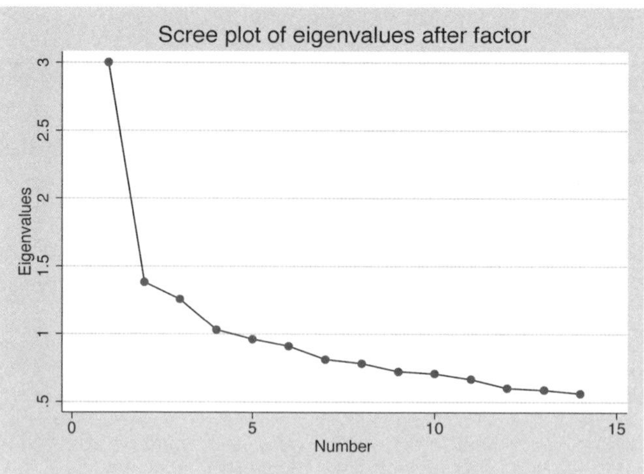

Figure 12.1. Screeplot: National priorities

This scree plot starts leveling off (going across more quickly than it is going down) with the fourth factor. This would suggest that we need 3 factors, and we could drop the rest of them. Deciding on the number of factors when doing a PCF analysis is not an easy task. We had 4 factors that had an eigenvalue of more than 1, although the fourth factor was just barely over 1. The scree plot suggests we might just focus on the first three factors, and the scree plot makes it clear that the first factor is dominant because there is such a big drop-off in the size of the eigenvalue between the first and second factors.

When there is more than one factor, it is usually useful to rotate the initial solution to see if a different solution is easier to interpret. We have two ways of doing this. One is an orthogonal rotation where we force the factors to be uncorrelated with each other. The other is an oblique rotation in which we allow the factors to be correlated. There are multiple options for both types of rotation (15 for orthogonal rotation and 8 more for oblique rotation). We will illustrate just one approach for each type.

12.6.1 Orthogonal rotation: varimax

Stata has kept the factor analysis solution in memory, so to do an orthogonal rotation all we need to do is enter the command `rotate` (this is the default rotation). If you forget this command, you can find it under the Postestimation menu. Here is what we obtain:

```
. rotate
Factor analysis/correlation                      Number of obs    =      1082
    Method: principal-component factors          Retained factors =         4
    Rotation: orthogonal varimax (Kaiser off)    Number of params =        50

    --------------------------------------------------------------------------
         Factor  |     Variance   Difference        Proportion   Cumulative
    -------------+------------------------------------------------------------
        Factor1  |      1.96564      0.29211            0.1404       0.1404
        Factor2  |      1.67353      0.04093            0.1195       0.2599
        Factor3  |      1.63260      0.22876            0.1166       0.3766
        Factor4  |      1.40384            .            0.1003       0.4768
    --------------------------------------------------------------------------
    LR test: independent vs. saturated:  chi2(91) = 1701.47 Prob>chi2 = 0.0000

Rotated factor loadings (pattern matrix) and unique variances

    ----------------------------------------------------------------
        Variable |  Factor1   Factor2   Factor3   Factor4 | Uniqueness
    -------------+----------------------------------------+-----------
        rnatspac |  -0.1295   -0.0143   -0.1449    0.7360 |    0.4204
       rnatenvir |   0.4623    0.3085    0.0890    0.2471 |    0.6221
        rnatheal |   0.6328    0.1374    0.2186    0.0975 |    0.5235
        rnatcity |   0.0262    0.4831    0.5052    0.0839 |    0.5036
       rnatcrime |   0.0941   -0.0276    0.8066   -0.0599 |    0.3361
        rnatdrug |   0.2187    0.0910    0.6618    0.0296 |    0.5051
        rnateduc |   0.6899    0.1329    0.0932    0.0808 |    0.4912
        rnatrace |   0.2352    0.6335    0.2259    0.0389 |    0.4909
        rnatarms |   0.1103   -0.6491    0.1967    0.1763 |    0.4968
        rnatfare |   0.2255    0.5642    0.0480    0.1061 |    0.6172
        rnatroad |   0.0039   -0.1708    0.2889    0.4252 |    0.7065
         rnatsoc |   0.6718   -0.1136    0.0967   -0.2007 |    0.4862
        rnatchld |   0.4475    0.3393    0.1075    0.0559 |    0.6699
         rnatsci |   0.1777    0.0334    0.0832    0.7110 |    0.4549
    ----------------------------------------------------------------

Factor rotation matrix

    --------------------------------------------------
                 |  Factor1  Factor2  Factor3  Factor4
    -------------+------------------------------------
         Factor1 |   0.6858   0.4868   0.5116   0.1758
         Factor2 |  -0.0311  -0.4717   0.1951   0.8594
         Factor3 |   0.1676  -0.6530   0.5615  -0.4798
         Factor4 |   0.7075  -0.3379  -0.6204  -0.0190
    --------------------------------------------------
```

The second table shows the factor loadings. With a varimax rotation, we can think of the loadings being the estimated correlation between each item and each factor. For example, the last item, `rnatsci` has a high correlation with the fourth factor, 0.711. We hope each item has a loading of 0.40 or higher on one factor and a smaller loading on the other factors. We use the items that have relatively large loadings on each factor to decide on what the factor is to be called. Let's work backward, starting with the fourth and weakest factor. It has just two items with high loadings, `rnatspac` and `rnatsci`. This factor might be labeled liberal spending support for science as a national priority. `Factor3` has three items with loadings higher than 0.4, namely, `rnatcity`, `rnatcrime`, and `rnatdrug`. This might be called a priority on urban problems, although we need to acknowledge that crime and drugs are problems outside of urban places as well. We could call this liberal spending support for urban problems.

`Factor2` has four items with loadings higher than 0.4. Support for minorities (`rnatrace`) and support for welfare programs (`rnatfare`) both have strong positive loadings. By contrast, support for the military (`rnatarms`) has just as strong a loading, but it is negative. This factor seems to be pitting together support for the disadvantaged against support for the military. This might make sense considering the disproportional burden minorities pay for military engagements and how the military expenses can drain money from social welfare. The problem with this factor and the third factor is that `rnatcity` loads strongly on both factors. It is compounded. Although Stata distinguishes between these two factors, the public sees both of these as special concerns in urban areas. We could call this factor liberal spending support for the disadvantaged.

`Factor1` is the strongest, and it has five items with loadings greater than 0.4. These involve the environment, health, education, social security, and child care. One might think of this as representing a liberal-conservative factor, where liberals want to put a higher spending priority on each of these than do conservatives.

12.6.2 Oblique rotation: promax

Sometimes it is unreasonable to think that the factors are orthogonal. We could argue that the first factor, which we have labeled liberalism, would be correlated with the second factor that we've labeled support for the disadvantaged. It is possible to do an oblique rotation in which we allow the factors to be correlated. This is an option for the `rotate` command. Here we illustrate one of the seven options for oblique rotation. We will do a promax rotation by using the command `rotate, promax`. When we allow the factors to be correlated, it is useful to see how correlated they are. When we run any command in Stata, the program saves some statistics it does not report in the Results window. These can be obtained with the `estat` command. To obtain the correlations between the common factors in an oblique rotation, the command is `estat common`. Here are the results of running these two commands in sequence:

```
. rotate, promax
Factor analysis/correlation                        Number of obs    =     1082
    Method: principal-component factors            Retained factors =        4
    Rotation: oblique promax (Kaiser off)          Number of params =       50

    --------------------------------------------------------------------------
         Factor  |     Variance   Proportion    Rotated factors are correlated
    -------------+------------------------------------------------------------
        Factor1  |      2.46691       0.1762
        Factor2  |      1.99629       0.1426
        Factor3  |      1.97508       0.1411
        Factor4  |      1.53936       0.1100
    --------------------------------------------------------------------------
    LR test: independent vs. saturated:  chi2(91) = 1701.47 Prob>chi2 = 0.0000
```

```
Rotated factor loadings (pattern matrix) and unique variances
```

Variable	Factor1	Factor2	Factor3	Factor4	Uniqueness
rnatspac	-0.1749	-0.0294	-0.2002	0.7787	0.4204
rnatenvir	0.4216	0.2378	-0.0437	0.2156	0.6221
rnatheal	0.6293	0.0361	0.0830	0.0460	0.5235
rnatcity	-0.1558	0.4893	0.5021	0.0281	0.5036
rnatcrime	-0.0338	-0.0575	0.8526	-0.1366	0.3361
rnatdrug	0.1026	0.0462	0.6555	-0.0410	0.5051
rnateduc	0.7179	0.0258	-0.0616	0.0370	0.4912
rnatrace	0.1043	0.6229	0.1546	-0.0083	0.4909
rnatarms	0.1864	-0.7131	0.2047	0.1762	0.4968
rnatfare	0.1329	0.5519	-0.0371	0.0790	0.6172
rnatroad	-0.0535	-0.2080	0.2760	0.4176	0.7065
rnatsoc	0.7637	-0.2163	-0.0129	-0.2463	0.4862
rnatchld	0.4121	0.2818	-0.0035	0.0167	0.6699
rnatsci	0.1175	-0.0307	-0.0209	0.7162	0.4549

```
Factor rotation matrix
```

	Factor1	Factor2	Factor3	Factor4
Factor1	0.8518	0.6478	0.6627	0.3341
Factor2	-0.0024	-0.4155	0.2418	0.8357
Factor3	0.1455	-0.5818	0.5015	-0.4341
Factor4	0.5032	-0.2632	-0.5008	-0.0396

```
. estat common
Correlation matrix of the promax(3) rotated common factors
```

Factors	Factor1	Factor2	Factor3	Factor4
Factor1	1			
Factor2	.3358	1		
Factor3	.3849	.1688	1	
Factor4	.1995	.1322	.2256	1

We can see that the factors are all positively correlated. The factor loadings can no longer be interpreted as simple correlations like before, but they tend to be "cleaner" than with the oblique rotation. By this I mean that the big loadings tend to be even bigger, and the small loadings tend to be even smaller. This sometimes simplifies the interpretation.

12.7 But we wanted one scale, not four scales

The rotations helped us gain insight into how people cluster different items and helped us recognize that there are some differences between people who would support each of these dimensions. However, we are trying to develop one scale, and we need to select items that represent one dimension. For this, we go back to our original PCF analysis results.

We clearly had one dominant factor as the first principal component, and most of the items loaded strongly on that first component. Indeed, 10 of the 14 items had loadings over 0.4. The items that failed to meet this standard are support for the space program, support for spending on science, support for roads, and support for the military. Space and science are special areas, as are roads and the military. Both liberals and conservatives may support or not support more spending for roads, space, and science. Some liberals may not support spending on the military in general, but many liberals and conservatives support spending to support the soldiers. By contrast, the 10 items with strong loadings on the first factor clearly differentiate liberals and conservatives. We could think of the 10 items as representing a general liberal-to-conservative dimension for federal spending policies.

We need to rerun our PCF analysis, including just the 10 items. Try this as an exercise. When we do this, the eigenvalue for the first factor is 2.941, explaining almost 30% of the variance in the set of 10 items. There are a second and third factor, but the eigenvalue for each of them is just barely over 1. All of the items load over 0.4 on the first factor.

Not all researchers use factor analysis when they are constructing a scale. However, if you use this approach and obtain a set of 10 items that have loadings over 0.4 on a first factor, and the first factor has a much larger eigenvalue than the other factors, your results of doing a reliability analysis are almost certainly going to be positive. For these 10 items, the alpha is 0.72 and dropping any of the items would reduce alpha.

12.7.1 Scoring our variable

We have used factor analysis to identify a set of items that load highly on one dimension. We can use the 10 items to construct our scale in one of two ways. One of these is exactly what we did before by using the `egen` command to generate the `rowmean` of the 10 items. We could do that and then do an alpha for the 10 items to estimate the reliability of our scale.

There is a more complex alternative, which is to estimate what is called a factor score. This is a standardized value that has a mean of 0 and a standard deviation of 1. If our variable is normally distributed, about two-thirds of the observations will have a score between -1.0 and $+1.0$ and about 95% will have a score between -2.0 and $+2.0$.

An advantage of a factor score over a mean or total score is that the factor score weights each of the items differently, based on how central it is to the first factor. For example, education has a loading of 0.6, and health has a loading of 0.63. These should be weighted more in the factor score than social security, which has a loading of 0.43. By contrast, when we generate a mean or total for the set of items, each item counts as if it were equally central to the concept.

It is easy to estimate a factor score. Using the default options for the command just after doing the PCF analysis, we enter the command `predict libfscore, norotate`, where `libfscore` is the name we are giving the factor score.

Which method is better? There is some controversy about how to best estimate a factor score, so I have simply used the default option, which is a regression-based approach. Although there is some disagreement about the best method, there is rarely a substantive difference in the result.

We should not forget that there is an advantage of a factor score over a mean or total score in that the factor score weights less-central items less and more-central items more, and this makes sense. However, this distinction rarely makes a lot of practical difference. The factor score may make a difference if there are some items with very large loadings, say, 0.9, and others with very small loadings, say, 0.2. But we would probably drop the weakest items. When the loadings do not vary a great deal, computing a factor score or a mean/total score will produce comparable results. Here are the commands you would use to compute a factor score and a mean score:

```
. factor rnatenvir rnatheal rnatcity rnatcrime rnatdrug rnateduc rnatrace
> rnatfare rnatsoc rnatchld, pcf
. predict libfscore, norotate
. egen libmean = rowmean(rnatenvir rnatheal rnatcity rnatcrime rnatdrug
> rnateduc rnatrace rnatfare rnatsoc rnatchld)
```

12.8 Summary

This chapter introduced some of the classical procedures for developing a scale. This is an area of social research that often is given too little attention. Without measures that are both reliable and valid, it is impossible for social science to develop. The intent of this chapter was to introduce the most basic ideas for developing a scale and how to execute basic Stata commands that help demonstrate the reliability and validity of a scale. We covered

- How to use the `egen` command to generate a scale score.
- Test–retest reliability.
- How to estimate alpha and how to use it to evaluate the internal consistency standard for a reliable scale. We saw that alpha depends on the average correlation between the items and the number of items in the scale; too few items makes it hard to obtain an adequate alpha, and too many items makes a questionnaire needlessly long. We also covered the Kuder–Richardson reliability coefficient for dichotomous items.
- Kappa as a measure of interrater reliability. We briefly discussed weighted kappa and how kappa can be used when there are three raters.
- Various types of validity including expert judges, face validity, criterion-related validity, predictive validity, and construct validity.
- Principal factor (PF) analysis and principal-component factor (PCF) analysis, with an example of how to use PCF analysis.

- Key terminology for understanding factor analysis, along with orthogonal and oblique rotations of the PCF analysis solution.
- How to construct a factor score and how this compares to generating a mean score.

12.9 Exercises

1. Use the `gss2002 and 2006_chapter12.dta` dataset. Compute a split-half correlation for the 10 items that load most strongly on the first factor in the PCF analysis (section 12.6). Correlate the mean of the first 5 items and the mean of the second 5 items. Why does this correlation underestimate the reliability of the 10-item scale?

2. Using the data from above, compute the alpha reliability for this scale. Interpret alpha to decide if this scale is adequate. Interpret the item-test correlations, item-rest correlations, and what happens to alpha if any of the items are dropped. Why is alpha a better measure of reliability for this 10-item scale than the split-half correlation?

3. Using the data from above, do an alpha analysis including all 14 items. Use the values of alpha if deleted column to drop the worst item and repeat alpha for the remaining 13 items. Repeat this process, dropping an item each time, until there is no item to drop that would increase alpha. Compare this set of items to what the factor analysis indicated was the best set of items.

4. Use the `gss2002 and 2006_chapter12.dta` dataset. Do a PCF analysis of the 10 items that load on the first factor for the spending items (section 12.6). Interpret the results.

5. Using the results from the last exercise, create the factor score for the 10 items and use `egen` to create a mean score for the 10 items. Compare the means and standard deviations of both scores, and explain the differences. What is the correlation between the two scale scores? What does it mean that it is so high? When would it not be so high?

6. Use the `gss2002 and 2006_chapter12.dta` dataset. Do a PCF analysis of the 14 items, and construct a factor score. Do a PCF analysis for the 10 items as done above, and construct a factor score. Correlate these two scores, and explain why they are so highly correlated even though they are based on different items.

13 Appendix: What's next?

13.1 Introduction to the appendix

The goal I had in writing this book was to help you learn how to use Stata, including creating a dataset, managing the dataset, changing variables, creating graphs and tables, and doing basic data analysis. There is much more to learn about each of these, as well as entirely new topics. At this point, you are ready to pursue these more advanced resources.

The purpose of this concluding chapter is to give you some guidance on what material is most useful as you develop greater expertise using Stata. What will give you a quick start? What requires a lot of statistical background? What are the most accessible resources? A word of caution as we start: new supporting resources appear regularly, so the suggestions here are current only at the time this book is published.

13.2 Resources

Many resources can help you expand your knowledge of Stata. More importantly, many of these are absolutely free, and they can be obtained over the Internet, including some online "movies" about specialized Stata techniques. Other resources include books about Stata and online courses.

13.2.1 Web resources

The premier web resource about Stata is at UCLA. They have an extraordinary web page at http://www.ats.ucla.edu/stat/stata/Library/. It has movies and extensions beyond what I have included. For example, I only briefly mentioned how to work with complex samples that involve clusters, stratification, and weighting. The UCLA web page has a link to a pair of movies about how to work with complex surveys (http://www.ats.ucla.edu/stat/stata/seminars/svy_stata_intro/default.htm). Some of the content is based on earlier versions of Stata, but even that content can be helpful. Stata maintains remarkable consistency between versions with new versions simply adding capabilities. The UCLA web page also has many links to statistics and data-analysis courses where you can obtain lecture information. It has many Stata programs that you can install on your machine to simplify your work. It even has Stata do-files that match examples in a few of the standard statistics textbooks. Much of what this web page does not have itself appears in the links it provides to other sources.

If you have some experience with SAS, the University of North Carolina's Population Center has a tutorial that was written for Stata 9 (at the time of this writing), and it has many useful examples of how to do different tasks and analyses. It is useful if you want to learn more about data management, and it is located at http://www.cpc.unc.edu/services/computer/presentations/statatutorial/.

Stata maintains a web page on getting started with Stata at http://www.stata.com/links/resources1.html, and this is an excellent place to start. This web page shows you links to a wide variety of sources that can help you learn Stata and extend your knowledge of Stata. Stata has a searchable frequently-asked-questions (FAQs) web page at http://www.stata.com/support/faqs/. This page includes many examples of how to do different procedures. Although some of these examples are extremely technical, many of them are accessible to a person who has completed this book. The way these pages work through examples and help you interpret results is way ahead of what we usually see in FAQs support.

We have installed a few commands from a web page at Boston College. You can check out the full list of available programs at http://ideas.repec.org/s/boc/bocode.html. This is the largest collection of user-written Stata commands and is maintained by Christopher F. Baum. For example, go down the list of programs for 2005, and find a program called `optifact`. This web page has a brief description of the program, and when you click on the link, it takes you to a more-detailed description. From there, you can click on the name of the author, Paul Millar, to get a list of papers he has written about this procedure. All of these commands are made available at no cost, and collectively they represent a considerable extension of the basic capabilities of Stata itself. The highly developed, user-driven extensibility of Stata is a feature that sets it above competing statistical analysis software.

There is a Stata newsgroup, Statalist, to which people submit questions and anybody who wants to provides an answer. It is hosted at the Harvard School of Public Health and has more than 2,000 subscribers. You can subscribe at http://www.stata.com/statalist/.

You probably want to subscribe as a digest, which gives you one or two messages per day, each of which contains many questions and answers. If you do not pick the digest option, you might get 20–50 or more messages per day. Much of the content on Statalist is for professional programmers and statisticians. However, the subscribers are often willing to answer questions from beginners, and many of these answers would cost you hundreds of dollars if you were paying for a consultant. The list will also keep you aware of new commands that you can install.

Stata is a completely web-aware package, and web-based resources are constantly expanding. If you enter the command `findit` followed by a keyword, Stata will search the web for relevant information. You might try it with a keyword, such as `findit missing`, if you were concerned about having a lot of missing values in your dataset. Stata will give you a list of relevant resources followed by a list of commands you can install that help you work with missing values.

We have discussed only a few of the available resources on the web. By the time you have checked these out, you will see more resources for your special interests and needs. Fortunately, many of these web pages have links to other web pages.

13.2.2 Books about Stata

Stata has a collection of reference manuals that are available from the Stata Bookstore web site, http://www.stata.com/bookstore/documentation.html. Many universities participate in a special program that Stata offers, the GradPlan, which lets you buy these at a reduced price. Even at the full price, however, they are inexpensive compared with other books about statistics and software. The manuals are among the best provided by a software company. In my experience, the price of the books that Stata Press publishes is less expensive if you buy them directly from Stata than if you buy them online from other sources. Before buying them elsewhere, check the Stata Press pricing.

The *Base Reference Manual* provides a description of the most commonly used commands, along with an example using data you can download. These examples typically include a brief guide on how to interpret the results. Most of the examples are at an appropriate level, being complicated enough to show you how to work with real data but not so esoteric that only a senior statistician would find them informative.

The *Stata Journal* is published quarterly. The editors are H. Joseph Newton and Nicholas J. Cox. It includes articles on commands people have written that you might want to install; tutorials on how to do different tasks by using Stata; articles on statistical analysis and interpretation, data management, and graphics; and book reviews. You can find out more about it at http://www.stata-journal.com/.

There are several excellent books on Stata. Ulrich Kohler and Frauke Kreuter published an English-language version of their German book *Data Analysis Using Stata* in 2009. The publisher is Stata Press. This book builds on what I have done in *Gentle Introduction to Stata* and covers some more-advanced data-analysis techniques. It goes much further than I did on programming Stata. Lawrence Hamilton published a book on Stata 9 in 2006, *Statistics with Stata*. The publisher is Brooks/Cole. This is a useful reference book, and you can quickly find commands that do specific tasks. It has a simple organization and many interesting empirical examples.

You have seen how graph commands are the longest and most complicated. The *Stata Graphics Reference Manual* is competently written, but many find it hard to follow. Remember, if a picture is worth a thousand words, it takes a lot of words to tell a computer how to make a good picture of your data. Michael Mitchell, who was previously associated with the UCLA Stata Portal, published a book in 2004 with Stata Press called *A Visual Guide to Stata Graphics* (now in its second edition); see Mitchell (2008). This book shows a few graphs on each page, and you scan the book until you see something similar to what you want to do. Next to the picture, you find the Stata command that produced the graph. This does not make use of the dialog system, but you can use his examples to enhance the graphs you make using the dialogs.

At the time this revision of *A Gentle Introduction to Stata* was written, Scott Long was writing a book on *Workflow in Data Analysis Using Stata* for Stata Press. This book will cover data management both conceptually and practically. It will introduce a set of commands that will be of assistance to managing a Stata dataset. It is scheduled for a 2008 publication by Stata Press. I have presented two chapters related to entering and managing data, but if you have a complex dataset to build or manage, Long's book should be a useful resource. If you develop strong data-management skills, you will be a valued member of any research team.

Scott Long and Jeremy Freese wrote *Regression Models for Categorical Dependent Variables Using Stata, Second Edition* and this builds on our chapter on logistic regression providing extensions to multinomial regression, ordered regression, Poisson regression, and zero-inflated models. It is extremely accessible given the complexity of the topic. One of the most important areas of development in Stata is multilevel analysis. Although this extends beyond the scope of my book, this type of analysis is an important extension to what was covered. Sophia Rabe-Hesketh and Anders Skrondal wrote *Multilevel and Longitudinal Modeling Using Stata, Second Edition*. This is a must-have book for anybody who needs to learn how to work with multilevel analysis and growth models.

Consult the Stata Bookstore, http://www.stata.com/bookstore/statabooks.html, to see a list of all the books about using Stata. These include books on many special topics.

13.2.3 Short courses

Stata provides NetCourses to help you learn to use Stata more effectively. These courses include two that are designed to introduce you to data management, analysis, and programming with Stata, and a third, more advanced programming course. You can enroll in a regularly scheduled NetCourse or in NetCourseNow, which allows you to take the course at your own pace. These courses may be a little too advanced for a true beginner, but those who have completed this book will find them useful, as will those who have expertise in a competing statistical package. These courses are useful if you already know the statistical content and if you are switching to Stata from some other package such as SYSTAT, SAS, or SPSS. Now that you have completed this book, you might want to pursue these NetCourses. You will be given weekly readings and assignments and have a web location where you can ask for clarification and interact with other students. For information, consult the Stata NetCourse web site at http://www.stata.com/netcourse/.

Many individual instructors provide supporting documentation for their own students, and they have placed it on the web where anybody has access to it. One example is Richard Williams, at Notre Dame, who has data and programs that cover a one-year graduate course in social statistics. Williams's page is especially useful for people who had been using SPSS and are switching to Stata.

13.2.4 Acquiring data

So you want to use Stata. Where can you get data? National funding organizations that support social science research expect researchers to make the data they collect available to others. This is a professional and ethical obligation that researchers have. As a result, you can acquire many national datasets at little or no cost. When you read an article that uses a dataset, you can do a web search on the name of the dataset with a search engine, such as Google. For example, I used a subset of variables from `NLSY97` (National Longitudinal Survey of Youth, 1997). Enter `NLSY97` in your search bar, and it will show you a link to a home page for the U.S. Department of Labor (http://www.bls.gov/nls/nlsy97.htm). Here you can download the documentation and data for all waves of this panel survey. They even provide an extraction-software program that will help you select variables and generate a Stata dataset. There is no charge to do this, unless you want a hard copy of documentation, and that charge is minimal. Not all datasets are free, but the cost is usually minimal.

There are several clearinghouses that archive datasets. One of the best is the Inter-University Consortium for Political and Social Research (ICPSR), located at the University of Michigan. Many universities pay an annual fee for membership in this organization. If you are affiliated with an organization that is a member, your organization has a set of numbers that identifies each computer on its network, and this range of numbers is recorded at ICPSR. If you are eligible, do a search for ICPSR, and go to their web page (http://www.icpsr.umich.edu/). Enter a keyword, and do a search. You might enter `recidivism`, for example. This will give you a list of all the surveys that involve recidivism or include questions about it. You can then download the dataset and documentation.

Although ICPSR is putting new datasets in the Stata format and updating some of the more widely used older datasets to Stata format, this is certainly not true for all the older datasets. SAS and SPSS had been used for statistical analysis for many years before

Stata was developed, and the dataset you want may be in one of those two formats. There is a simple solution. Stat/Transfer is a program you can buy from Stata or from Circle Systems, the publisher of Stat/Transfer (http://www.stattransfer.com/), which converts datasets from one format to another. We discussed this in an early chapter, and you should make use of it.

When working with national datasets, you can run into some limitations in Stata. One is the number of variables you can have in a dataset. With Stata/IC, you are limited to 2,047 variables, and with Stata/SE, you are limited to 32,767 variables. Both of these numbers sound huge for any study, but some large datasets can exceed these numbers.

13.3 Summary

At the beginning of this book, I assumed that you had no experience using a computer program to create and manage data or to do data analysis and graphics. If you already had experience using another software program and had a strong statistical background, our gentle approach may have been a bit too slow at times. I deliberately focused on making few assumptions about your background and decided to build from the simplest applications to the more complex. Now you can reflect on what you have learned. The book covered creating graphs, charts, tables, statistical inferences for one or two variables, analyses of variance, correlations, regressions, multiple regressions, and logistic regressions; entering data; labeling variables and values; managing datasets; generating new variables; and recoding variables. This is a lot of material. At this point, you know enough to do your own study or to manage a study for a research team. Although this book has been designed more for learning Stata than as a reference manual, the extended index can help you use it in the future as a reference manual. You now have a valuable set of skills! Congratulations!

References

Altman, D. G. 1991. *Practical Statistics for Medical Research.* London: Chapman & Hall.

Bureau of Labor Statistics. 2005. NLSY97 Questionnaires and Codebooks. http://www.bls.gov/nls/quex/y97quexcbks.htm.

Hamilton, L. C. 2006. *Statistics with Stata (Updated for Version 9).* Belmont, CA: Brooks/Cole.

Kohler, U., and F. Kreuter. 2009. *Data Analysis Using Stata.* 2nd ed. College Station, TX: Stata Press.

Landis, J. R., and G. G. Koch. 1977. The measurement of observer agreement for categorical data. *Biometrics* 33: 159–174.

Lawshe, C. H. 1975. A quantitative approach to content validity. *Personnel Psychology* 28: 563–575.

Long, J. S., and J. Freese. 2006. *Regression Models for Categorical Dependent Variables Using Stata.* 2nd ed. College Station, TX: Stata Press.

Mitchell, M. 2004. *A Visual Guide to Stata Graphics.* College Station, TX: Stata Press.

———. 2008. *A Visual Guide to Stata Graphics.* 2nd ed. College Station, TX: Stata Press.

Shultz, K. S., and D. J. Whitney. 2005. *Measurement Theory in Action: Case Studies and Exercises.* Thousand Oaks, CA: Sage.

Author index

Subject index